EXCEPTIONAL FOSSIL PRESERVATION

Critical Moments and Perspectives in Earth History and Paleobiology

DAVID J. BOTTJER
RICHARD K. BAMBACH
Editors

Critical Moments and Perspectives in Earth History and Paleobiology
David J. Bottjer and Richard K. Bambach, Editors

EXCEPTIONAL FOSSIL PRESERVATION

A Unique View on the Evolution of Marine Life

Edited by
DAVID J. BOTTJER, WALTER ETTER,
JAMES W. HAGADORN, AND CAROL M. TANG

COLUMBIA UNIVERSITY PRESS
New York

Columbia University Press
Publishers Since 1893
New York   Chichester, West Sussex

Library of Congress Cataloging-in-Publication Data
 Exceptional fossil preservation: a unique view on the evolution of marine life /
    edited by David J. Bottjer . . . [et al.].
        p. cm. — (Critical moments and perspectives in Earth history and
        paleobiology)
    Includes bibliographical references and index.
    ISBN 0-231-10254-2 (cloth : acid-free paper)
    ISBN 0-231-10255-0 (pbk. : acid-free paper)
    1. Marine animals, Fossil.   2. Taphonomy.   I. Bottjer, David J.
    II. Critical moments in earth history and paleobiology series.

QE766.E93 2001
560′.457–dc21   2001042434

CONTENTS

FOREWORD

FOSSILS, REMAINS OF ANCIENT LIFE, HAVE PLAYED A LARGE role in our exploration of the Earth's past. In antiquity they led to the recognition of sea-level oscillations. During the Renaissance, they raised questions about the tales of Genesis. In the nineteenth century, when paleontology became a science of its own, they served to develop an Earth history and demonstrated the great complexity of the tree of life. In the twentieth century, these concepts were greatly refined.

Part of what makes fossils so useful is their enormous abundance. In many outcrops, shells of clams or snails or brachiopods weather out by the hundreds, even thousands, and many limestone cliffs are replete with corals and bryozoans. A teaspoon of clay or chalk may contain hundreds of foraminifera or millions of coccoliths.

But such simply armored organisms constitute only a fraction of any living fauna. Carcasses with compound skeletons, such as echinoderms and vertebrates, are normally disarticulated, as they are torn apart by scavengers or rotted by bacteria and fungi, and few are the bones or ossicles that are not dragged off by scavengers; distributed by rivers, currents, and waves; or scattered by sediment burrowers. Animals with fairly standardized body plans, such as vertebrates, can still generally be assembled from scattered bones, but in the case of echinoderms, commonly having thousands to tens of thousands of individual ossicles and a baffling variety of body plans, the task is hopeless.

Beyond this, faunas include many animals that have little or nothing in the way of mineralized skeletons that escape bacterial decomposition. How and when did these groups evolve, and what role, if any, did they play in fossil faunas of the past?

Furthermore, extinct groups of animals often pose problems of relationship. Trilobites, for example, clearly were arthropods, but were they crustaceans, merostomes, or an independent and now wholly extinct class of their own? Arthropod groups are classified largely on the basis of their appendages—legs, gills, and antennae—which in trilobites were not covered by the mineralized armor.

Thus, destructive physical processes of wave and current transport, the dissolution by rains and marine waters, the activity of scavengers, borers, and burrowers, and decomposition by bacteria and fungi ensure that generally only the most resistant structures—mineralized skeletons—mostly battered and scattered, get preserved.

But here and there, in the space-time continuum of history, those destructive forces were briefly suppressed, allowing more of the record to slip through: fishes or dinosaurs that did not get scattered down the riverbed or over the seafloor; populations of trilobites or brittle stars overwhelmed and buried by a turbidite or mudflow; a seafloor where lack of oxygen or presence of poisons kept the sediment-eaters at bay; and, very rarely, bacterial decomposition proceeding hand-in-hand with replacement of tissues, generally by phosphate or pyrite, to leave a record of delicate appendages, hairs and organs, and of the outline of the body. It is these deposits that have preserved complete skeletons, more rarely some of the organs, and, in very rare cases, the form of worms and jellyfish, of parasites and of larvae.

Thus, while the metazoan paleontologist works mostly with the bits and pieces that pass through the normal gauntlet of destruction, much of the deeper fabric of the science depends on these special localities, these fossil-Lagerstätten. These, where so much more of the organism and of the fauna or flora comes to view, are our Rosetta stones—serving to reconstruct otherwise disassembled bodies, to clarify biological relationships of extinct forms, and to cast light on the soft-bodied members of ancient faunas.

In the Proterozoic and Cambrian, extraordinary patterns of preservation were not so localized but widely distributed, suggesting that in those beginning days of metazoan life the destructive processes did not yet comb the sedimentary accumulations as thoroughly as they did later.

Monte Bolca, with the world's largest known fossil fish fauna, has been worked for collectors since approximately 1500. The Solnhofen quarries have provided lithographic and decorative stone since the eighteenth century, and such spectacular fossils as the famous bird *Archaeopteryx,* horseshoe crabs, jellyfishes, and pterodactyls are a rare byproduct of industry.

Many other sites have been discovered, and every year brings to light new examples of extraordinary preservation. These fossils either are in distant areas where science was slow to arrive, as in the case of Chengjiang, or are microscopic and long overlooked, as in the orsten of the Swedish Cambro-Ordovician.

In this volume, David J. Bottjer and colleagues have provided a sampling of marine Lagerstätten from the Precambrian Ediacara to the Eocene Monte Bolca sites. Each case considers the taxa preserved, with appropriate illustrations. They also survey the environmental setting, the sedimentology with special focus on the preservational (taphonomic) aspects, and the nature of the biological community.

The book walks us through time, through the treasure troves of marine paleontology, from one spectacular fauna to another. We are exposed not only to an enormous diversity of ancient life, but to diverse preservational processes, which, in many cases, remain problematic. Much is yet to be learned from and about each of these sites, from others not dealt with here, and from the many that remain to be discovered.

Alfred G. Fischer

ACKNOWLEDGMENTS

OUR INTEREST IN PRODUCING A BOOK COVERING THE broad variety of marine conservation Lagerstätten was kindled through mutual fascination with how the mechanisms that produce the fossil record at times lead to cases of amazing preservation. As this endeavor progressed, it involved many in the broad community who are research specialists on one or more fossil Lagerstätten. These individuals provided the photographs that we have used, and many were involved in reviewing chapters. Production of this book has involved the ongoing encouragement and efforts of Robin Smith, Alessandro Angelini, and Irene Pavitt of Columbia University Press. Encouragement to pursue this project to completion was provided by Dick Bambach, Ed Lugenbeel, and Holly Hodder. Chris Maples and Sally Walker read all the chapters and provided valuable input on improving our presentation. Bill Ormerod has been invaluable in providing suitable copies of photos from disparate sources. To everyone who helped in assembling the many parts that make this book, we offer our heartfelt thanks!

EXCEPTIONAL FOSSIL PRESERVATION

1

Fossil-Lagerstätten: Jewels of the Fossil Record

David J. Bottjer, Walter Etter, James W. Hagadorn,
and Carol M. Tang

THE FOSSIL RECORD IS THE REPOSITORY OF THE HISTORY of life and the processes that have governed it. Most nonscientists are usually aware of fossils, but it is commonly believed that they are extremely rare. This seems to be because people typically do not know how or where to look for fossils—and even what a fossil would look like if they were to find one. But, in fact, fossils are exceptionally common in many sedimentary rocks and are used extensively in geology for such things as age dating, interpreting ancient environments, and exploring for natural resources. Similarly, most typical fossils preserve aspects of the skeletons of once-living organisms and thus have provided us with the evidence for much of what we know about life's history, and the broad trends that have played out during this history (Bambach 1977, 1983; Bottjer and Ausich 1986; Sepkoski 1993).

However, there is another type of fossil deposit that *is* truly rare. These rare fossil deposits preserve the remains of soft tissues or the articulated nature of skeletal elements. These deposits with exceptional preservation have been known for centuries, and because they have long been popular with the public, it is not unusual to find museums associated with such deposits.

These deposits of exceptional fossil preservation were typically considered to be scientific curiosities, where, because of unusual preservation, paleontologists could decipher usually unobtainable aspects of organism morphology, behavior, function, and evolution. These deposits have long had significance as the place where soft tissues fundamental to

2 • *Fossil-Lagerstätten*

our understanding of life history, such as the feathers of *Archaeopteryx* (Chapter 18) or of dinosaurs (Qiang et al. 1998), are found.

However, it is only over the past 25 years that a concerted effort has been made to understand the overall significance of these rare and remarkable fossil deposits. This modern era in the study of such deposits began with the work of Adolf Seilacher and colleagues in Germany, where several deposits exhibiting exquisite preservation, such as the Hunsrück Slate, Solnhofen, and the Posidonia Shale, are located. Perhaps the founding paper in our modern understanding of these deposits is that by Seilacher (1970), in which he first popularized the general use of the word "Fossil-Lagerstätten" (singular, "Lagerstätte"). In German, a Lagerstätte is a geological deposit of economic interest (Shields 1998), and a fossil-Lagerstätte is a deposit containing fossils that are so exceptionally preserved or abundant that it warrants special exploitation, if only for scientific purposes (Seilacher 1990). Seilacher, Reif, and Westphal (1985) defined two types of fossil-Lagerstätten: conservation Lagerstätten, which contain exceptionally preserved fossils such as the impressions of soft parts of organisms; and concentration Lagerstätten, which contain an abundance of fossils that are typically not exceptionally preserved (Seilacher 1990). The term "fossil-Lagerstätten" is now so widely accepted that paleontologists commonly shorten it to "Lagerstätten" (Brett, Baird, and Speyer 1997; but see Shields 1998). Significant strides in our overall understanding of conservation Lagerstätten have been made in the past two decades, particularly by Derek Briggs, Peter Allison, Simon Conway Morris, and James Gehling. Their work was initially driven in large part by the need to understand the nature and preservation of early metazoan animals preserved in such deposits as the Burgess Shale of Canada and Ediacaran fossil sites around the world.

SELECTED EXAMPLES OF LAGERSTÄTTEN

Much of the literature on Lagerstätten is widely scattered, and information on many of these deposits, especially those outside of North America and Europe, is commonly published in sources that are not easily accessible to the global community. This book is an overview of conservation Lagerstätten that contain a significant proportion of marine metazoan fossils. We have not been able to include all such deposits, but have attempted to provide a representative selection through time from the Neoproterozoic into the Cenozoic (Table 1.1). The examples offered here are mostly from Europe and North America, which have received more paleontological exploration and study than other parts of the world, and thus have more well-known Lagerstätten sites. In large part, our goal is to present an easily accessible casebook of marine Lagerstätten with exceptional macrofossil preservation, outlining and illustrating the great variety in styles of preservation, the types of organisms pre-

TABLE 1.1 Age Distribution of Lagerstätten Treated in This Volume with Geologic Ages in Millions of Years (Ma)

Era	Period/Epoch	Lagerstätten	Chapter
Cenozoic (0–65 Ma)	Eocene (34–53 Ma)	Monte Bolca	20
Mesozoic (65–250 Ma)	Cretaceous (65–145 Ma)	Smoky Hill Chalk	19
	Jurassic (145–200 Ma)	Solnhofen	18
		Oxford Clay	17
		La Voulte-sur-Rhône	16
		Posidonia Shale	15
		Osteno	14
	Triassic (200–251 Ma)	Berlin- Ichthyosaur	13
		Monte San Giorgio	12
		Grès à Voltzia	11
Paleozoic (251–543 Ma)	Carboniferous (295–355 Ma)	Mazon Creek	10
		Bear Gulch	9
	Devonian (355–410 Ma)	Hunsrück Slate	8
	Ordovician (440–495 Ma)	Beecher's Trilobite	7
	Cambrian (495–543 Ma)	Orsten	6
		Burgess Shale	4, 5
		Chengjiang	3
Neoproterozoic (543–1000 Ma)	Vendian (543–600 Ma)	Ediacaran	2

served, the mechanisms that led to such preservation, and the lessons to be learned from this type of fossil deposit. For each example, we discuss the condition and context of its status as a Lagerstätte, its depositional conditions, and its specific paleobiological significance.

TAPHONOMY AND THE PRESERVATION OF FOSSILS

Taphonomy is a fundamental branch of paleontology that aims to understand the processes behind the preservation of all fossils ranging from the most common shells to the rare, soft-body remains found in many Lagerstätten (Allison and Briggs 1991a; Donovan 1991; Kidwell and Behrensmeyer 1993). As such, taphonomy involves the study of the ecological, biogeochemical, and sedimentary processes that occur in the environment before and after burial of organisms. Thus, taphonomy, and the study of the formation of Lagerstätten, is necessarily a science that integrates biology, chemistry, and geology.

Under typical taphonomic conditions where oxygen is present in marine settings, scavenging and microbial decay rapidly remove soft tissue from mineralized skeletal elements such as shells and bones. These elements are also subject to scattering by carnivores and scavengers, degradation by agents such as boring microorganisms, chemical dissolution, and physical erosion by waves and currents. Thus, biological remains are typically destroyed before they can be buried by sediment.

However, a small proportion of organic remains do become buried below the seafloor. If the sediment pore waters are undersaturated in dissolved calcium carbonate (the mineral of which most shells are made) or calcium phosphate (of which bones are made), chemical dissolution will occur. If they are not dissolved, continued deposition of sediment can bury organic remains to the point where they are no longer in the *taphonomically active zone* (TAZ) (Davies, Powell, and Stanton 1989; Walker and Goldstein 1999), and become immune to reexposure by erosion and damage by organisms that burrow through the sediment surface. It is these biological, chemical, and sedimentary processes that almost all fossils must pass through in order to become preserved.

Very rarely, these processes of decay and destruction are inhibited, producing a conservation Lagerstätte with exceptional preservation of biological remains. If, for example, a carcass is buried before it can be disturbed, the quality of preservation may be very high. One mechanism for this is *obrution* (Seilacher, Reif, and Westphal 1985; Brett, Baird, and Speyer 1997), by which the organism (either alive or recently dead) is smothered and immediately buried by a pulse of sedimentation, such as a storm or turbidite. Once the remains are buried below the TAZ, they are protected from further disturbance. Another way to decrease the chance of being disturbed is to introduce the organism into an environment where dissolved oxygen concentrations are so low that scavengers and aerobic microorganisms are not present; conservation Lagerstätten resulting from this process are termed *stagnation deposits* (Seilacher, Reif, and Westphal 1985).

Due to the lack of disturbance under both obrution and stagnation conditions, skeletal articulations will be maintained but soft tissues will commonly still not be preserved (Allison and Briggs 1991b, 1991c). So how is soft-tissue preservation accomplished in marine environments? Usually it is through the presence of unique geochemical conditions, commonly mediated by the presence of microbial activity, which leads either to the preservation of original organic material or to the early precipitation of pyrite, calcium carbonate, or phosphate, which produces a mineral replica of the soft tissues (Allison and Briggs 1991b, 1991c).

Thus, conservation Lagerstätten are products of the fortuitous co-occurrence of unique biological, chemical, and sedimentary conditions and are rare exceptions to the usual fate of organic remains as they proceed to possible inclusion in the fossil record. Although additional Lagerstätten are sure to be discovered in the future, fewer than 100 marine Lagerstätten with soft-tissue preservation are presently known (Conway Morris 1986; Allison and Briggs 1993a), whereas the number of more typical fossil deposits is several orders of magnitude greater. Clearly, the formation of marine Lagerstätten represents some of the more unusual events in the Earth's biological and geological history, and their uniqueness is best exemplified by their scattered occurrence in the sedimentary record of the past 700 million years (Table 1.1).

UNIFORMITARIANISM AND LAGERSTÄTTEN

Most paleontological and geological work in the twentieth century has been done following the rubric of uniformitarianism, the idea that "the present is the key to the past" (Gould 1965; Bottjer et al. 1995; Bottjer 1998). In fact, the conditions for formation of marine Lagerstätten with exceptional preservation have not been particularly well studied in modern environments. Although it will undoubtedly be a rapidly growing field in the future (Allison and Pye 1994), most of what we know about the formation of marine Lagerstätten is from the study of ancient examples.

Research on these examples has revealed a great deal of secular variation in the temporal and paleoenvironmental distribution of marine Lagerstätten, particularly those that preserve evidence of soft tissues (Allison and Briggs 1991b, 1991c, 1993a; Butterfield 1995). Although general sedimentological processes such as obrution and stagnation have operated since at least the Neoproterozoic, for preservation of soft tissue, it appears that the present will not in fact be the key to the past, and that many intervals of geological time represent somewhat different taphonomic windows for Lagerstätten development.

For example, perhaps the oldest distinct group of Lagerstätten (not treated in this volume) is a set of Late Riphean (Proterozoic) age deposits that preserve organic-walled microorganisms (Butterfield 1995). In addition, newly discovered late Neoproterozoic (early Vendian) and Early Cambrian age deposits that contain phosphatized animal soft tissues including eggs and embryos (Bengtson and Zhao 1997; Li, Chen, and Hua 1998; Xiao, Zhang, and Knoll 1998; Chen et al. 2000), may also represent the presence of unique and temporally restricted taphonomic conditions. Similarly, later Vendian deposits are characterized by the preservation of soft-body impressions in shallow-water sandstones, termed the Ediacaran fauna (Chapter 2). Preservation conditions for the Ediacaran fauna seem to have been best in the Vendian, but such fossils are also more rarely found preserved in the Cambrian (Chapter 2), indicating the closing of a taphonomic window during the early Paleozoic.

The next youngest group of conservation Lagerstätten includes the Burgess Shale–type faunas of the Early and Middle Cambrian, which generally show preservation of soft metazoan tissues as carbonaceous compressions similar to the preservation of microorganisms at Late Riphean age sites (Chapters 3, 4, 5; Butterfield 1995). However, the Late Cambrian orsten faunas of southern Sweden (Chapter 6), characterized by three-dimensional calcium phosphate preservation of tiny soft-bodied organisms within calcium carbonate nodules, exhibit a preservational style found in other Paleozoic and Mesozoic examples (Allison and Briggs 1991b, 1991c) and thus have less temporal restriction.

Between the Cambrian and the Devonian, the occurrence of marine Lagerstätten with soft-bodied faunas is relatively rare (Briggs and Siveter 1996). Perhaps the most significant Ordovician marine Lagerstätte is Beecher's Trilobite Bed of northern New York (Chapter 7), where trilobites with pyritized soft tissues have been found at the base of a deep-water microturbidite. Silurian marine Lagerstätten with soft-tissue preservation are also rather sparse, although a new example of calcite-infilled, three-dimensional soft-bodied organisms from British volcaniclastic deposits has recently been discovered (Briggs and Siveter 1996). Since there does not appear to be a significant difference in the preservation potential of Cambrian, Ordovician, and Silurian organisms, the relative abundance of Cambrian Lagerstätten with soft-tissue preservation is more likely due to temporal differences in preservation processes themselves than to biological differences through time (Allison and Briggs 1991b, 1991c, 1993a; Butterfield 1995; Briggs and Siveter 1996).

An apparent exception to this generality is the Devonian Hunsrück Slate of Germany (Chapter 8), which exhibits preservation much like Beecher's Trilobite Bed. Here soft tissues of a wide variety of organisms were deposited in a deep-marine environment and were also preserved as pyrite. But preservation by pyrite mineralization is rare in the rest of the Phanerozoic and never plays as prominent a role as it did in Beecher's and the Hunsrück Slate—a puzzling fact that only highlights the nonuniformitarian nature of such preservation processes (Allison and Briggs 1991b, 1991c).

The occurrence of soft-tissue preservation shifted gears again in the succeeding time interval, the Carboniferous, where major Lagerstätten are found deposited in deltaic environments, with both marine and nonmarine organisms preserved (Allison and Briggs 1991b, 1991c). The most prominent of these Lagerstätten is the Mazon Creek biota of northern Illinois (Chapter 10). Interestingly, this mode of development of Lagerstätten in deltaic environments continued into the early Mesozoic, with the best-known example being the Triassic Grès à Voltzia of northeastern France (Chapter 11). The Carboniferous also marks the advent of another major style of exceptional preservation, the Plattenkalk, a fine-grained limestone facies (lithographic limestone) deposited in basinal environments under a stratified water column (Allison and Briggs 1991b, 1991c). This type of preservation is typified by the fish-bearing Lagerstätte at Bear Gulch in Montana (Chapter 9).

Many Mesozoic Lagerstätten that preserve articulation of skeletal components and/or preservation of soft tissues could be called stagnation deposits and were deposited in marine environments characterized by restricted circulation and/or a stratified water column. Some of these, such as the Jurassic Solnhofen of Germany (Chapter 18), are like the Carboniferous Bear Gulch and preserve soft tissues within a Plattenkalk lithofacies. Others, such as the Triassic Monte San Giorgio from along the Swiss-Italian border, as well as the Jurassic Holzmaden of Germany, La Voulte-sur-Rhône in France, and Oxford Clay of England (Chapters 12, 15, 16, 17), exhibit preservation of soft tissues within a black shale lithofacies. In contrast, the Lagerstätte with soft-tissue preservation at Osteno in northern Italy (Chapter 14) is considered to be an intermediate between Solnhofen and Holzmaden (Allison and Briggs 1991c).

Some Mesozoic marine Lagerstätten, which were deposited under restricted circulation conditions, do not exhibit soft-tissue preservation, but do contain specimens with remarkably well articulated skeletal elements; such deposits include the Triassic Berlin-Ichthyosaur site of Nevada (Chapter 13) and the Cretaceous Smoky Hill Chalk of Kansas (Chapter 19). Although Lagerstätten deposited in paleoenvironments with restricted circulation appear concentrated in the Mesozoic, there are other Paleozoic examples, such as Bear Gulch, and Cenozoic examples, such as the Eocene Plattenkalk site at Monte Bolca in Italy (Chapter 20), as well as the Miocene Monterey faunas in diatomites of California.

Along with these secular variations in taphonomic pathways toward soft-tissue preservation, there is some evidence that Lagerstätten with soft-body preservation are overrepresented in the Cambrian and the Jurassic, when compared with relative outcrop area (Allison and Briggs 1993a). The overrepresentation of Lagerstätten in the Jurassic may be explained by the abundance of stagnation deposits resulting from sluggish oceanic circulation related to the breakup of Pangaea and the formation of the Atlantic Ocean (Allison and Briggs 1993a). However, the causes of Burgess Shale–type preservation, and why it appears limited to the Cambrian, are hotly debated (Chapters 3, 4, 5; Aronson 1992, 1993; Allison and Briggs 1993a, 1993b, 1994; Butterfield 1995, 1996; Towe 1996). Investigation of the processes that may have produced this pattern of preservational variation through time is at only an initial stage, and this promises to be an exciting area of future research on marine Lagerstätten.

GROWING IMPORTANCE OF LAGERSTÄTTEN STUDIES

Whereas in the past, Lagerstätten were considered novelties that exhibited extraordinary preservation, modern workers on Lagerstätten are increasingly addressing significant paleoenvironmental and paleobiological questions. With advanced image and geochemical analyses not available just a few decades ago, sophisticated and detailed analyses of the paleoenvironmental, sedimentary, and geochemical conditions necessary for the preservation of soft tissues and organic-rich sediments are now possible (Chapters 4, 5, 6, 7, 8, 17). These studies, in turn, have introduced new techniques and protocols with which to search for undiscovered Lagerstätten and provide information on ancient paleoceanographic conditions and the formation of economically important natural resources such as coal and oil.

In addition to providing new geological information, studies of fossils from Lagerstätten have made enormous contributions to paleobiology. A century ago, fossils from the Monte Bolca Lagerstätte (Chapter 20) were used in the first monograph on the evolutionary relationships between fish. Today, exceptionally preserved fossils continue to play a large role in phylogenetics and the clarification of the basic evolutionary relationships among animal groups (Chapters 2, 3, 4, 5, 6, 10, 14, 18) because fossils provide a fundamental test for evolutionary hypotheses (Smith 1994). In addition, Lagerstätten preserving soft-bodied organisms provide key evidence needed for understanding critical biological phenomena such as the nature of the earliest larger metazoans (Chapter 2) and the Cambrian explosion (Chapters 3, 4, 5). Perhaps the most inviting new field in the study of conservation Lagerstätten will be analyses of the recently discovered phosphatized metazoan embryos and larvae in late Neoproterozoic and Cambrian age deposits, and what can be learned from them about evolution of early metazoans (Bengtson and Zhao 1997; Li, Chen, and Hua 1998; Xiao, Zhang, and Knoll 1998; Xiao and Knoll 1999; Chen et al. 2000). In large part, it is this use of Lagerstätten in answering some of the most basic questions about life on Earth that has dramatically spurred the pace of studies of Lagerstätten and generated attention to the intrinsic utility of such deposits.

It is likely that in the future, as the potential significance of Lagerstätten becomes more widely known and the ability to search for Lagerstätten grows, many more fascinating examples will be found from around the world, especially from relatively poorly explored regions. Thus, there is great potential for new and dazzling discoveries of Lagerstätten of all ages, much like the discoveries of additional deposits containing typical Ediacaran, Burgess Shale, and Orsten faunas that have been found in the past decade (Chapters 2, 3, 4, 5, 6).

Although the contents of this book are a clear indication that paleontologists have already learned much from these remarkable deposits, we have very likely only scratched the surface of the record of marine conservation Lagerstätten. As these scientific inquiries continue, paleontologists can provide further understanding of why such deposits occur, how they have varied since the advent of marine metazoan life, and how their presence affects our overall understanding of the ways that life has evolved in the Earth's oceans. In this way, the study of Lagerstätten continues to move toward the mainstream of paleobiological, biological, and geological research, and away from its former status as the examination of mere curiosities.

REFERENCES

Allison, P. A., and D. E. G. Briggs, eds. 1991a. *Taphonomy: Releasing the Data Locked in the Fossil Record*. New York: Plenum.

Allison, P. A., and D. E. G. Briggs. 1991b. The taphonomy of soft-bodied animals. In S. K. Donovan, ed., *The Processes of Fossilization*, pp. 120-140. New York: Columbia University Press.

Allison, P. A., and D. E. G. Briggs. 1991c. Taphonomy of nonmineralized tissues. In P. A. Allison and D. E. G. Briggs, eds., *Taphonomy: Releasing the Data Locked in the Fossil Record*, pp. 25–70. New York: Plenum.

Allison, P. A., and D. E. G. Briggs. 1993a. Exceptional fossil record: Distribution of soft-tissue preservation through the Phanerozoic. *Geology* 21:527–530.

Allison, P. A., and D. E. G. Briggs. 1993b. Burgess Shale biotas burrowed away? *Lethaia* 26:184–185.

Allison, P. A., and D. E. G. Briggs. 1994. Exceptional fossil record: Distribution of soft-tissue preservation through the Phanerozoic: Comment and reply. Reply to comment by R. K. Pickerill. *Geology* 22:184.

Allison, P. A., and K. Pye. 1994. Early diagenetic mineralization and fossil preservation in modern carbonate concretions. *Palaios* 9:561–575.

Aronson, R. B. 1992. Decline of the Burgess Shale fauna: Ecologic or taphonomic restriction? *Lethaia* 25:225–229.

Aronson, R. B. 1993. Burgess Shale–type biotas were not just burrowed away: Reply. *Lethaia* 26:185.

Bambach, R. K. 1977. Species richness in marine benthic habitats through the Phanerozoic. *Paleobiology* 3:152–167.

Bambach, R. K. 1983. Ecospace utilization and guilds in marine communities through the Phanerozoic. In M. J. S. McCall and P. L. Tevesz, eds., *Biotic Interactions in Recent and Fossil Benthic Communities*, pp. 719–746. New York: Plenum.

Bengtson, S., and Y. Zhao. 1997. Fossilized metazoan embryos from the earliest Cambrian. *Science* 277:1645–1648.

Bottjer, D. J. 1998. Phanerozoic non-actualistic paleoecology. *Geobios* 30:885–893.

Bottjer, D. J., and W. I. Ausich. 1986. Phanerozoic development of tiering in soft substrata suspension-feeding communities. *Paleobiology* 12:400–420.

Bottjer, D. J., K. A. Campbell, J. K. Schubert, and M. L. Droser. 1995. Palaeoecological models, non-uniformitarianism, and tracking the changing ecology of the past. In D. W. J. Bosence and P. A. Allison, eds., *Marine Palaeoenvironmental Analysis from Fossils*, pp. 7–26. Special Publication, no. 83. London: Geological Society.

Brett, C. E., G. C. Baird, and S. E. Speyer. 1997. Fossil Lagerstätten: Stratigraphic record of paleontological and taphonomic events. In C. E. Brett and G. C. Baird, eds., *Paleontological Events: Stratigraphic, Ecological, and Evolutionary Implications*, pp. 3–40. New York: Columbia University Press.

Briggs, D. E. G., and D. J. Siveter. 1996. Soft-bodied fossils from a Silurian volcaniclastic deposit. *Nature* 382:248–250.

Butterfield, N. J. 1995. Secular distribution of Burgess-Shale–type preservation. *Lethaia* 28:1–13.

Butterfield, N. J. 1996. Fossil preservation in the Burgess Shale: Reply. *Lethaia* 29:109–112.

Chen, J. Y., P. Oliveri, C. W. Li, G. Q. Zhou, F. Gao, J. W. Hagadorn, K. J. Peterson, and E. J. Davidson. 2000. Precambrian animal diversity: New evidence from high resolution phosphatized embryos. *Proceedings of the National Academy of Sciences* 97:4457–4462.

Conway Morris, S. 1986. The community structure of the Middle Cambrian phyllopod bed (Burgess Shale). *Palaeontology* 29:423–467.

Davies, D. J., E. N. Powell, and R. J. Stanton Jr. 1989. Taphonomic signature as a function of environmental process: Shells and shell beds in a hurricane-influenced inlet on the Texas coast. *Palaeogeography, Palaeoclimatology, Palaeoecology* 72:317–356.

Donovan, S. K., ed. 1991. *The Processes of Fossilization.* New York: Columbia University Press.

Gould, S. J. 1965. Is uniformitarianism necessary? *American Journal of Science* 263:223–228.

Kidwell, S. M., and A. K. Behrensmeyer, eds. 1993. *Taphonomic Approaches to Time Resolution in Fossil Assemblages.* Short Courses in Paleontology, no. 6. Paleontological Society. Knoxville: Department of Geological Sciences, University of Tennessee.

Li, C. W., J. Y. Chen, and T. E. Hua. 1998. Precambrian sponges with cellular structures. *Science* 279:879–882.

Qiang, J., P. J. Currie, M. A. Norell, and J. Shu-An. 1998. Two feathered dinosaurs from northeastern China. *Nature* 393:753–761.

Seilacher, A. 1970. Begriff und bedeutung der Fossil-Lagerstätten. *Neues Jahrbuch für Geologie und Paläontologie, Abhandlungen* 1970:34–39.

Seilacher, A. 1990. Taphonomy of Fossil-Lagerstätten. In D. E. G. Briggs and P. R. Crowther, eds., *Palaeobiology: A Synthesis,* pp. 266–270. Oxford: Blackwell Science.

Seilacher, A., W. E. Reif, and F. Westphal. 1985. Sedimentological, ecological and temporal patterns of Fossil Lagerstätten. *Philosophical Transactions of the Royal Society of London, B* 311:5–23.

Sepkoski, J. J., Jr. 1993. Ten years in the library: New data confirm paleontological patterns. *Paleobiology* 19:43–51.

Shields, G. 1998. What are Lagerstätten? *Lethaia* 31:124.

Smith, A. B. 1994. *Systematics and the Fossil Record.* Oxford: Blackwell Science.

Towe, K. M. 1996. Fossil preservation in the Burgess Shale. *Lethaia* 29:107–108.

Walker, S. E., and S. T. Goldstein. 1999. Taphonomic tiering: Experimental field taphonomy of molluscs and foraminifera above and below the sediment-water interface. *Palaeogeography, Palaeoclimatology, Palaeoecology* 149:227–244.

Xiao, S., and A. H. Knoll. 1999. Fossil preservation in the Neoproterozoic Doushantuo phosphorite Lagerstätte, South China. *Lethaia* 32:219–240.

Xiao, S., Y. Zhang, and A. H. Knoll. 1998. Three-dimensional preservation of algae and animal embryos in a Neoproterozoic phosphorite. *Nature* 391:553–558.

2
Enigmatic Ediacara Fossils: Ancestors or Aliens?

David J. Bottjer

THE EDIACARA FOSSILS, OF LATE PRECAMBRIAN (VENDIAN) through Cambrian age, are among the most remarkable fossil biotas known from the stratigraphic record. This stems from the fact that this biota is thought to include fossils of some of the earliest larger organisms, whose nature has been much debated: Are they ancient representatives of still extant metazoan phyla, do they represent phyla or a kingdom now extinct on Earth, or could they even be colonial procaryotes or fossil lichens? Furthermore, when compared with younger deposits, this biota is in general a taphonomic anomaly. The Ediacara fossils represent remains of completely soft-bodied organisms, and yet they are commonly preserved in coarser-grained siliciclastics deposited in relatively well oxygenated marine environments, a seemingly improbable phenomenon not known elsewhere from the marine fossil record. Because Ediacara fossil preservation is commonly associated with some sort of event bed, varying from tidal sandstones to storm beds, to turbidites and subaqueous ash falls, their taphonomic context is best thought of as obrution deposits.

Fossils we now recognize as Ediacara were discovered as early as the nineteenth century in England at the Charnwood Forest locality (Hill and Bonney 1877) and in the early twentieth century in Namibia (Gürich 1930). However, their importance was not internationally recognized until the 1940s when R. C. Sprigg, an assistant government geologist of South Australia, discovered fossils of late Precambrian soft-bodied organisms in the Ediacara Hills of the Flinders Range, 600 km north of Adelaide. Sprigg's (1947, 1949) discoveries led to the extensive work of Glaessner (1961, 1969, 1983, 1984; Glaessner and Wade 1966) and

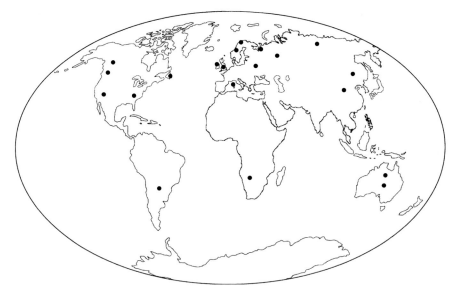

FIGURE 2.1 Global location, marked by dots, of the Ediacara biota; see text for indicated sites.

Wade (1972a, 1972b), who systematically documented this fauna and its preservation. This work in South Australia led to the appellation "Ediacara fauna" and fostered recognition that these fossils of soft-bodied organisms correspond to those in Charnwood Forest and Namibia and have a worldwide distribution, with other occurrences including the United States (California, Nevada, North Carolina), Canada (British Columbia, Yukon, Northwest Territories, Newfoundland), South America, Wales, Ireland, Sardinia, Norway, Finnmark, Russia (White Sea area, Urals, Siberia), Ukraine, central Australia, and China (Liao-Dun Peninsula, Heilongjiang Province, Yangtze Gorges) (Fedonkin 1992; Waggoner 1999) (Figure 2.1). The entire Ediacara biota was once thought to have become extinct well before the beginning of the Cambrian, but recent research indicates that at least portions of this biota survived into the Cambrian (Conway Morris 1993; Crimes, Insole, and Williams 1995; Grotzinger et al. 1995; Jensen, Gehling, and Droser 1998; Hagadorn, Fedo, and Waggoner 2000).

GEOLOGICAL CONTEXT

Among the numerous localities where Ediacara fossils are found, paleo-environmental reconstructions indicate that these organisms lived in a variety of shallow- to deep-marine environments (Conway Morris 1990; Narbonne and Aitken 1990; Runnegar 1992; Seilacher 1992; Crimes, Insole, and Williams 1995; Narbonne 1998). Detailed accounts in this chapter will concentrate on two examples: the Flinders Ranges fauna, which is interpreted to have lived in nearshore to shallow shelf environments, and the fauna found on the Avalon Peninsula of Newfoundland, which has been interpreted as deep marine in origin (Gehling 1999; Narbonne, Dalrymple, and Gehling 2001; Wood et al. 2001).

Folded and faulted outcrops of upper Proterozoic strata occur discontinuously in the Flinders Ranges (Figure 2.2). Ediacara fossils are found in the Ediacara Member of the Rawnsley Quartzite (Pound Subgroup), which occurs in a thick sequence of sedimentary rocks of late Proterozoic age (Figure 2.3). The Pound Subgroup is overlain by Cambrian strata with definite Cambrian trace fossils, although the exact

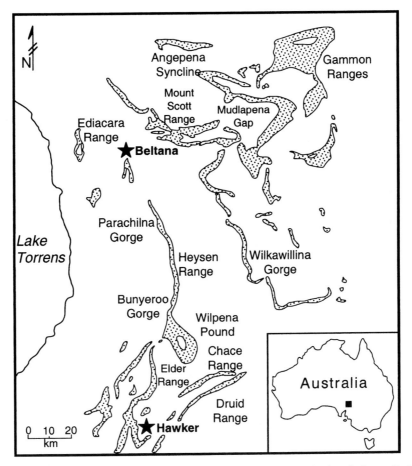

FIGURE 2.2 Generalized geologic map of the Flinders Ranges in South Australia illustrating the distribution of the upper Proterozoic Pound Subgroup (stippled). (Modified from Mount 1989)

relationships at the contact are of considerable controversy (Mount 1989, 1991; Nedin and Jenkins 1991).

Upper Proterozoic rocks are a prominent component of the Avalon Peninsula in Newfoundland, and the Mistaken Point area represents a classic locality for Ediacara fossils (Figure 2.4). Ediacara fossils from the Avalon Peninsula are found in the upper part of the Conception Group (Briscal and Mistaken Point Formations) and the overlying lower part of the St. John's Group (Trepassey and Fermeuse Formations) (Figure 2.5). The stratigraphic interval that bears Ediacara fossils contains mainly deep-marine slope turbidites with interbedded graded ash layers (Narbonne, Dalrymple, and Gehling 2001; Wood et al. 2001). In the Mistaken Point Formation, there is a tuff band that has been dated at 565 ± 3 my (Benus 1988).

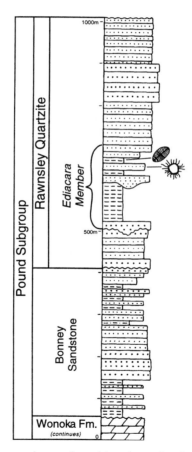

FIGURE 2.3 Generalized, composite stratigraphic column for the upper Proterozoic Pound Subgroup of the Flinders Ranges in South Australia; thicknesses and lithostratigraphy are approximate. The stratigraphic interval that contains the Ediacara biota is indicated by schematic fossils. (Modified from Mount 1989)

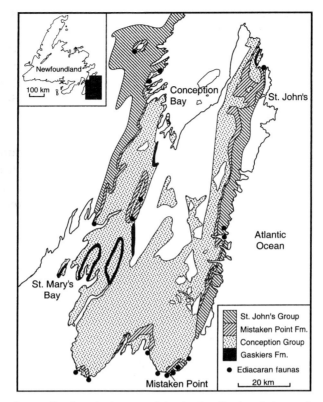

FIGURE 2.4 Generalized geologic map of the Avalon Peninsula in southeastern Newfoundland showing the distribution of upper Proterozoic St. John's Group and Conception Group (which includes the Mistaken Point and Gaskiers Formations) and of sites at which Ediacara faunas are found. (Modified from Jenkins 1992)

PALEOENVIRONMENTAL SETTING

The Rawnsley Quartzite contains thick, clean feldspathic sandstones that Gehling (1982, 1983, 1999) and Jenkins, Ford, and Gehling (1983) have interpreted as having accumulated in an environmental gradient ranging from shallow marine to intertidal sand flats. A sequence of thin, wavy-bedded sandstones; massive, channelized sandstones; and siltstones is found in the Rawnsley Quartzite in the central and southern Flinders Ranges (Mount 1989). The wavy-bedded sandstone interval is the lithofacies in which the Ediacara fossils are found, and this has been termed the Ediacara Member (Figure 2.3). Fossils are found only on the soles of flaggy sandstone beds (Glaessner 1984).

A number of sedimentological studies have been made of the flaggy sandstones of the Ediacara Member. Sprigg (1947) originally postulated that organisms of the Ediacara soft-bodied fauna were preserved on intertidal flats or along the strandline. Deposition in a tidally influenced

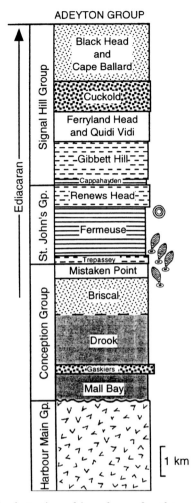

FIGURE 2.5 Generalized stratigraphic column for the upper Proterozoic of the Avalon Peninsula, Newfoundland, with stratigraphic intervals where Ediacara faunas are found indicated by schematic fossils. (Modified from Jenkins 1992)

environment was also supported by the studies of Glaessner (1961) and Jenkins, Ford, and Gehling (1983). However, Goldring and Curnow (1967) made a detailed study of the fossiliferous interval and interpreted it as having been deposited in an offshore neritic environment. Similarly, Gehling (1982, 1983, 1999) has concluded that the upper fossiliferous portion of the Ediacara Member (Figure 2.3) represents shallow subtidal, storm-dominated shelf sandstones and siltstones (Mount 1991). Thus, the most likely interpretation for the origin of the flaggy sandstone beds is that they are of storm origin.

Seilacher (1989) has observed that although many of the thinner sandstone beds show various characteristics of storm beds (grading of grain sizes and bedforms, oscillation-rippled tops), their soles lack the tool marks and other erosional features that are usually characteristic of Phanerozoic storm beds. Seilacher (1989) has also noted that when these sandstone beds directly overlie each other, soles mold the ripple marks of the underlying bed without any of the erosion typically found in storm beds. This variety of evidence thus led him to conclude that the sands were covered with extensive cyanobacterial mats that inhibited erosion. In the Ediacara Member, Seilacher (1989) also observed flat pebbles of coarse sand that were bent like a piece of leather during transport, and concluded that this is further evidence for the presence of extensive mats. The presence of these microbial mats very likely strongly influenced the preservation of these Ediacara fossils (Gehling 1986, 1999).

In Newfoundland, turbidites in the Mistaken Point Formation are relatively fine-grained (Landing et al. 1988; Jenkins 1992). In the Mistaken Point area (Figure 2.4), which is a well-known fossil site, these turbidites (thicknesses ranging from 10 to 80 cm) thicken and thin over a vertical distance of 6 to 8 m and were deposited in a deep-marine, most likely slope, environment (Jenkins 1992; Narbonne, Dalrymple, and Gehling 2001; Wood et al. 2001). The Trepassey Formation outcropping at Mistaken Point also contains deep-marine slope thin- to medium-bedded turbidites (Narbonne, Dalrymple, and Gehling 2001; Wood et al. 2001). Outcrops of the Fermeuse Formation have also been interpreted as deep marine in origin (Jenkins 1992).

TAPHONOMY

In the Ediacara Member (Figure 2.3), fossils of soft-bodied organisms are typically preserved in part-and-counterpart preservation as casts or molds on the soles of storm event beds with complimentary casts or molds formed on the top surfaces of underlying beds (Jenkins, Ford, and Gehling 1983; Gehling 1999) (Figure 2.6). Using trace fossil preservation terminology, fossils found on soles of sandstones as concave impressions are negative hyporeliefs, and casts in convex relief are positive hyporeliefs (Glaessner 1984; Seilacher 1989); preservation on the underlying sandstone beds thus occurs as epirelief counterpart casts and molds (Gehling 1999) (Figure 2.6).

Fossils that have been traditionally termed medusoids (Glaessner 1961, 1984) sometimes occur in an overlapping position or are pressed close together so they show apparent tears that may have occurred during physical battering (Wade 1972b; Jenkins, Ford, and Gehling 1983). Discoidal fossil forms are also preserved as mass kills, with some bedding planes covered with numerous individuals (Jenkins, Ford, and Gehling 1983). Frond-like fossils that have traditionally been identified possibly as sea pens are commonly torn free of their anchoring structures (Figure

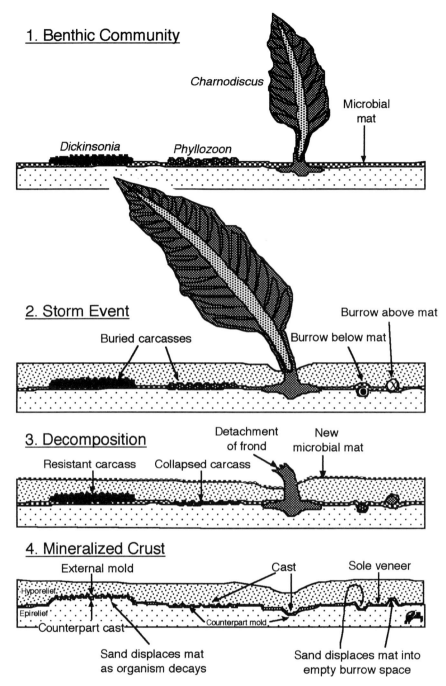

1. Benthic Community

Charnodiscus

Microbial mat

Dickinsonia

Phyllozoon

2. Storm Event

Buried carcasses

Burrow above mat

Burrow below mat

3. Decomposition

Detachment of frond

New microbial mat

Resistant carcass

Collapsed carcass

4. Mineralized Crust

External mold

Cast

Sole veneer

Hyporelief

Epirelief

Counterpart cast

Counterpart mold

Sand displaces mat as organism decays

Sand displaces mat into empty burrow space

FIGURE 2.6 A schematic model for Ediacara taphonomy: (1) Benthic community of Ediacara biota on a microbial mat; prostrate *Dickinsonia* and *Phylozoon;* upright frond, *Charniodiscus.* (2) Storm event buries Ediacara organisms and mat; frond survives with buried holdfast; infaunal burrows made above mat and below mat. (3) Decomposition of organisms begins with rapid collapse of *Phylozoon* and slow decay of *Dickinsonia;* storm detaches frond, but new mat prevents erosion. Bacterial reduction of iron in the sole veneer. (4) Mineralized crust forms as a death mask in the sole veneer; after complete decay of organic material, the underlying sand forms epirelief counterpart casts and molds. (Modified from Gehling 1999)

2.6), which in a few instances can be seen to correspond with the discoidal forms, and some show evidence of decomposition (Jenkins, Ford, and Gehling 1983). None of the fossils in the Ediacara Member show any evidence of predation (Glaessner 1979).

The presence of actual trace fossils of infaunal bilaterians in the Ediacara Member (Glaessner 1969; Narbonne 1998) indicates that surface sediments and the overlying seawater were well oxygenated. Glaessner (1984) has postulated that the bodies of medusoids, although composed of 96 to 98 percent water, are sufficiently tough that they would not break apart with compaction, with continued decomposition and compaction eventually forming the molds and casts that are now found.

In the Briscal, Mistaken Point, and Trepassey Formations (Figure 2.5), fossils are preserved on the upper surfaces of beds that are overlain by graded volcanic ash layers, which range from a few millimeters to 0.5 m thick (Jenkins 1992; Narbonne 1998). These ash layers are interpreted to have originated from large nearby phreatomagmatic explosions that caused steam-buoyed ash to move in an apron across the sea (Jenkins 1992). Ash crystals would have settled quickly to the seafloor, and since the density of unskeletonized marine organisms is about that of seawater, these crystals would have settled faster than any Ediacara organism living in the water column (Anderson 1978; Jenkins 1992). Thus, it is generally agreed that this preserved Ediacara assemblage represents organisms that lived on the seafloor (Jenkins 1992). The preservation of these organisms involved their being pushed down into the underlying sand bed by the weight of the overlying ash (Jenkins 1992). As the organisms decayed, ash then filled the mold as decomposition proceeded (Jenkins 1992). In some examples, composite molds were produced; in other cases, the crystals of the tuff are coarse enough to obscure morphological details, so that morphological aspects of the fossils are commonly difficult to distinguish (Jenkins 1992), particularly where they have also suffered tectonic deformation.

Of utmost importance to understanding the Ediacara biota is that soft-bodied animals, such as jellyfishes, are not preserved in any younger rocks the same way as they are at sites with Ediacara fossils (Seilacher 1984). Seilacher (1984:161) has thus posed the question "Why did the Ediacaran mode of preservation become 'extinct'?" Possibly, Ediacara organisms had a flexible cuticle that was not digestible by contemporaneous microorganisms (Seilacher 1984), as is that of modern soft-bodied organisms. In a similar vein, Norris (1989) has concluded from experimental studies that Ediacara organisms had a stiff cuticle that is not as easily torn or folded as proposed modern analogues. Taphonomic studies by Crimes, Insole, and Williams (1995) on Late Cambrian examples of the Ediacara biota also led them to conclude that members of the Ediacara biota were not truly soft-bodied and that they had a rigid outer wall. Even though burrows made by bilaterians are found in the Ediacara Member (Glaessner 1969), they apparently did not scavenge the remains of buried Ediacara organisms, as would happen in modern well-oxygenated environments. Similarly, in a study analyzing three-dimensional preservation of the Ediacara fossil *Ernietta* from classic fossil localities in Namibia, Dzik (1999) concluded that the presence of the Ediacara biota in the fossil record is in large part because decomposers, which could consume collagen, had not yet evolved. From this analysis, Dzik (1999) further postulated that most Ediacara fossils do not represent the complete anatomy of the original organisms, but typically only the preservable internal hydraulic skeletons of these animals.

Although many studies on taphonomy of the Ediacara biota have concentrated on proposed degradational properties of Ediacara soft tissues, several recent studies have focused on the effects of microbial mats on the preservation of these fossils (Gehling 1986, 1999; Narbonne and Dalrymple 1992; Narbonne, Dalrymple, and MacNaughton 1997). Evidence is beginning to develop that, just as for carbonate substrates, Neoproterozoic siliciclastic seafloors were typically covered with microbial mats (Pflüger and Sarkar 1996; Hagadorn and Bottjer 1997, 1999; Bottjer, Hagadorn, and Dornbos 2000). Gehling (1999) has proposed that when event beds covered siliciclastic seafloors on which Ediacara organisms lived, the smothered microbial mats inhibited the vertical movement of pore fluids, hence promoting rapid cementation of a sole veneer in the overlying sand (Figure 2.6). In this way, the microbial mats may have acted as "death masks" for buried Ediacara organisms (Gehling 1999).

Thus, we still do not completely understand all the causes that led to preservation of Ediacara organisms and the specific contributions of the degradational behavior of Ediacara soft tissues versus characteristics of the environment, particularly the effects of microbial mats. However, since it now appears that some Ediacara fossils occur in the Cambrian, future studies will certainly focus on how this taphonomic window closed in the Proterozoic–Phanerozoic transition (Hagadorn and Bottjer 1997; Seilacher 1997; Jensen, Gehling, and Drosser 1998; Hagadorn, Fedo, and Waggoner 2000).

PALEOBIOLOGY AND PALEOECOLOGY

Fossils of soft-bodied organisms from the Ediacara Member have traditionally been referred to as still-extant phyla (Glaessner 1961, 1983; Wade 1972a, 1972b; Jenkins 1992) (Figure 2.7). Using this approach, members of the Cnidaria are most commonly identified, including forms that were solitary medusoids or polypoids (e.g., *Cyclomedusa, Ediacara*), frond-like fossils that are considered to probably be sea pens (e.g., *Charniodiscus*), as well as hydrozoans (e.g., *Ovatoscutum*), cubozoans (e.g.,

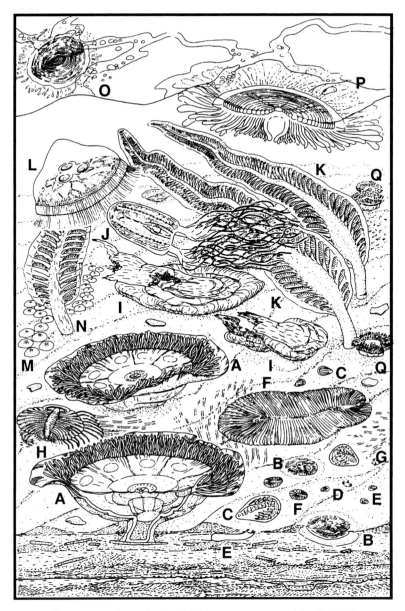

FIGURE 2.7 Reconstruction of subtidal biotas represented in the Ediacara assemblage of the Flinders Ranges, South Australia: (A) sectioned *Ediacaria flindersi* individual; (B) *Tribrachidium heraldicum;* (C) *Paravancorina minchami;* (D) *Praecambridium sigillium;* (E) burrowing organism with three sclerotized elements on anterior; (F) *Dickinsonia costata,* juveniles and mature examples; (G) scratch marks made by feeding epibenthos; (H) soft-bodied trilobite; (I) dead and torn specimens of *Cyclomedusa* partly exhumed from the substrate; (J) *Kimberella quadrata;* (K) *Charniodiscus longus;* (L) *Rugoconites enigmaticus;* (M) *Nemiana simplex;* (N) *Charniodiscus oppositus;* (O) *Ovatoscutum concentricum;* (P) *Eoporpita medusa;* and (Q) *Medusinites asteroides,* an anemone-like creature. Organisms are shown at varying scales to true natural size. (Modified from Jenkins 1992)

Kimberella), and scyphozoans (e.g., *Rugoconites*) (Figure 2.7). Other common fossils include those that are interpreted as polychaete worms (e.g., *Dickinsonia* [Figure 2.8], *Spriggina* [Figure 2.9]) and a representative (*Tribrachidium*) of a phylum, Tribrachidia, that is thought to occur only in the Ediacara biota and has been compared with the edrioasteroids (Glaessner 1984) (Figure 2.7).

Other than forms recognized from other sites, none of the members of the Ediacara fauna on the Avalon Peninsula in Newfoundland have been formally described (Jenkins 1992). However, initial assessment of the fauna indicates that it contains about 20 genera (Anderson 1978) with at least 30 species (Anderson and Conway Morris 1982). Jenkins (1992) has reconstructed some of these fossils as they may have existed in life (Figure 2.10). The following five main groups of fossils occur in decreasing order of approximate abundance (Jenkins 1992): (1) spindle-shaped forms (Figures 2.10–2.12) in the upper parts of the Mistaken Point Formation; (2) complexly branched forms (Figure 2.10), including the bush-like form, probably the Pectinate forms, and animals with numerous fronds joined by branching or zigzag connections, in the upper parts of the Mistaken Point Formation, the Trepassey Formation, and the lower Fermeuse Formation; (3) strongly frondose forms with stalks expanded basally or terminating in a disk (Figure 2.10), traditionally interpreted as sea pens, including *Charnia masoni* and about four

FIGURE 2.8 *Dickinsonia costata* from the Ediacara fauna of South Australia. Length of larger specimen is 13 cm. (Photo courtesy of B. Runnegar, University of California, Los Angeles)

FIGURE 2.9 *Spriggina flindersi* from the Ediacara fauna of South Australia. Length of specimen is 4 cm. (Photo courtesy of B. Runnegar, University of California, Los Angeles)

species that may be loosely grouped in *Charniodiscus* (Figures 2.11 and 2.12), from the upper part of the Mistaken Point Formation, the Trepassey Formation, and the lower Fermeuse Formation; (4) discoidal organisms (Figure 2.10), traditionally interpreted as medusoids, with either radial lobes or irregular lobes, found in the Briscal, Mistaken Point, and Trepassey Formations; and (5) discoidal organisms with strongly developed annulations, from the mid-Fermeuse Formation.

Seilacher (1984, 1989, 1992) revolutionized the study of Ediacara fossils by disputing the assignment of most of the Ediacara biota to still-extant phyla. He argued that the morphology of the Ediacara fossils interpreted as medusoids does not match that of Phanerozoic jellyfish fossils or of extant jellyfish. Bruton (1991) has studied preservation of modern jellyfish in the field and the laboratory, and believes his results show that casts and molds of modern jellyfish do not resemble the Ediacara fossils interpreted as medusoids. Similarly, Crimes, Insole, and Williams (1995) dispute the medusoid origin of certain Ediacara fossils. However, Norris (1989) studied preservation of modern jellyfish in the laboratory, compared his results with Ediacara fossils interpreted as medusoids, and concluded that one could not reject the contention that the fossils are in fact molds and casts of medusoids.

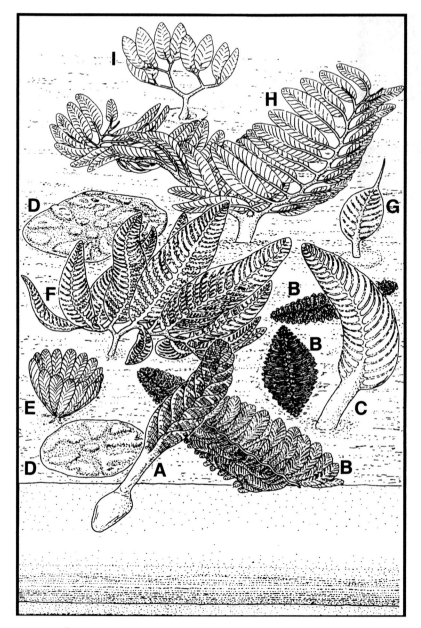

FIGURE 2.10 Reconstruction of Ediacara fauna from the upper part of the Mistaken Point Formation, Avalon Peninsula, Newfoundland: (A) *Charnia masoni;* (B) spindle-shaped fossils; (C) *Charniodiscus concentricus;* (D) lobate discoidal remains; (E) bush-like fossils; (F) branched frondose fossil; (G) *Charniodiscus* sp.; (H) pectinate organism; (I) dichotomously branched frondose organism. *Charnia* is approximately 18 cm long; all other organisms are shown at the same scale. (Modified from Jenkins 1992)

FIGURE 2.11 *Charniodiscus,* a frond-like fossil with a disk-shaped holdfast (*right*), with portion of a spindle-shaped fossil, from the Mistaken Point Formation, Avalon Peninsula, Newfoundland. *Charniodiscus* is 6.4 cm tall. (Photo courtesy of A. Seilacher, Tübingen University and Yale University)

Seilacher (1989) also questioned the assignment of the frondose forms to the pennatulaceans, or sea pens. He maintained that many of the supposed Vendian sea pens lack an axial stem (possessed by modern sea pens) and that all the fronds of the Ediacara sea pens are leaf-like structures without branch separation. Modern sea pens have branch separation so that currents can pass through, thus allowing polyps on the branches to feed (Seilacher 1989). Similarly, Seilacher (1989) contested the assignment of *Dickinsonia* to the polychaete worms.

As an alternative model for understanding the biology of many Ediacara soft-bodied organisms, Seilacher (1989, 1992) proposed that they shared a serially or fractally quilted pneu structure, similar to that of an air mattress, that allowed them to build a relatively rigid, broad flat organism. Because no real organs have been identified in Ediacara fossils, Seilacher postulated that these organisms may have operated their metabolic processes (nutrient uptake, respiration, excretion) through the body surface, in which case maximizing surface area through such a pneu structure would be advantageous (Seilacher 1989). Analysis of several specimens of *Dickinsonia* has led Seilacher (1989) to conclude that the quiltings were attached together by rigid internal struts rather than a continuous sheet, such as is found in an air mattress.

FIGURE 2.12 Four *Charniodiscus* and at least 12 spindle-shaped fossils from the Mistaken Point Formation, Avalon Peninsula, New-foundland. Slab surface is 70 cm across. (Photo courtesy of A. Seilacher, Tübingen University and Yale University)

Animals that employ this quilted structure were termed the Vendobionta by Seilacher (1992), and they represent a phylogenetically distinct animal group of kingdom or phylum rank (Seilacher 1992; Buss and Seilacher 1994). In this interpretation, Seilacher (1989, 1992) acknowledged that other metazoans coexisted with the Vendobionta and that they have left a fossil record in the form of trace fossils. Seilacher's (1984) earlier work on the Ediacara biota suggested that their large surface areas would be advantageous for organisms that were photosymbiotic. However, the presence of Ediacara fossils in strata deposited below the photic zone (e.g., Mackenzie Mountains of northwestern Canada) indicates that at least some Ediacara organisms did not need sunlight to live, and hence were probably not photosymbiotic (Seilacher 1992).

McMenamin (1986) and McMenamin and McMenamin (1990) proposed that members of the Ediacara biota may have been well adapted for chemosymbiosis, using as an analogy the chemosymbiotic organisms that live at modern hydrothermal vents and cold seeps. McMenamin and McMenamin (1990) acknowledge that there is no evidence for hydrothermal activity in deeper-water strata in which the Ediacara biota is found, nor is there evidence for fossil cold seeps. McMenamin and McMenamin (1990), however, do suggest that these organisms may have been trapping a diffuse methane or hydrogen sulfide flow from the seafloor to fuel their chemosymbiotic activity. Seilacher (1989) also proposed that members of the Ediacara biota may have been chemosymbiotic and that the microbial mats covering Proterozoic seafloors produced a sharp boundary between oxygenated bottom-water and reducing pore waters, allowing broad, flat Ediacara organisms to adsorb hydrogen sulfide or methane through their bottom surface and oxygen through the exposed upper surface. In a study of the Mackenzie Mountains Ediacara fossil locality from northwestern Canada, Narbonne and Dalrymple (1992) interpreted the depositional environment as deep-sea, and found that the Ediacara fossils were associated with pyritic intervals that also show evidence of microbial mats. Narbonne and Dalrymple (1992) suggested that the microbial mats aided preservation of these fossils (Narbonne, Dalrymple, and MacNaughton 1997), much as has been proposed for shallow-water Ediacara fossils by Gehling (1986, 1999). Narbonne and Dalrymple (1992) also postulated that these particular Ediacaran organisms lived under exaerobic conditions (Savrda and Bottjer 1987), which potentially supply additional support for the chemosymbiosis hypothesis.

Some members of the Ediacara biota attained relatively large sizes— up to 1 m long for some specimens of *Dickinsonia*. Although these large organisms would represent a ready food source for predators, no evidence of predation on them or any other Ediacara organisms has been found. Thus, McMenamin (1986) has termed this "The Garden of Ediacara" because it was a world with no apparent predators of Ediacara organisms.

Until recently, it was believed that the Ediacara biota became extinct before the beginning of the Cambrian. This disappearance of the Ediacara biota before the Cambrian was thus interpreted as a mass extinction due to the evolution of the first predators, which presumably were non-Ediacara metazoans (McMenamin 1986; Seilacher 1989; McMenamin and McMenamin 1990). The discovery of members of the Ediacara biota through the late Proterozoic and into the Cambrian (Conway Morris 1993; Crimes, Insole, and Williams 1995; Grotzinger et al. 1995; Jensen, Gehling, and Drosser 1998; Hagadorn, Fedo, and Waggoner 2000), however, indicates that the history of the Ediacara biota cannot be easily separated from that of other Phanerozoic organisms.

A diverse and stimulating array of data and ideas continues to be generated on the nature of these fossils. For example, based on a comparative taphonomic analysis, Retallack (1994) proposed that the Ediacara biota may actually represent fossil lichens, an idea that has generated much discussion (Retallack 1995; Waggoner 1995). Taking a different view, Steiner and Reitner (2001) have postulated that some Ediacara organisms were procaryotic colonies or symbiotic organisms involving procaryotes. In contrast, however, many students of this biota continue to support conclusions that although a number of Ediacara fossils certainly appear to be strange, they are genuinely ancestors of metazoan groups we know today. Thus, Valentine (1992) has interpreted *Dickinsonia* as a benthic polypoid that is very likely of cnidarian affinity, and Fedonkin and Waggoner (1996) have postulated that *Kimberella,* originally described as a cubozoan medusa, actually is the fossil of a benthic mollusc-like organism. Similarly, Buss and Seilacher (1994) have suggested that the Vendobionta are a monophyletic sister group to the Eumetazoa and were cnidarian-like organisms that lacked cnidae. Dzik's (1999) hypothesis that the complete anatomy of Ediacara organisms may not be preserved as fossils, as well as the interpretation by Grazhdankin and Seilacher (2002) that some members of the Ediacara biota were infaunal, may also have important implications for ultimately understanding the biological affinities of this fauna.

As we learn more about the phylogenetic affinities and functional morphology of the Ediacara fossils, we will be able to reconstruct the paleoecology of these organisms in more detail. A beginning has already been made by Seilacher (1997; Seilacher and Pflüger 1994), who has postulated that benthic Ediacara organisms were specifically adapted for a microbial mat–covered seafloor. Thus, various Ediacara organisms (such as the mollusc *Kimberella*) fed on the mat surface and hence were *mat scratchers,* while others (such as *Tribrachidium*) were firmly attached to the mat surface and were *mat encrusters.* Narbonne (1998) has proposed that the Ediacara fauna represents the initiation of complex ecological tiering, with three feeding levels: an elevated level in the water column occupied by fronds; a seafloor level with a variety of organisms, including mat encrusters and scratchers; and a subsurface level represented by trace fossils made by bilaterians (Ausich and Bottjer 2001).

CONCLUSIONS

The Ediacara biota represents more than just another example of exceptional fossil preservation. Younger marine Lagerstätten are usually in fine-grained sediments and are recognized because they contain (1) soft-bodied members of a fauna that also contain taxa with mineralized skeletons, (2) preservation of soft tissues of organisms that also have mineralized skeletons, or (3) articulated skeletons of organisms that are usually preserved disarticulated. Other than evidence on sponges (Brasier, Green, and Shields 1997; Li, Chen, and Hua 1998) and bilaterian trace fossils, as well as initial results on cnidarians (Chen et al. 2000; Xiao, Yuan, and Knoll 2000) and bilaterian embryos and larvae (Chen et al. 2000), the Ediacara biota represents most of the currently known evidence of metazoan life that existed during the Neoproterozoic. Thus, it does not provide supplemental information, but much of the information on possible metazoan life at this time. Because we are only just beginning to understand how to investigate Lagerstätten, as compared with the typical Phanerozoic assemblage with mineralized skeletons, study of the meaning of the Ediacara biota is doubly difficult.

ACKNOWLEDGMENTS

I would like to thank A. Seilacher and R. J. F. Jenkins for their thorough review of this chapter.

REFERENCES

Anderson, M. M. 1978. Ediacaran fauna. In *McGraw-Hill Yearbook of Science and Technology,* pp. 146–149. New York: McGraw-Hill.

Anderson, M. M., and S. Conway Morris. 1982. A review, with descriptions of four unusual forms, of the soft-bodied fauna of the Conception and St. John's Groups (late Precambrian), Avalon Peninsula, Newfoundland. In B. Mamet and M. J. Copeland, eds., *Proceedings: Third North American Paleontological Convention,* vol. 1, pp. 1–8. Toronto: Business and Economic Service.

Ausich, W. I., and D. J. Bottjer. 2001. Sessile invertebrates. In D. E. G. Briggs and P. R. Crowther, eds., *Palaeobiology II,* pp. 388–391. Oxford: Blackwell Science.

Benus, A. P. 1988. Sedimentological context of a deep-water Ediacaran fauna (Mistaken Point Formation, Avalon Zone, eastern Newfoundland). *Bulletin of the New York State Museum* 463:8–9.

Bottjer, D. J., J. W. Hagadorn, and S. Q. Dornbos. 2000. The Cambrian substrate revolution. *GSA Today* 10:1–7.

Brasier, M., O. Green, and G. Shields. 1997. Ediacarian sponge spicule clusters from southwestern Mongolia and the origins of the Cambrian fauna. *Geology* 25:303–306.

Bruton, D. L. 1991. Beach and laboratory experiments with the jellyfish *Aurelia* and remarks on some fossil "medusoid" traces. In A. M. Simonetta and S. Conway Morris, eds., *The Early Evolution of Metazoa and the Significance of Problematic Taxa*, pp. 125–129. Cambridge: Cambridge University Press.

Buss, L. W., and A. Seilacher. 1994. The phylum Vendobionta: A sister group of the Eumetazoa? *Paleobiology* 20:1–4.

Chen, J. Y., P. Oliveri, C. W. Li, G. Q. Zhou, F. Gao, J. W. Hagadorn, K. J. Peterson, and E. J. Davidson. 2000. Precambrian animal diversity: New evidence from high resolution phosphatized embryos. *Proceedings of the National Academy of Sciences* 97:4457–4462.

Conway Morris, S. 1990. Late Precambrian–Early Cambrian metazoan diversification. In D. E. G. Briggs and P. R. Crowther, eds., *Palaeobiology: A Synthesis*, pp. 30–36. Oxford: Blackwell Science.

Conway Morris, S. 1993. Ediacaran-like fossils in Cambrian Burgess Shale–type faunas of North America. *Palaeontology* 36:593–635.

Crimes, T. P., A. Insole, and B. P. J. Williams. 1995. A rigid-bodied Ediacaran biota from Upper Cambrian strata in Co. Wexford, Eire. *Geological Journal* 30:89–109.

Dzik, J. 1999. Organic membranous skeleton of the Precambrian metazoans from Namibia. *Geology* 27:519–522.

Fedonkin, M. A. 1992. Vendian faunas and the early evolution of metazoa. In J. H. Lipps and P. W. Signor, eds., *Origin and Early Evolution of the Metazoa*, pp. 87–129. New York: Plenum.

Fedonkin, M. A., and B. M. Waggoner. 1996. The later Precambrian fossil *Kimberella* is a mollusc-like bilaterian organism. *Nature* 388: 868–871.

Gehling, J. G. 1982. The sedimentology and stratigraphy of the late Precambrian Pound Subgroup, Central Flinders Ranges, South Australia. Master's thesis, University of Adelaide.

Gehling, J. G. 1983. The Ediacara Member: A shallowing-upward submarine fan sequence, within the Pound Subgroup. *Geological Society of Australia Abstracts* 10:52–54.

Gehling, J. G. 1986. Algal binding of siliciclastic sediments: A mechanism in the preservation of Ediacaran fossils. *Twelfth International Sedimentology Congress, Abstracts*, p. 117.

Gehling, J. G. 1999. Microbial mats in terminal Proterozoic siliciclastics: Ediacaran death masks. *Palaios* 14:40–57.

Glaessner, M. F. 1961. Pre-Cambrian animals. *Scientific American* 204:72–78.

Glaessner, M. F. 1969. Trace fossils from the Precambrian and basal Cambrian. *Lethaia* 2:369–393.

Glaessner, M. F. 1979. Precambrian. In R. A. Robison and C. Teichert, eds., Introduction: Fossilization (taphonomy), biogeography and biostratigraphy. *Treatise on Invertebrate Paleontology, Part A*, pp. A79–A118. Lawrence: Geological Society of America and University Press of Kansas.

Glaessner, M. 1983. The emergence of metazoa in the early history of life. *Precambrian Research* 290:427–441.

Glaessner, M. 1984. *The Dawn of Animal Life: A Biohistorical Study*. Cambridge: Cambridge University Press.

Glaessner, M. F., and M. Wade. 1966. The late Precambrian fossils from Ediacara, South Australia. *Palaeontology* 9:599–628.

Goldring, R., and C. N. Curnow. 1967. The stratigraphy and facies of the late Precambrian at Ediacara, South Australia. *Journal of the Geological Society of Australia* 14:195–214.

Grazhdankin, D., and A. Seilacher. 2002. Underground Vendobionta from Namibia. *Paleontology* 45:57–78.

Grotzinger, J. P., S. A. Bowring, B. Z. Saylor, and A. J. Kaufman. 1995. Biostratigraphic and geochronologic constraints on early animal evolution. *Science* 270:598–604.

Gürich, G. 1930. Die bislang ältesten Spuren von Organismen in Südafrika. *International Geological Congress, Comptes Rendus* 15:670–680.

Hagadorn, J. W., and D. J. Bottjer. 1997. Wrinkle structures: Microbially mediated sedimentary structures common in subtidal siliciclastic settings at the Proterozoic–Phanerozoic transition. *Geology* 25:1047–1050.

Hagadorn, J. W., and D. J. Bottjer. 1999. Restriction of a late Neoproterozoic biotope: Suspect-microbial structures and trace fossils at the Vendian–Cambrian transition. *Palaios* 14:73–85.

Hagadorn, J. W., C. M. Fedo, and B. M. Waggoner. 2000. Early Cambrian Ediacaran-type fossils from California. *Journal of Paleontology* 74:731–740.

Hill, E., and T. G. Bonney. 1877. The Precarboniferous rocks of Charnwood Forest. *Quarterly Journal of the Geological Society of London* 33:754–789.

Jenkins, R. J. F. 1992. Functional and ecological aspects of Ediacaran assemblages. In J. H. Lipps and P. W. Signor, eds., *Origin and Early Evolution of the Metazoa*, pp. 131–176. New York: Plenum.

Jenkins, R. J. F., C. H. Ford, and J. G. Gehling. 1983. The Ediacara Member of the Rawnsley Quartzite: The context of the Ediacara assemblage (late Precambrian, Flinders Ranges). *Journal of the Geological Society of Australia* 30:101–119.

Jenkins, R. J. F., and J. G. Gehling. 1978. A review of the frond-like fossils of the Ediacara assemblages. *Records of the South Australian Museum* 17:347–359.

Jensen, S., J. G. Gehling, and M. L. Droser. 1998. Ediacara-type fossils in Cambrian sediments. *Nature* 393:567–569.

Landing, E., G. M. Narbonne, P. Myrow, A. P. Benus, and M. M. Anderson. 1988. Faunas and depositional environments of the upper Precambrian through Lower Cambrian, southeastern Newfoundland. *Bulletin of the New York State Museum* 463:18–52.

Li, C. W., J. Y. Chen, and T. E. Hua. 1998. Precambrian sponges with cellular structures. *Science* 279:879–882.

McMenamin, M. A. S. 1986. The garden of Ediacara. *Palaios* 1:178–182.

McMenamin, M. A. S., and D. L. S. McMenamin. 1990. *The Emergence of Animals: The Cambrian Breakthrough*. New York: Columbia University Press.

Mount, J. F. 1989. Re-evaluation of unconformities separating the "Ediacaran" and Cambrian systems, South Australia. *Palaios* 4:366–373.

Mount, J. F. 1991. Re-evaluation of unconformities separating the "Ediacaran" and Cambrian systems, South Australia: Reply. *Palaios* 6:105–108.

Narbonne, G. M. 1998. The Ediacara biota: A terminal Neoproterozoic experiment in the evolution of life. *GSA Today* 8:1–6.

Narbonne, G. M., and J. D. Aitken. 1990. Ediacaran fossils from the Sekwi Brook area, Mackenzie Mountains, northwestern Canada. *Palaeontology* 33:945–980.

Narbonne, G. M., and R. W. Dalrymple. 1992. Taphonomy and ecology of deep-water Ediacaran organisms from northwestern Canada. In S. Lidgard and P. R. Crane, eds., *Fifth North American Paleontological Convention: Abstracts and Program,* p. 219. Paleontological Society Special Publication, no. 6. Knoxville: Department of Geological Sciences, University of Tennessee.

Narbonne, G. M., R. W. Dalrymple, and J. G. Gehling. 2001. *Trip B5: Neoproterozoic Fossils and Environments of the Avalon Peninsula, Newfoundland.* Geological Association of Canada–Mineralogical Association of Canada Field Trip Guidebook. St. John's, Newfoundland: Geological Association of Canada.

Narbonne, G. M., R. W. Dalrymple, and R. B. MacNaughton. 1997. Deep-water microbialites and Ediacara-type fossils from northwestern Canada. *Geological Society of America Abstracts with Programs* 29:A-193.

Nedin, C., and R. J. F. Jenkins. 1991. Re-evaluation of unconformities separating the "Ediacaran" and Cambrian systems, South Australia: Comment. *Palaios* 6:102–105.

Norris, R. D. 1989. Cnidarian taphonomy and affinities of the Ediacaran biota. *Lethaia* 22:381–393.

Pflüger, F., and S. Sarkar. 1996. Precambrian bedding planes—Bound to remain. *Geological Society of America Abstracts with Programs* 28:491.

Retallack, G. J. 1994. Were the Ediacaran fossils lichens? *Paleobiology* 20:523–544.

Retallack, G. J. 1995. Ediacaran lichens—A reply to Waggoner. *Paleobiology* 21:398–399.

Runnegar, B. 1992. Evolution of the earliest animals. In J. W. Schopf, ed., *Major Events in the History of Life,* pp. 65–93. Boston: Jones and Bartlett.

Savrda, C. E., and D. J. Bottjer. 1987. The exaerobic zone: A new oxygen-deficient marine biofacies. *Nature* 327:54–56.

Seilacher, A. 1984. Late Precambrian and Early Cambrian metazoa: Preservation or real extinctions? In H. D. Holland and A. F. Trendal, eds., *Patterns of Change in Earth Evolution,* pp. 159–168. Heidelberg: Springer-Verlag.

Seilacher, A. 1989. Vendozoa: Organismic construction in the Proterozoic biosphere. *Lethaia* 22:229–239.

Seilacher, A. 1992. Vendobionta and Psammocorallia: Lost constructions of Precambrian evolution. *Journal of the Geological Society* 149:607–613.

Seilacher, A. 1997. Precambrian life styles related to biomats. *Geological Society of America Abstracts with Programs* 29:A-193.

Seilacher, A., and F. Pflüger. 1994. From biomats to benthic agriculture: A biohistoric revolution. In W. E. Krumbein et al., eds., *Biostabilization of Sediments,* pp. 97–105. Oldenburg: Universität Oldenburg, Bibliotheksund Informationssystem.

Sprigg, R. C. 1947. Early Cambrian (?) jellyfishes from the Flinders Ranges, South Australia. *Transactions of the Royal Society of South Australia* 71:212–224.

Sprigg, R. C. 1949. Early Cambrian "jellyfishes" of Ediacara, South Australia and Mount John, Kimberley District, Western Australia. *Transactions of the Royal Society of South Australia* 73:72–99.

Steiner, M., and J. Reitner. 2001. Evidence of organic structures in Ediacara-type fossils and associated microbial mats. *Geology* 29:1119–1122.

Valentine, J. W. 1992. *Dickinsonia* as a polypoid organism. *Paleobiology* 18:378–382.

Wade, M. 1972a. *Dickinsonia:* Polychaete worms from the late Precambrian Ediacara fauna, South Australia. *Memoirs of the Queensland Museum* 16:171–190.

Wade, M. 1972b. Hydrozoa and scyphozoa and other medusoids from the Precambrian Ediacara fauna, South Australia. *Palaeontology* 15:197–225.

Waggoner, B. M. 1995. Ediacaran lichens: A critique. *Paleobiology* 21:393–397.

Waggoner, B. M. 1999. Biogeographic analyses of the Ediacara biota: A conflict with paleotectonic reconstructions. *Paleobiology* 25:440–458.

Wood, D. A., R. W. Dalrymple, G. M. Narbonne, M. E. Clapham, and J. G. Gehling. 2001. Sedimentology and taphonomy of Ediacaran fossils at Mistaken Point, southeastern Newfoundland. *Geological Association of Canada–Mineralogical Association of Canada Joint Annual Meeting, Abstracts* 26:165.

Xiao, S., X. Yuan, and A. H. Knoll. 2000. Eumetazoan fossils in terminal Proterozoic phosphorites? *Proceedings of the National Academy of Sciences* 97:13684–13689.

3
Chengjiang:
Early Record of the Cambrian Explosion

James W. Hagadorn

When I found the first fossil . . . I knew right away that it was an arthro-pod with paired appendages, extending forward, as if it was swimming on the moistened surface of a mudstone. But I realized that you could see the impression of the soft body parts. That night I put the fossils under my bed. But because I was so excited, I couldn't sleep very well. I got up often and pulled out the fossils just to look at them.

THIS IS HOW CHINESE PALEONTOLOGIST HOU XIANGUANG (personal communication, 2000) describes his initial reaction to the discovery of the Chengjiang Lagerstätte. In 1984, Hou was working on bradoriid-rich deposits at Maotian Hill, near the town of Chengjiang in the Yunnan Province of China (Figure 3.1). He split open a rock on the west face of the mountain to reveal an unusual arthropod: the soft-bodied trilobite *Naraoia* (Gore 1993; Monastersky 1993). Al-though a few obscure soft-bodied fossils from the Chengjiang deposit had been published (Mansuy 1912; Pan 1957), Hou was the first to rec-ognize both Chengjiang's status as a Lagerstätte and its link to other soft-bodied deposits like the Burgess Shale. Considering the impact this de-posit has had on the paleontological community, his enthusiastic reac-tion was more than justified.

The Chengjiang deposit is a conservation Lagerstätte that contains a variety of soft-bodied and biomineralized metazoans representing one of the earliest records of the Cambrian explosion in metazoan diversity. The Chengjiang biota is quite diverse, including algae, acritarchs, sponges, chancellorids, anemones, ctenophores, hyoliths, inarticulate

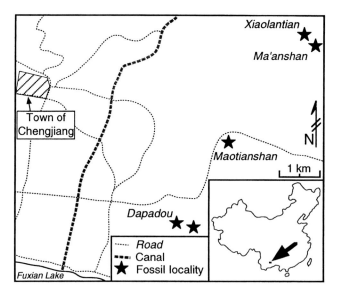

FIGURE 3.1 Chengjiang fossil localities. (Modified from Chen and Erdtmann 1991)

and linguloid brachiopods, paleoscolecids, priapulids, echinoderms, trilobites, primitive chordates including fish, trace fossils, several types of armored lobopods, a varied assemblage of arthropods, and a variety of taxonomically enigmatic forms. The abundance of specimens and prolific advance of recent fossil discoveries in this deposit stems partly from the spectacular soft-tissue preservation in the deposit, and partly from the pervasive labor-intensive collecting efforts that have been focused on outcrops in the Chengjiang area. Before the recent burst of activity on the Chengjiang Lagerstätte, the earliest known metazoan fauna with such great diversity and strong resemblance to extant phyla was the Burgess Shale fauna of the Canadian Rockies, which came to prominence 75 years earlier (Chapter 4).

In addition to expanding the temporal and geographic distribution of Burgess Shale–type faunas, the Chengjiang deposit is notable because it contains a variety of taxa that have radically affected our understanding of the deep history of many fossil groups (e.g., the agnathan-like *Haikouichthys*), as well as articulated forms, which elsewhere were previously known from only fragmentary remains (e.g., the net-like plates of *Microdictyon*). The presence of possible stem-group fossils in the deposit has also provided the opportunity for further speculation about early metazoan phylogenies and the origin of body plans, by providing a morphological groundtruthing for molecular-based phylogenies. In addition to its importance for individual clades, the entire Chengjiang community is significant, simply because it has greatly reduced estimates of the time during which metazoans diversified and colonized marine habitats shortly after the onset of the Cambrian. Before its discovery and the recalibration of the Cambrian timescale (Bowring et al. 1993; Landing et al. 1998), paleontologists' main view of Cambrian diversity was dominated by the Burgess Shale, which reflects a snapshot of Cambrian life nearly 10 Ma later.

GEOLOGICAL CONTEXT

At the time of this writing, the Lower Cambrian stratigraphic nomenclature for this region is in a state of flux. For the purposes of this chapter, the framework established by Luo et al. (1984) and modified by Chen et al. (1996) and Chen and Zhou (1997) is used.

The Chengjiang soft-bodied fossils occur primarily within the Maotianshan Shale, which is one of four proposed units within the Yuanshan Member of the Qiongzhusi Formation (Lu 1941; Ho 1942; Luo et al. 1984; Chen et al. 1996; Chen and Zhou 1997) (Figure 3.2). The Qiongzhusi Formation is underlain by the lowest Lower Cambrian Meishucun Formation, which is dominated by a 15 to 20 m thick condensed sequence of phosphatic hardgrounds, dolomitic and stromatolitic phosphorites, and metabentonites (Zhang et al. 1997; Zhu 1997). The Meischucun is capped by a glauconite-bearing phosphatic conglomerate, the Dahai Member, and is thought to record a shallowing-upward sequence cut by a hiatus of unknown duration. Unconformably overlying the Meishucun Formation is the 100 to 150 m thick Shiyantou Member of the Qiongzhusi Formation, which consists of four units: a thin phosphatic conglomerate overlain by a thicker clay member, a black silty shale member, and a siltstone member (Chen et al. 1996; Chen and Zhou 1997; Zhu 1997). The base of the Shiyantou records a rapid deepening event, followed by a shoaling sequence from dysaerobic black shale facies into more aerobic dark gray siltstones intercalated with gray sharp-based dolomitic silty tempestites, including well-developed slump and loading structures (Zhu 1997). Like the Shiyantou Member, the Yuanshan Member is 100 to 150 m thick and composed of four units representing a shoaling-upward sequence. Basal units are dominated by black siltstones and concretion-laden black shales, overlain by thinbedded clays and shales of the fossiliferous Maotianshan Shale unit, and interbedded mudstones and dolomitic siltstones of the upper unit. The 50 m thick Maotianshan Shale unit has numerous storm-dominated features, such as graded bedding, wave ripples, tool marks, and flute casts, suggesting formation through deposition of distal mud-tempestites. The Yuanshan Member grades upward into sandstones, siltstones, and mudstones of the Guansan Member of the Canglangpu Formation, which represents the shallower wave-dominated subtidal–intertidal component of the upward-shoaling sequence recorded in the Yuanshan.

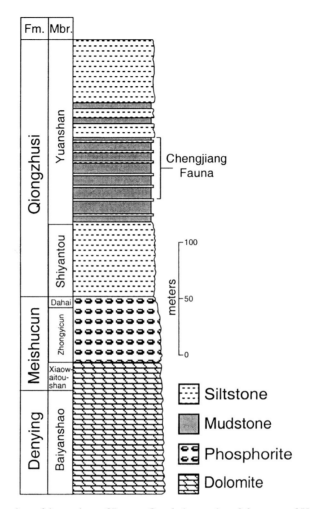

FIGURE 3.2 Stratigraphic section of Lower Cambrian units of the central Yunnan region. (Modified from Chen and Erdtmann 1991)

Although soft-bodied fossils extend as far upward as the Canglangpu Formation, the lowest and most abundant occurrences are found in the Maotianshan Shale, which corresponds to the *Eoredlichia* trilobite biozone and is underlain by the earliest trilobites (*Abadiella* zone) and bradoriids. These are, in turn, underlain by three small shelly fossil zones, the uppermost of which is indicative of a late Atdabanian age (Qian 1977; Zhou and Yuan 1982; Zhang and Hou 1985; Hou 1987a, 1987d; Chen, Hou, and Lu 1989c; Conway Morris 1989; Qian and Bengtson 1989). Although reliable chronostratigraphic age estimates are not available for the Chengjiang section, carbon isotope stratigraphy and acritarch biostratigraphy corroborate trilobite biostratigraphic data that suggest a post-Tommotian age for the deposit (Zang 1992; Zhang et al. 1997).

The Chengjiang fauna has been documented from a number of localities in southern China, most notably at localities near the town of Chengjiang, located about 50 km southeast of Kunming, in Yunnan Province (Figure 3.1). In addition to the original and most well documented locality at Maotianshan, a number of fossiliferous localities have been documented from exposures of the Yuanshan Member in east-central Yunnan (e.g., Dapotou, Fengkoushao, Haikou, Meishucun, and Xiaolantian) (Hou and Sun 1988; Hou and Bergström 1997; Luo et al. 1997). Fossils occur in at least 10 horizons within this unit, and because the mudstones are very gently dipping, outcrops spanning the fossiliferous interval are exposed laterally for as much as 100 km (Chen and Erdtmann 1991; Hou and Bergström 1991).

PALEOENVIRONMENTAL SETTING

Fossiliferous units are thought to have been deposited in a detrital belt located along the margin of the Yangtze Platform (Hou and Sun 1988; Chen and Erdtmann 1991). Before the terminal Neoproterozoic, the Yangtze Platform was part of a landmass that was likely separated from the two other landmasses (North China–Tarim and South Tibet) that later formed China (Jiang 1992). During the Cambrian, widespread evidence suggests, the Yangtze Platform was a shallow tropical sea bordered by relatively high-relief landmasses on three sides that was open (and deepened) to the east. Chengjiang was likely located in a bay that was in this seaway (Hou 1987a). During terminal Neoproterozoic–Early Cambrian time, this basin experienced several shallowing events, likely resulting from larger-scale eustatic shifts as well as local tectonism within the basin (Luo et al. 1984; Liang, Fang, and van de Voo 1990; Zhang et al. 1997; Zhu 1997).

Soft-bodied fossils of the Maotianshan unit are preserved in finely laminated 1 to 3 mm thick graded mudstones that have sharp-bottomed laminated silty bases overlain by nonlaminated clays. These beds are intercalated with 1 to 50 cm thick fine-grained nonfossiliferous sandstones. The presence of flute marks, storm wave ripples, and rare hummocky cross-stratification indicates rapid deposition of beds below or near storm wave base (Shu, Geyer, et al. 1995; Zhu 1997). Grading within mudstone and siltstone layers suggests that fossils were likely entrained in and/or buried by a series of microturbidites deposited in a relatively quiescent setting; fine-grained sandstones likely reflect the influence of sporadic strong storm events.

Based on regional paleogeography, the Maotianshan Shale was likely deposited around 60 to 70 km east of the paleoshoreline of the Yangtze seaway and about 600 to 700 km west of the continental slope (Zhang 1987). Flute and groove casts support an easterly current direction for microturbidite layers in the Maotianshan, and lithofacies stacking patterns suggest that the entire depositional sequence bounding the Chengjiang fauna reflects eastward progradation of a delta, with the fossiliferous Maotianshan Shale reflecting episodic turbid transport of distal marine muds at the foot of the delta front (Chen and Lindström 1991; Lindström 1995; Chen and Zhou 1997). Detailed paleoenvironmental analyses of the Chengjiang are still under way, with a recent study (Babcock, Zhang, and Leslie 2001) suggesting the operation of tidal processes in this environment.

TAPHONOMY

Fossils are typically preserved in beds consisting of sharp-based, 1 to 2 mm thick graded mud layers overlain by a thin nonlaminated claystone. The lower mudstone appears black to grayish black, and the middle to upper portions of these beds are typically weathered grayish yellow to green. Soft-bodied fossils are preserved as aluminosilicate films (Zhu 1997). Biomineralized skeletons are similarly preserved, but in a decalcified state. In organic-walled fossils, proteins have been replaced by hematite and iron-rich clay minerals through early diagenetic alteration (Jin, Wang, and Wang 1991). However, in a study of bradoriid mineralogy, Leslie et al. (1996) noted that nonmineralized arthropod cuticle contained higher concentrations of phosphorus than the surrounding rock matrix, suggesting that preservation of soft parts may have been mediated by early precipitation of phosphatic minerals. Many of the fossils at Chengjiang localities are severely weathered, appearing as reddish, ferric oxide–stained films on a yellow-weathered matrix. This reddish color is thought to result from oxidation of finely dispersed framboidal pyrite on bedding and fossil surfaces. This pyrite may have been produced by bacterial activities during early fossil diagenesis (Chen and Erdtmann 1991).

Fossils are moderately compressed, as indicated by wrinkling of convex and cylindrical shapes (Hou 1987b), and the appendages of arthropods are commonly visible through the body or carapace. Such appendages are expressed as shallow furrows and reflect decay and collapse of the underlying appendage and compression of the overlying exoskeleton into the vacant space (Chen, Zhou, and Ramsköld 1995a; Hou and Bergström 1997). Organisms characterized by exoskeleton-like carcasses are typically preserved roughly parallel to the bedding, with their appendages and related features penetrating several laminae in both upward and downward directions (Hou, Ramsköld, and Bergström 1991). Thus, rocks tend to cleave along the exoskeletal dorsal surface or along the upper margin of the valve (e.g., bradoriids, trilobites), and although some of the appendages are visible on split slabs, more commonly they are buried beneath sediment laminae.

In general, fossils appear to be exquisitely preserved because they have undergone minimal transport and sustained minimal postburial taphonomic overprinting (Hou, Chen, and Lu 1989). For example, even in transported specimens, delicate features such as exopod setae are preserved in three dimensions (Hou and Bergström 1997). Some forms may have been buried *in situ,* including lingulid brachiopods preserved with their pedicle traces extending obliquely downward into underlying layers (Jin, Hou, and Wang 1993) or infaunal priapulid worms, such as *Maotianshania,* buried in their burrows (Sun and Hou 1987b).

Although the majority of the fossils were likely transported, many were probably buried while still alive or perhaps very shortly after death. Examples of live transport followed by burial include epibiotic forms preserved in attached position, such as the onychophoran-like *Microdictyon,* which was apparently buried while still attached to its shelled host *Eldonia* (Chen, Zhou, and Ramsköld 1995a, 1995b). Although the mere occurrence of soft-tissue preservation is often used as evidence to support minimal transport, Chengjiang includes several examples that are particularly convincing, including the probable cnidarian *Cambrorhytium* preserved still attached to a lingulid shell in life position (Chen et al. 1996). Evidence for catastrophic burial also includes the presence of sediment in the pharynx of mobile vertebrates found in the deposit (Shu, Luo, et al. 1999), as well as indirect evidence such as the overall rarity of mobile swimmers in these deposits (but see Vannier and Chen 2001), suggesting that mobile animals may have been able to avoid benthic sediment flows. Other forms, such as the predator *Anomalocaris,* seem to have undergone significant decay before burial (Hou and Bergström 1997).

Death of preserved organisms is interpreted to have occurred as a result of asphyxia, although it is unclear whether asphyxia is related to suffocation by repeated benthic sediment flows or by incursion of oxygen-depleted waters. Evidence for entrainment of faunas by abundant microturbidite flows has been mentioned. Hou and Bergström (1991) suggest that shelfward movement of the oceanic oxygen minimum zone or periodic upwelling of oxygen-deficient waters (mediated by transgressive conditions) over the outer portions of the central Yunnan continental shelf caused frequent anoxic poisoning of the fauna. Absence of escape burrows in the deposit lends support to this anoxia-poisoning hypothesis. In either case, evidence for asphyxia includes everted worm proboscises, coiling in soft-bodied worms, and extended mantle setae in inarticulate brachiopods (Sun and Hou 1987b; Hou and Sun 1988; Jin and Wang 1992). Other specimens exhibit features indicative of metabolic stasis, such as an extruded esophagus (Chen and Erdtmann 1991).

Relevant to the oxygenation debate is a detailed taphonomic study by Babcock and Chang (1997), in which they examined disarticulation patterns of the relatively common soft-bodied arthropod *Naraoia* and compared them with decay of a modern limulid. Like most arthropods in the deposit, naraoiid skeletons became pliable several hours after death, as evidenced by wrinkling patterns within dorsal and posterior shields. More important, however, is evidence suggesting that in oxygenated water, postmortem naraoiid appendages could remain articulated for weeks, and their two exoskeletal shields could remain articulated for months. In anoxic water, such features might even remain articulated for years. Together, this evidence suggests that *Naraoia,* and perhaps many of the other components of this arthropod-dominated Lagerstätte, could reflect dead skeletons swept into amalgamated storm deposits.

Regardless of tenuous evidence for possible reduction of benthic oxygen levels, the mere presence of a diverse biota (described later) certainly suggests that the water column in this setting was sufficiently oxygenated to allow colonization by a rich pelagic and benthic marine community. Sediments in this environment may have been characterized by low oxygen levels or fluctuating salinity (Babcock, Zhang, and Leslie 2001), thus inhibiting the scavenging that would have ordinarily damaged the many nonmineralized carcasses buried in the sediment (Babcock and Chang 1997).

Decay of organisms may also have been inhibited through postmortem sealing by microbial mats. Evidence for postmortem coating of sediment-covered carcasses by bacterial sheaths is suggested by a bluish surface stain that occurs on many of the segmented worms, brachiopods, and sea anemones (Chen and Erdtmann 1991). In addition, low pyrite levels and the presence of sulphide mineral coatings on carcasses suggest that sulfate reduction was minimal during sediment deposition (Chen and Erdtmann 1991).

Although some of the aforementioned information is incomplete, it is worth noting that taphonomic study of this deposit is still in a nascent state, largely because the majority of the focus thus far has been on the paleobiology of Chengjiang organisms. Taphonomic work has mostly focused on isolated fossils. Lithologic, geochemical, and petrographic data, as well as related taphonomic information, have rarely been collected in direct association with the fossils (but see Zhu 1992; Leslie, Babcock, and Chang 1996; Leslie et al. 1996; Babcock and Chang 1997).

PALEOBIOLOGY

Given the recent discovery of Chengjiang, it is not surprising that the vast majority of work on the deposit has focused on the systematics of the faunas and their evolutionary significance. Because intense collecting efforts at multiple Chengjiang quarries seem to yield new taxa almost monthly, the following is (at best) only a cursory overview of the more notable organisms described from this Lagerstätte. Readers are directed to Chen, Cheng, and Iten (1997) and Hou et al. (1999) for richly illustrated perspectives on this deposit, and, for a systematic review of the arthropods, to Hou and Bergström (1997).

Megascopic algae are common members of the Chengjiang biota, and many forms consist of only algal thalli, like *Yuknesia* sp. More elaborate forms also occur and include delicate, long, slender looping forms such as *Sinocylindra yunnanensis* or the beautiful helically coiled *Megaspirellus houi* (Chen and Erdtmann 1991). The most common alga is *Fuxianospira gyrata*—a tightly coiled form commonly found splayed in large looping strings on bedding planes (Chen and Zhou 1997).

Chengjiang contains one of the oldest and most spectacularly preserved articulated sponge communities (but see Steiner et al. 1993; Mehl and Erdtmann 1994; Gehling and Rigby 1996; Brasier, Green, and Shields 1997). Sponges are typically preserved articulated or aligned on bedding planes, along with coarse algal debris. Sponges have the second-highest diversity of the Chengjiang metazoan groups (Chen, Hou, and Lu 1989c; Chen, Hou, and Li 1990; Rigby and Hou 1995), with more than 15 genera and 30 species represented among over 1,000 specimens. Sponges are dominated by the tubular Demospongea, including thin-walled balloon- or fan-shaped forms like *Leptomitus teretiusculus, Leptomitella conica,* and *Paraleptomitella dictyodroma* (Chen, Hou, and Lu 1989c), and spinose funnel-shaped forms like *Choiaella radiata* (Rigby and Hou 1995). Although rare, hexactinellids also occur, and are represented by the large globose reticulate *Quadrolaminiella diagonalis* (Chen, Hou, and Li 1990; Reitner and Mehl 1995), the sac-shaped *Crumillospongia* sp., and the elongate fan-shaped *Halichondrites* sp. (Chen and Zhou 1997). Taxonomically enigmatic sponge-like forms are also known, including several undescribed species of the sac-like spiculate *Chancelloria* sp. (Chapter 4; Chen and Zhou 1997).

Rare anemone-like forms suggest the presence of cnidarians on the Chengjiang seafloor, and include forms with sac-shaped bodies capped by distal tentacles, such as *Xianguangia sinica* (Chen and Erdtmann 1991). These specimens have as many as 16 long flexible tentacles, each of which is lined with exquisitely preserved feathery setae along the medial ridge. A funnel-shaped solitary form described as *Cambrorhytium* is also known and, although rare, occurs attached to *Lingulepis* valves or to other *Cambrorhytium* (Chen and Zhou 1997).

Ctenophores have recently been reported from the deposit, and include the smooth lobed form *Sinoascus papillatus* and the spectacular globose form *Maotianoascus octonatius. Maotianoascus* has eight petaloid lobes, and each comb-row preserves impressions of delicate cilia and ciliary support structures (Chen and Zhou 1997).

As in the Burgess Shale, there is a diverse assemblage of priapulid worms in the Chengjiang Lagerstätte, including elongate forms such as

Maotianshania, ornamented forms such as *Palaeoscolex,* segmented forms such as *Cricocosmia,* and possible U-shaped forms such as *Acosmia maotiania* (Sun and Hou 1987b; Hou and Sun 1988; Hou and Bergström 1994; Chen and Zhou 1997). Worms are typically preserved flattened either parallel or nearly parallel to bedding in the shales and mudstones (Jiang 1992). The most common priapulid is *Maotianshania cylindrica,* which may constitute as much as 5 percent of the fauna (Figure 3.3). Like many of the worms, *Maotianshania* exhibits nearly perfect preservation—not only are its proboscis, trunk annulations, setae, and intestinal canal preserved, but even the spiny papillae on its everted proboscis are visible.

Tentaculates are also present, represented by the two spectacularly preserved forms *Eldonia eumorpha* and *Rotadiscus grandis.* Each of these disk-shaped lophophorates contains a U-shaped intestine, concentrically arranged growth lines on the convex surfaces of their disks, and a tentacle-shaped lophophore (Sun and Hou 1987a; Dzik 1991; Chen, Zhu, and Zhou 1995; Dzik, Zhao, and Zhu 1997). These presumably benthic epifaunal forms are typically compressed into thin disks or flat films between shale and mudstone bedding planes (Sun and Hou 1987a) and are often covered with epibionts, such as the linguloid brachiopods *Lingulella* and *Lingulepis,* the lobopod *Microdictyon,* as well as a number of other taxonomically enigmatic epizoans (Chen, Zhou, and Ramsköld 1995a, 1995b; Dzik, Zhao, and Zhu 1997).

Among the more typical Cambrian skeletonized invertebrates, Chengjiang's lingulate brachiopods are of particular interest because they are among a handful of brachiopods that contain fossilized pedicles (Liu 1979; Jin, Hou, and Wang 1993). Some of the more elaborate forms are *Lingulepis malongensis* (Figure 3.4) and *Lingulella chengjianensis,* each

FIGURE 3.3 *Maotianshania cylindrica.* Lower specimen length is approximately 1.7 cm. (Photo courtesy of Hou Xianguang, Nanjing Institute of Geology and Paleontology, China)

FIGURE 3.4 *Lingulepis malongensis.* Length of specimen is approximately 4 cm. (Photo courtesy of Hou Xianguang, Nanjing Institute of Geology and Palaeontology, China)

of which has looping vermiform pedicles with well-preserved cuticular fibers. Less elaborate nonpediculate brachiopods are also present, such as the very common nonmineralized form *Heliomedusa orienta*. It is characterized by a biconvex shape and delicate features, including setae and apparent nerves (Jin and Wang 1992), but because its closely spaced setae are distributed along its entire shell margin, it was inadvertently identified as a hydrozoan in its initial description (Sun and Hou 1987a). Subsequent interpretations of its distinctive setal arrangement suggest that in life position, *H. orienta* lay on the seafloor with its shell open (Jin and Wang 1992).

Other typical Cambrian organisms include trilobites (discussed later) and hyoliths. Hyolithids are quite common at some Chengjiang localities, where they typically occur as concentrations aligned in apparent alimentary canals and in coprolites (Chen et al. 1996). Many of the hyoliths, such as *Ambrolinevitus* sp., exhibit growth lines and are preserved with opercula and helens intact.

Arthropods are the most common group at Chengjiang, in both number of specimens and number of species (Hou and Bergström 1991, 1997). Although the most common arthropods (constituting approximately 80 percent of the total number of individuals) are the tiny bivalved bradoriids, arthropods exhibit a wide size range in the deposit (from less than 1 mm up to 0.5 m) and, overall, are the largest organisms among the Chengjiang fauna. More than half of the over 80 described taxa from Chengjiang are arthropods, including a number of forms also known from the Burgess Shale, such as *Anomalocaris, Canadaspis, Leanchoilia, Naraoia, Waptia*-like forms, and members of the family Helmetiidae (Zhang and Hou 1985; Hou 1987b, 1987d; Hou and Bergström 1991; Chen, Ramsköld, and Zhou 1994; Luo et al. 1997; Shu, Luo, et al. 1999). Like the nonarthropod taxa, nonmineralized arthropods exhibit spectacular preservation, including well-preserved eyes, digestive organs, appendages, and fine surface textures.

Despite their status as the most common fossil in Chengjiang, only a few (of the many thousands) of the ostracod-like bradoriids are preserved associated with soft-tissue preservation. Among the more spectacular of these forms is *Kunmingella douvillei*, which often exhibits well-preserved stalked eyes, uniramous antennae, appendages of thoracic segments, and a long telson with furcal rami (Hou and Bergström 1997; Hou 1999). *Isoxys auritus* and *I. paradoxus* are another common component of the arthropod fauna, and are characterized by a body enclosed in a large bivalved carapace (Hou 1987c; Shu, Zhang, and Geyer 1995). Although the abdomen of *Isoxys* is poorly defined due to enclosure by the bivalved carapace, a number of specimens exhibit soft-part preservation of stalked eyes and trunk limbs bearing pairs of foliaceous exopods fringed with setae.

Among the soft-bodied arthropods, *Naraoia* predominates—comprising nearly 7 percent of the total number of Chengjiang specimens and 10 percent of the arthropods—and is preserved in all manner of orientations within beds (Zhang and Hou 1985; Leslie, Babcock, and Chang 1996). Like *Naraoia* from the Burgess Shale (Chapter 4), two species are present (*Naraoia spinosa* and *N. longicaudata*) and are characterized by relatively large anterior and dorsal shields and elongate antennae directed antero-laterally. These shields were joined together so that the animal could enroll—as indicated by the many enrolled and partially enrolled specimens at Chengjiang. Where preserved fully extended on bedding surfaces, hundreds of specimens exhibit near-perfect preservation of segmented uniramous antennae, biramous limbs, mouth, esophagus, appendages, delicate appendage setae, and other organs (Zhang and Hou 1985; Chen, Edgecombe, and Ramsköld 1997).

A number of the other bivalved arthropods, such as *Canadaspis laevigata* and *Chuandianella ovata,* are preserved with extruded abdomens, antennae, biramous appendages bearing spine-tipped podomeres, alimentary tract, and stalked eyes (Hou and Bergström 1991, 1997). Among these, *Vetulicola cuneatus* is notable because of its large partially fused carapace, which bears a pair of lateral slits; its paddle-shaped abdomen; and its lack of locomotive appendages (Hou 1987c; Hou and Bergström 1991).

Fuxianhuia protensa is one of the more spectacularly preserved of the megathoracic arthropods (Figure 3.5), and is characterized by a short head shield bearing bulbous eyes and raptorial uniramous appendages, a wide segmented trunk bearing biramous limbs, an elongated legless abdomen, and posterior fins bearing telson-like spines (Hou 1987b; Hou and Bergström 1991; Chen, Edgecombe, et al. 1995). This unusual suite of characteristics has led to debate about its taxonomic affinity, with arguments for a chelicerate (Wills 1996) or a primitive euarthropod affinity suggested (Chen, Edgecombe, et al. 1995; Hou and Bergström 1997).

The small, but delicately preserved jawless arthropod *Jianfengia multisegmentalis* (Figure 3.6) is also of note, largely because of the exquisite preservation of nearly all its appendages, including features of its head, segmented trunk, and paddle-like telson. These include a pair of giant pre-oral appendages capped by four distal spines, possible stalked eyes, a streamlined body, and 22 post-cephalic segments bearing biramous endopods, each of which is characterized by spinose podomeres and broad teardrop-shaped exopods (Hou 1987a; Hou and Bergström 1991).

Trilobites are a relatively minor component of the fauna, in terms of both diversity and abundance. Although most of the trilobites are preserved as complete carapaces (rather than disarticulated exuviae), soft-part preservation is extremely rare among the calcified forms (Zhang and Hou 1985; Zhang 1987; Shu, Geyer, et al. 1995). Careful removal of

FIGURE 3.5 *Fuxianhuia protensa.* Length of specimen is approximately 10 cm. (Photo courtesy of Hou Xianguang, Nanjing Institute of Geology and Palaeontology, China)

exoskeletal elements from rock surfaces allowed Shu, Geyer, et al. (1995) and Hou and Bergström (1997) to document spectacularly preserved antennae, biramous appendages, and digestive tracts in the many Redlichiacean trilobites of the deposit, such as *Eoredlichia intermedia* and *Kuanyangia* sp. A variety of soft-bodied trilobite-like arthropods is also known, including forms like *Kuamaia lata* and *Retifacies abnormis*. These taxa are similar to forms from the Burgess Shale (Chapter 4) in that they are characterized by a wide, flat, thin carapace composed of a head shield, thorax, and long tail shield (Hou 1987b; Hou, Chen, and Lu

FIGURE 3.6 *Jianfengia multisegmentalis.* Length of specimen is 1.7 cm. (Photo courtesy of Hou Xianguang, Nanjing Institute of Geology and Palaeontology, China)

1989), but differ in that some specimens preserve stalked eyes, biramous appendages, and jointed antennae covered with setae.

Perhaps the most exciting recent discoveries at Chengjiang are of putative early chordates, including a tunicate (Shu et al. 2001), an agnathan-like fish (Shu, Luo, et al. 1999), and pipiscid-like forms (Shu, Conway Morris, et al. 1999). For example, the extremely rare eel-like *Cathaymyrus diadexus* is thought to be preserved with its pharyangeal gill slits, myotomes, and notochord-like structures intact (Shu, Conway Morris, and Zhang 1996). The lamprey-like *Haikouichthys ercaicuensis* possesses a probable branchial basket and dorsal fin with prominent fin-radials, and the hag-fish-like *Myllokunmingia fengjiaoa* has well-developed gill pouches and probable hemibranchs (Shu, Luo, et al. 1999). Both forms are notable because they contain distinctive chordate features, including complex myomeres, a pericardial cavity, paired ventral fin-folds, and probable imprints of a cartilagenous skull. Not only do these agnathans extend the range of fish by 45 Ma, but they suggest the presence of a Neoproterozoic ancestry for chordates (Shu, Luo, et al. 1999). A possible precursor to the agnathans is also preserved, in the probable pipiscid *Xidazoon stephanus* (Shu, Conway Morris, et al. 1999). This taxon, known from only two specimens on a single bedding plane, has a segmented sac-like body with an exquisitely preserved feeding apparatus characterized by a circlet of plates. *Xidazoon* is notable because it extends the stratigraphic range of this group from the upper Carboniferous to the Cambrian, and because it may reflect a primitive stem-group deuterostome (Shu, Conway Morris, et al. 1999). Perhaps related to *Xidazoon* is *Yunnanozoon lividum* (Figure 3.7), a blade-shaped or-

FIGURE 3.7 *Yunnaonzoon lividum.* Length of specimen is approximately 3.5 cm. (Photo courtesy of Hou Xianguang, Nanjing Institute of Geology and Palaeontology, China)

ganism with segmented musculature (Hou, Ramsköld, and Bergström 1991) and perhaps a circum-oral set of plates (Dzik 1995). However, there is some debate as to the affinity of this form, with early work suggesting that it constituted the earliest chordate (Chen, Dzik, et al. 1995; Dzik 1995; Chen and Li 1997), and subsequent work suggesting a hemichordate affinity (Shu, Zhang, and Chen 1996), or possibly neither (Shu, Conway Morris, et al. 1996; Bergström et al. 1998; Conway Morris 1998). Discovery of *Haikouella,* similar in many ways to *Yunnanozoon,* has furthered the debate with the suggestion that they both be considered as early craniates (Chen, Huang, and Li 1999).

Among the more enigmatic fossils at Chengjiang is *Anomalocaris,* which is also known from Burgess Shale–type localities (Chapters 4 and 5), a large predatory animal that exhibits an unusual combination of arthropod- and aschelminth-like features, including a disk-like mouth, anterior crustacean-shaped frontal appendages, lateral flaps along its torso, stalked eyes, and segmented walking legs (Chen, Ramsköld, and Zhou 1994; Hou, Bergström, and Ahlberg 1995) (Figure 3.8). Although the number of anomalocariid taxa present in this deposit is debated (Minicucci 1999), key forms include *A. saron, Cucumericrus decoratus,* and *Parapeytoia yunnanensis.* These taxa are characterized by transverse rows of scales, walking legs on each appendage, a fan-shaped tail, and a backward-facing inverted mouth (Hou, Bergström, and Ahlberg 1995). In addition to the articulated anomalocariid specimens, a number of isolated, very large putative anomalocariid mouths have been found, each characterized by several circlets bearing numerous sharp, inwardly pointing teeth (Chen, Erdtmann, and Steiner 1992); presumably such teeth were employed in macrophagous predation (Nedin 1999).

Perhaps some of the most famous forms from Chengjiang are the so-called lobopodians, a group of onychophoran-like, small segmented animals bearing a dorsal or dorso-lateral series of plates, spines, or sclerites.

FIGURE 3.8 *Anomalocaris saron.* Length of specimen is 8 cm. (Photo courtesy of Junyuan Chen, Nanjing Institute of Geology and Palaeontology, China)

FIGURE 3.9 *Microdictyon sinicum.* Width of specimen is approximately 2 cm. (Photo courtesy of Hou Xianguang, Nanjing Institute of Geology and Palaeontology, China)

At least six lobopod taxa are known, including *Cardiodictyon catenulum, Hallucigenia fortis, Luolishania longicruris, Microdictyon sinicum, Onychodictyon ferox,* and *Paucipodia inermis* (Chen, Hou, and Lu 1989a; Hou and Chen 1989b; Hou, Ramsköld, and Bergström 1991; Ramsköld and Hou 1991; Chen, Zhou, and Ramsköld 1995a; Hou and Bergström 1995). In some forms, the mouth is visible at the anterior end of the head region, and an apparent anal extension is present in other forms. Of these taxa, *Luolishania* and *Microdictyon* are mentioned shortly. *Hallucigenia,* perhaps the most famous lobopod from Chengjiang, is discussed in Chapter 4 (Ramsköld and Hou 1991; Ramsköld 1992; Hou and Bergström 1995).

Microdictyon is one of the more heavily armored lobopods, bearing 10 pairs of netted phosphatic plates on its lateral margins, just above its 10 pairs of elongate soft legs (Chen, Hou, and Lu 1989b) (Figure 3.9). The plates on *Microdictyon* are unusual in that they may not have grown by accretion of mineralized tissue, but by molting (Chen, Hou, and Lu 1989c). Thus, despite the similarity of *Microdictyon* to other lobopodians, the taxonomic assignment of *Microdictyon,* and hence the placement of its enigmatic mode of biomineralization into an evolutionary context, is unclear. However, *Microdictyon*'s fame among the Chengjiang lobopods does not stem merely from the evolutionary significance of its biomineralization mechanism. Before the discovery of articulated specimens at Chengjiang, *Microdictyon*'s rounded, oval, and polygonal phosphatic plates were well-known microfossil components of Lower Cambrian deposits all over the world (Bengtson, Matthews, and Missarzhevsky 1986; Chen, Zhou, and

Ramsköld 1995b). Until the discovery of articulated specimens of *Micro-dictyon* at Chengjiang, the origin of these plates was unknown. In fact, the function of the plates is still not well understood, with some authors suggesting a defensive function, and others noting a resemblance to schizochroal trilobite eyes (Dzik 1993). Regardless, the mere occurrence of this taxon in the Early Cambrian suggests the possible early development of protective armor and may give clues to the sources of many of the other enigmatic small shelly fossils found in other Cambrian deposits (Bengtson, Matthews, and Missarzhevsky 1986).

Luolishania is another peculiar Chengjiang lobopod, which, together with *Microdictyon*, is characterized by a pair of terminal legs and a small anal extension (although the anterior–posterior interpretation of this taxon is controversial) (Chen, Hou, and Lu 1989b; Hou and Chen 1989b; Chen, Zhou, and Ramsköld 1995b; Hou and Bergström 1995). This taxon is of interest because it has very small plates (Hou and Bergström 1995) and has segmentation similar to that of the lobopod *Xenusion auerswaldae* from the Lower Cambrian of the Baltic (Chapter 5; Dzik and Krumbiegel 1989). In particular, some authors (Chen and Erdtmann 1991) have suggested that it may be intermediate between annelids and arthropods, perhaps providing an ancestral link among the onycophorans, hexapods, and myriapods.

Another problematic taxon perhaps related to extant entoprocts or echinoderms is *Dinomischus venustus,* known from only 13 specimens. These stalked organisms may have been attached to the seafloor and had a cup-shaped body circled by petal-like rays, each of which bears radial canals and corrugated banding (Chen, Hou, and Lu 1989a; Chen and Erdtmann 1991). Based on these features, a solitary sessile passive suspension-feeding mode of life is inferred (Conway Morris 1977).

Another enigmatic taxon is the worm-like *Facivermis yunnanicus,* which is characterized by an elongated cylindrical body capped by a tapered papillated head bearing five pairs of finely annulated tentacles. This form, known from only five specimens, has been variously allied with the annelids (Hou and Chen 1989b), the lobopodians (Hou, Ramsköld, and Bergström 1991), and the lophophorates (Chen, Zhu, et al. 1995). Obviously, more specimens must be collected before its taxonomic affinity can be fully resolved.

In addition to the taxonomically enigmatic taxa mentioned earlier, the Chengjiang assemblage contains numerous other enigmatic faunal elements whose paleobiological importance is certain, but are currently known from only a few specimens. These include the flattened oval-shaped arthropod *Saperion glumaceum,* the eurypterid-like arthropod *Xandarella spectaculum* (Hou, Ramsköld, and Bergström 1991), the *Jianfengia*-like arthropod *Alalcomenaeus illecebrosus* (Hou 1987a), possible chelicerates (Babcock and Chang 1996), possible lophophorates (Babcock and Chang 1995), and echinoderm-like forms (Chen and Zhou 1997; Hou et al. 1999).

Finally, the Qiongzhusi Formation contains a well-preserved suite of trace fossils. Although a detailed analysis of traces occurring in individual units (e.g., within the mudstones containing soft-bodied fossils) has not yet been published, Zhu (1997) notes that overall bioturbation within the Maotianshan Shale is weak, and documents trace fossils from only two intervals within a 60 m thick section. Zhu (1997) does note, however, that arthropod-style traces such as *Monomorphichnus* and *Diplichnites* are common through this interval, as well as bed-parallel traces such as *Planolites* and *Palaeophycus,* and trails with a vertical component, including *Arenicolites, Chondrites,* and *Treptichnus.* Sun and Hou (1987b) corroborate these observations by noting the presence of tubular burrows and grazing trails on several bedding planes.

PALEOECOLOGY

Although a systematic paleoecologic analysis of Chengjiang has not yet been published (but see Leslie, Babcock, and Chang 1996), fossiliferous horizons at Chengjiang are much like those at other Burgess Shale–type deposits (Chapters 4 and 5) because they are dominated by arthropods and algae, and over 97 percent of the fauna are nonmineralized forms. Unlike these other deposits, Chengjiang also seems to have a small, but significant number of lobopodians. Dominant members of the arthropod component are bivalved forms such as bradoriids, which comprise nearly 70 percent of available specimens. There are a variety of life habits represented by the Chengjiang biota, including mobile benthic infaunal taxa, mobile and sessile benthic epifaunal taxa, as well as actively and passively mobile pelagic taxa.

In terms of trophic structure, the abundance of algae and acritarchs in the deposit, coupled with the organic-rich nature of the strata in the Qiongzhusi Formation, suggests that the base of the food chain was likely occupied by phytoplankton and zooplankton, as well as detritus from these sources. The infaunal benthic community was dominated by a number of suspension- or deposit-feeding forms, such as lingulate brachiopods, as well as some primary consumers, such as the priapulid worm *Maotianshania* and the worm-like *Facivermis.* Epifaunal benthic habits were characterized by suspension feeders (such as *Choiaella, Dinomischus, Eldonia, Heliomedusa, Rotadiscus,* and *Xianguangia*) and deposit feeders (such as *Clypecaris* and *Naraoia*). Among these, the poorly preserved bradoriids have the largest number of individuals and the highest biomass and are thought to have been a major food source for most carnivores, as they are commonly found concentrated in ovate flattened coprolites (Leslie, Babcock, and Chang 1996; Chen and Zhou 1997). The presence of such coprolites, as well as similar hyolithid concentrations, denotes the existence of numerous predators (such as *Luolishania* and *Anomalocaris*) that may have occupied both epifaunal and pelagic habitats. Floating pelagic forms are also well known, such as the ctenophore *Maotianoascus*. Swimming pelagic forms may include a number of the smaller arthropods such as *Isoxys*, as well as chordates such as *Cathaymyrus, Haikouichthys,* and *Myllokunmingia,* which represent primary or secondary consumers. Commensal or possibly scavenging modes of life are also suggested by the presence of a variety of epizoans attached to or associated with *Eldonia* and *Rotadiscus.*

Although the amalgamation of transported and *in situ* faunas at Chengjiang complicates interpretation of distinct paleocommunities, when considered together these faunas suggest that Early Cambrian shallow-marine environments were inhabited by a relatively diverse and well-developed marine community. Among the life habits discussed earlier, the epifaunal benthic habit seems to be the best represented—including a variety of suspension- and deposit-feeding organisms reflecting development of several tiering levels rising up to about 30 cm above the seafloor, and perhaps several centimeters beneath it (Chen, Hou, and Lu 1989c).

CONCLUSIONS

Exquisite soft-body preservation of diverse benthic and pelagic faunas makes Chengjiang one of the most promising of early Paleozoic Lagerstätten. In addition to being well preserved, specimens are abundant, and fossil-bearing units can be traced laterally for approximately 100 km, allowing for intensive collection of fossils from Chengjiang outcrops in the Kunming region.

One of the major contributions of the Chengjiang biota is that it has provided data on the most ancient, diverse group of fossils from which paleontologists can attempt to constrain the evolutionary relationships between and among major animal clades. Although much of the fauna bears most directly on the vagaries of early arthropod evolution (Hou and Bergström 1997; Bergström and Hou 1998; Ramsköld and Chen 1998), the collection of additional nonarthropod taxa promises to extend rigorous phylogenetic analyses to rarer taxa such as the various palaeoscolecid worms (Hou and Bergström 1994) and the lobopodians (Ramsköld and Chen 1998).

Because of its early occurrence in geologic time, the Chengjiang fauna has been instrumental in advancing our understanding of poorly preserved or enigmatic groups from younger Cambrian Lagerstätten. Furthermore, some Chengjiang taxa (e.g., sponges) have direct analogues in both Ediacaran and Burgess Shale Lagerstätten. Thus, in addition to providing clues about the rates and mechanisms of the Cambrian explosion, the Chengjiang fauna may lead to a greater understanding of the connection between organisms that inhabited both the late Proterozoic (Chapter 2) and early Paleozoic (Chapters 4, 5, 6, 7) worlds. Although thousands of specimens have been systematically collected from Chengjiang, the deposit still contains much untapped paleoecologic and taphonomic information–thus, revelation of the impact of this deposit on our understanding of paleobiology and early animal evolution has just begun.

ACKNOWLEDGMENTS

I would like to thank X. Hou, L. Ramsköld, and J. Bergström for their constructive reviews of the original manuscript, and J. Y. Chen, L. Babcock, S. Bengtson, and S. Leslie for subsequent discussions and insights that helped improve this contribution.

REFERENCES

Babcock, L. E., and W. T. Chang. 1995. New Early Cambrian animal from China, and its implications for the early radiations of lophophorates. *Geological Society of America Abstracts with Programs* 27:366.

Babcock, L. E., and W. T. Chang. 1996. Early Cambrian chelicerate arthropod from China. In J. E. Repetski, ed., *Sixth North American Paleontological Convention: Abstracts of Papers,* p. 8. Paleontological Society Special Publication, no. 8. Washington, D.C.: Paleontological Society.

Babcock, L. E., and W. Chang. 1997. Comparative taphonomy of two nonmineralized arthropods: *Naraoia* (Nektaspida; Early Cambrian, Chengjiang biota, China) and *Limulus* (Xiphosurida; Holocene, Atlantic Ocean). *Bulletin of the National Museum of Natural Science* 10:233–250.

Babcock, L. E., W. Zhang, and S. A. Leslie. 2001. The Chengjiang biota: Record of the Early Cambrian diversification of life and clues to exceptional preservation of fossils. *GSA Today* 11:4–9.

Bengtson, S., S. C. Matthews, and V. V. Missarzhevsky. 1986. The Cambrian netlike fossil *Microdictyon.* In A. Hoffman and M. H. Nitecki, eds., *Problematic Fossil Taxa,* pp. 97–115. New York: Oxford University Press.

Bergström, J., and X. Hou. 1998. Chengjiang arthropods and their bearing on early arthropod evolution. In G. D. Edgecombe, ed., *Arthropod Fossils and Phylogeny,* pp. 151–184. New York: Columbia University Press.

Bergström, J., W. W. Naumann, J. Viehweg, and M. Martí-Mus. 1998. Conodonts, calcichordates and the origin of vertebrates. *Mitteilungen des Museum für Naturkunde zu Berlin, Geowissenschaftliche* 1:81–92.

Bowring, S. A., J. P. Grotzinger, C. E. Isachsen, A. H. Knoll, S. M. Pelechaty, and P. Kolosov. 1993. Calibrating rates of Early Cambrian evolution. *Science* 261:1293–1298.

Brasier, M. D., O. Green, and G. Shields. 1997. Ediacaran sponge spicule clusters from southwestern Mongolia and the origins of the Cambrian fauna. *Geology* 25:303–306.

Chen, J. Y., J. Bergström, M. Lindström, and X. Hou. 1991. Fossilized soft-bodied fauna: The Chengjiang fauna—Oldest soft-bodied fauna on Earth. *National Geographic Research and Exploration* 7:8–19.

Chen, J. Y., Y. Cheng, and H. V. Iten, eds. 1997. *The Cambrian Explosion and the Fossil Record.* Taichung, Taiwan: National Museum of Natural Science.

Chen J. Y., J. Dzik, G. D. Edgecombe, L. Ramsköld, and G. Q. Zhou. 1995. A possible Early Cambrian chordate. *Nature* 377:720–722.

Chen, J. Y., G. D. Edgecombe, and L. Ramsköld. 1997. Morphological and ecological disparity in Naraoiids (Arthropoda) from the Early Cambrian Chengjiang fauna, China. *Records of the Australian Museum* 49:1–24.

Chen, J. Y., G. D. Edgecombe, L. Ramsköld, and G. Q. Zhou. 1995. Head segmentation in Early Cambrian *Fuxianhuia:* Implications for arthropod evolution. *Science* 268:1339–1343.

Chen J. Y., and B. D. Erdtmann. 1991. Lower Cambrian fossil Lagerstätte from Chengjiang, Yunnan, China: Insights for reconstructing early metazoan life. In A. M. Simonetta and S. Conway Morris, eds., *The Early Evolution of Metazoa and the Significance of Problematic Taxa,* pp. 57–76. Cambridge: Cambridge University Press.

Chen, J. Y., B. D. Erdtmann, and M. Steiner. 1992. Die unterkambrische Fossillagerstätte Chengjiang (China). *Fossilien* 5:273–282.

Chen, J. Y., X. Hou, and G. Li. 1990. New Lower Cambrian demosponge *Quadrolaminiella* (gen. nov.). *Acta Palaeontologica Sinica* 29:402–414.

Chen, J. Y., X. Hou, and H. Lu. 1989a. Early Cambrian hock glass-like rare sea animal *Dinomischus* (Entoprocta) and its ecological features. *Acta Palaeontologica Sinica* 28:58–71.

Chen, J. Y., X. Hou, and H. Lu. 1989b. Early Cambrian netted scale-bearing worm-like sea animal. *Acta Palaeontologica Sinica* 28:1–16.

Chen, J. Y., X. Hou, and H. Lu. 1989c. Lower Cambrian leptomitids (Demospongea), Chengjiang, Yunnan. *Acta Palaeontologica Sinica* 28:58–71.

Chen, J. Y., D. Y. Huang, and C. W. Li. 1999. An Early Cambrian craniate-like chordate. *Nature* 402: 518–522.

Chen, J. Y., and C. Li. 1997. Early Cambrian chordate from Chengjiang, China. *Bulletin of the National Museum of Natural Sciences of Taiwan* 10:257–273.

Chen, J. Y., and M. Lindström. 1991. Lower Cambrian non-mineralized fauna from Chengjiang, Yunnan, China. *Geologiska Föreningens I Stockholm Förhandlingar* 113:79–81.

Chen, J. Y., L. Ramsköld, and G. Q. Zhou. 1994. Evidence for monophyly and arthropod affinity of Cambrian giant predators. *Science* 264:1304–1308.

Chen, J. Y., and G. Zhou. 1997. Biology of the Chengjiang fauna. *Bulletin of the National Museum of Natural Sciences of Taiwan* 10:11–106.

Chen, J. Y., G. Zhou, and L. Ramsköld. 1995a. A new Early Cambrian onychophoran-like animal, *Paucipodia* gen. nov., from the Chengjiang fauna, China. *Transactions of the Royal Society of Edinburgh* 85:275–282.

Chen, J. Y., G. Q. Zhou, and L. Ramsköld. 1995b. The Cambrian lobopodian *Microdictyon sinicum* and its broader significance. *Bulletin of the National Museum of Natural Science* 5:1–93.

Chen, J. Y., G. Q. Zhou, M. Y. Zhu, and K. Y. Yeh. 1996. *The Chengjiang Biota—A Unique Window of the Cambrian Explosion.* Taichung, Taiwan: National Museum of Natural Science.

Chen, J. Y., M. Y. Zhu, and G. Q. Zhou. 1995. The Early Cambrian "medusoid" *Eldonia* from the Chengjiang fauna of China. *Acta Palaeontologica Polonica* 40:213–244.

Collins, D. 1996. The "evolution" of *Anomalocaris* and its classification in the arthropod class Dinocarida (Nov.) and order Radiodonta (Nov.). *Journal of Paleontology* 70:280–293.

Conway Morris, S. C. 1977. A new entoproct-like organism from the Burgess Shale of British Columbia. *Palaeontology* 20:833–845.

Conway Morris, S. 1989. The persistence of Burgess Shale–type faunas: Implications for the evolution of deeper-water faunas. *Transactions of the Royal Society of Edinburgh: Earth Sciences* 80:271–283.

Conway Morris, S. C. 1998. *The Crucible of Creation: The Burgess Shale and the Rise of Animals.* Oxford: Oxford University Press.

Dzik, J., 1991. Is fossil evidence consistent with traditional views of the early metazoan phylogeny? In A. M. Simonetta and S. Conway Morris, eds., *The Early Evolution of Metazoa and the Significance of Problematic Taxa,* pp. 47–56. Cambridge: Cambridge University Press.

Dzik, J. 1993. Early metazoan evolution and the meaning of its fossil record. *Evolutionary Biology* 27:339–386.

Dzik, J. 1995. *Yunnanozoon* and the ancestry of chordates. *Acta Palaeontologica Polonica* 40:341–360.

Dzik, J., and G. Krumbiegel. 1989. The oldest "onychopohoran" *Xenusion:* A link connecting phyla? *Lethaia* 22:29–38.

Dzik, J., Y. Zhao, and M. Zhu. 1997. Mode of life of the Middle Cambrian eldonioid lophophorate *Rotadiscus. Palaeontology* 40:385–396.

Gehling, J. G., and J. K. Rigby. 1996. Long expected sponges from the Neoproterozoic Ediacara fauna of South Australia. *Journal of Paleontology* 70:185–195.

Gore, R. 1993. Explosion of life: The Cambrian period. *National Geographic* 184:120–136.

Ho, C. 1942. Phosphate deposits of Tungshan, Chengjian, Yunnan. *Bulletin of the Geological Survey of China* 35:41–43.

Hou, X. 1987a. Two new arthropods from Lower Cambrian, Chengjiang, eastern Yunnan. *Acta Palaeontologica Sinica* 26:236–256.

Hou, X. 1987b. Three new large arthropods from Lower Cambrian, Chengjiang, eastern Yunnan. *Acta Palaeontologica Sinica* 26:272–285.

Hou, X. 1987c. Early Cambrian large bivalved arthropods from Chengjiang, eastern Yunnan. *Acta Palaeontologica Sinica* 26:286–298.

Hou, X. 1987d. Oldest Cambrian bradoriids from eastern Yunnan. In *Stratigraphy and Palaeontology of Systemic Boundaries in China: Precambrian–Cambrian Boundary,* vol. 1, pp. 535–547. Nanjing: Nanjing University Publishing House.

Hou, X. 1999. New rare bivalved arthropods from the Lower Cambrian Chengjiang fauna, Yunnan, China. *Journal of Paleontology* 73:102–116.

Hou, X., and J. Bergström. 1991. The arthropods of the Lower Cambrian Chengjiang fauna, with relationships and evolutionary significance. In A. M. Simonetta and S. Conway Morris, eds., *The Early Evolution of Metazoa and the Significance of Problematic Taxa,* pp. 11–17. Cambridge: Cambridge University Press.

Hou, X., and J. Bergström. 1994. Palaeoscolecid worms may be nematomorphs rather than annelids. *Lethaia* 27:11–17.

Hou, X., and J. Bergström. 1995. Cambrian lobopodians—Ancestors of extant onychophorans? *Zoological Journal of the Linnean Society* 114:3–19.

Hou, X., and J. Bergström. 1997. *Arthropods of the Lower Cambrian Chengjiang Fauna, Southwest China*. Fossils & Strata. no. 45. Oslo: Scandinavian University Press.

Hou, X., J. Bergström, and P. Ahlberg. 1995. *Anomalocaris* and other large animals in the Lower Cambrian Chengjiang fauna of southwest China. *Geologiska Foreningens I Stockholm Forhandlingar* 117:163–183.

Hou, X., J. Bergström, H. Wang, X. Feng, and A. Chen. 1999. *Exceptionally Well-Preserved Animals from 530 Million Years Ago*. Kunming: Yunnan Science and Technology Press.

Hou, X., and J. Chen. 1989a. Early Cambrian tentacled worm-like animals (*Facivermis* gen. nov.) from Chengjiang, Yunnan. *Acta Palaeontologica Sinica* 28:32–41.

Hou, X., and J. Chen. 1989b. Early Cambrian arthropod–annelid intermediate *Luolishania longicruris* (gen. et sp. nov.) from Chengjiang, eastern Yunnan. *Acta Palaeontologica Sinica* 28:42–57.

Hou, X., J. Chen, and H. Lu. 1989. Early Cambrian new arthropods from Chengjiang, Yunnan. *Acta Palaeontologica Sinica* 28:42–57.

Hou, X., L. Ramsköld, and J. Bergström. 1991. Composition and preservation of the Chengjiang fauna—Lower Cambrian soft-bodied fauna. *Zoologica Scripta* 20:395–411.

Hou, X., D. J. Siveter, M. Williams, D. Walossek, and J. Bergström. 1996. Appendages of the arthropod *Kunmingella* from the Early Cambrian of China: Its bearing on the systematic position of Bradoriida and the fossil record of Ostracoda. *Philosophical Transactions of the Royal Society of London, B* 351:1131–1145.

Hou, X., and W. Sun. 1988. Discovery of Chengjiang fauna at Meishucun, Jinning, Yunnan. *Acta Palaeontologica Sinica* 27:1–12.

Jiang, Z. 1992. The Lower Cambrian fossil record of China. In J. H. Lipps and P. W. Signor, eds., *Origin and Early Evolution of the Metazoa*, pp. 311–333. New York: Plenum.

Jin, Y., X. Hou, and H. Wang. 1993. Lower Cambrian pediculate lingulids from Yunnan, China. *Journal of Paleontology* 67:788–798.

Jin, Y. G., and H. Y. Wang. 1992. Revision of the Lower Cambrian brachiopod *Heliomedusa* Sun & Hou, 1987. *Lethaia* 25:35–49.

Jin, Y., H. Wang, and W. Wang. 1991. Palaeoecological aspects of brachiopods from Chingchussu Formation of Early Cambrian age, eastern Yunnan, China. In Y. Jin, J. Wang, and S. Xu, eds., *Palaeoecology of China*, vol. 1, pp. 25–47. Nanjing: Nanjing University Press.

Landing, E., S. A. Bowring, K. L. Davidek, S. R. Westrop, G. Geyer, and W. Heldmaier. 1998. Duration of the Early Cambrian: U-Pb ages of volcanic ashes from Avalon and Gondwana. *Canadian Journal of Earth Sciences* 35:329–338.

Leslie, S. A., L. E. Babcock, and W. T. Chang. 1996. Community composition and taphonomic overprint of the Chengjiang biota (Early Cambrian, China). In J. E. Repetski, ed., *Sixth North American Paleontological Convention: Abstracts of Papers,* p. 237. Paleontological Society Special Publication, no 8. Washington, D.C.: Paleontological Society.

Leslie, S. A., L. E. Babcock, J. C. Mitchell, and W. T. Chang. 1996. Phosphatization and its relationship to exceptional preservation of fossils in the

Chengjiang Lagerstätte (Lower Cambrian, China). *Geological Society of America Abstracts with Programs* 27:294.

Liang, Q. Z., W. Fang, and R. van de Voo. 1990. Further study on palaeomagnetism of the Precambrian–Cambrian boundary candidate stratotype section at Meishucun, Yunnan, China. *Acta Geologica Sinica* 64:264–274.

Lindström, M. 1995. The environment of the Early Cambrian Chengjiang fauna. In J. Y. Chen, G. Edgecombe, and L. Ramsköld, eds., *International Cambrian Explosion Symposium, Programme and Abstracts,* p. 17.

Liu, D. 1979. Earliest Cambrian brachiopods from southwest China. *Acta Palaeontologica Sinica* 18:505–511.

Lu, Y. 1941. Lower Cambrian stratigraphy and trilobite fauna of Kunming, Yunnan. *Bulletin of the Geological Society of China* 21:71–90.

Luo, H. L., S. Hu, S. Zhang, and Y. Tao. 1997. New occurrence of the Early Cambrian Chengjiang fauna from Haikou, Kunming, Yunnan province. *Acta Geologica Sinica* 71:97–104.

Luo, H. L., Z. W. Juan, X. C. Wu, X. L. Song, L. Ouyang, Y. S. Xing, G. Z. Liu, S. S. Zhang, and Y. H. Tao. 1984. *Sinian–Cambrian Boundary Stratotype Section in Meishucun, Jingling, Yunnan, China.* Beijing: People's Publishing House.

Mansuy, H. 1912. Etude géologique du Yunnan oriental. Part 2: Paléontologie. *Mémoires du Service Géologique de l'Indo-Chine* 1:1–146.

Mehl, D., and B. Erdtmann. 1994. *Sanhapentella dapingi* n. gen. N. sp.—A new hexactinellid sponge from the Early Cambrian (Tommotian) of China. *Berliner Geowissenschaftlich Abhandlungern, E* 13:315–319.

Minicucci, J. M. 1999. Forward to the Cambrian—Anomalocarid studies at the end of the millennium. *Palaeontological Association Newsletter* 41:23–32.

Monastersky, R. 1993. Mysteries of the Orient. *Discover* 14:38–49.

Nedin, C. 1999. *Anomalocaris* predation on nonmineralized and mineralized trilobites. *Geology* 27:987–990.

Pan, K. 1957. On the discovery of homopoda from South China. *Acta Palaeontologica Sinica* 5:523–6.

Qian, Y. 1977. Hyolitha and some problematica from the Lower Cambrian Meischucun Stage in central and s. w. China. *Acta Palaeontologica Sinica* 16:256–275.

Qian, Y., and S. Bengtson. 1989. *Palaeontology and Biostratigraphy of the Early Cambrian Meishucunian Stage in Yunnan Province, South China.* Fossils & Strata, no. 24. Oslo: Universitetsforlag.

Ramsköld, L. 1992. The second leg row of *Hallucigenia* discovered. *Lethaia* 25:443–460.

Ramsköld, L., and J. Chen. 1998. Cambrian lobopodians: Morphology and phylogeny. In G. D. Edgecombe, ed., *Arthropod Fossils and Phylogeny,* pp. 107–150. New York: Columbia University Press.

Ramsköld, L., and X. Hou. 1991. New Early Cambrian animal and onychophoran affinities of enigmatic metazoans. *Nature* 351:225–228.

Reitner, J., and D. Mehl. 1995. Early Palaeozoic diversification of sponges: New data and evidence. *Geologische und paläontologische Mitteilungen Innsbruck* 20:335–347.

Rigby, J. K., and X. Hou. 1995. Lower Cambrian demosponges and hexactinellid sponges from Yunnan, China. *Journal of Paleontology* 69:1009–1019.

Shu, D., L. Chen, J. Han, and X. L. Zhang. 2001. An Early Cambrian tunicate from China. *Nature* 411: 472–473.

Shu, D., S. Conway Morris, and X. L. Zhang. 1996. A *Pikaia*-like chordate from the Lower Cambrian of China. *Nature* 384:157–158.

Shu, D., S. Conway Morris, X. L. Zhang, L. Chen, Y. Li, and J. Han. 1999. A pipiscid-like fossil from the Lower Cambrian of south China. *Nature* 400:746–749.

Shu, D., G. Geyer, L. Chen, and X. Zhang. 1995. Redlichiacean trilobites with preserved soft-parts from the Lower Cambrian Chengjiang fauna (South China). *Beringeria Special Issue* 2:203–241.

Shu, D., H. L. Luo, S. Conway Morris, X. L. Zhang, S. X. Hu, L. Chen, J. Han, M. Zhu, Y. Li, and L. Z. Chen. 1999. Lower Cambrian vertebrates from south China. *Nature* 402:42–46.

Shu, D., X. L. Zhang, and L. Chen. 1996. Reinterpretation of *Yunnanozoon* as the earliest known hemichordate. *Nature* 380:428–430.

Shu, D., X. L. Zhang, and G. Geyer. 1995. Anatomy and systematic affinities of Lower Cambrian bivalved arthropod *Isoxys auritus*. *Alcheringa* 19:333–342.

Steiner, M., D. Mehl, J. Reitner, and B. Erdtmann. 1993. Oldest entirely preserved sponges and other fossils from the lowermost Cambrian and a new facies reconstruction of the Yangtze Platform (China). *Berliner Geowissenchaften, Abhandlungen* 9:293–329.

Sun, W., and X. Hou. 1987a. Early Cambrian medusae from Chengjiang, Yunnan, China. *Acta Palaeontologica Sinica* 26:257–271.

Sun, W., and X. Hou. 1987b. Early Cambrian worms from Chengjiang, Yunnan, China: *Maotianshania* gen. nov. *Acta Palaeontologica Sinica* 26:299–305.

Vannier, J., and J. Y. Chen. 2001. The Early Cambrian colonization of pelagic niches exemplified by Isoxys (Arthropoda). *Lethaia* 33:295–311.

Wills, M. A. 1996. Classification of the arthropod *Fuxianhuia*. *Science* 272:746–747.

Zang, W. 1992. Sinian and Early Cambrian floras and biostratigraphy on the South China platform. *Palaeontographica, B* 224:75–119.

Zhang, W. 1987. Early Cambrian Chengjiang fauna and its trilobites. *Acta Palaeontologica Sinica* 26:223–235.

Zhang, W., and X. Hou. 1985. Preliminary notes on the occurrence of the unusual trilobite *Naraoia* in Asia. *Acta Palaeontologica Sinica* 24:591–595.

Zhang, J., G. Li, C. Zhou, M. Zhu, and Z. Yu. 1997. Carbon isotope profiles and their correlation across the Neoproterozoic–Cambrian boundary interval on the Yangtze Platform, China. *Bulletin of the National Musuem of Natural Science* 10:107–116.

Zhou, Z., and J. Yuan. 1982. A tentative correlation of the Cambrian system in China with those in selected regions overseas. *Bulletin of the Nanjing Institute of Geology and Palaeontology, Academia Sinica* 5:289–306.

Zhu, M. Y. 1992. Taphonomy of the Chengjiang Lagerstätte, Yunnan. Ph.D. diss., Nanjing Institute of Geology and Paleontology.

Zhu, M. Y. 1997. Trace fossils of Yunnan. *Bulletin of the National Museum of Natural Science* 10:275–312.

4
Burgess Shale:
Cambrian Explosion in Full Bloom

James W. Hagadorn

THE MIDDLE CAMBRIAN BURGESS SHALE IS ONE OF THE world's best-known and best-studied fossil deposits. The story of the discovery of its fauna is a famous part of paleontological lore. While searching in 1909 for trilobites in the Burgess Shale Formation of the Canadian Rockies, Charles Walcott discovered a remarkable "phyllopod crustacean" on a shale slab (Yochelson 1967). Further searching revealed a diverse suite of soft-bodied fossils that would later be described as algae, sponges, cnidarians, ctenophores, brachiopods, hyoliths, priapulids, annelids, onychophorans, arthropods, echinoderms, hemichordates, chordates, cirripeds, and a variety of problematica. Many of these fossils came from a single horizon, in a lens of shale 2 to 3 m thick, that Walcott called the Phyllopod (leaf-foot) Bed. Subsequent collecting at and near this site by research teams led by Walcott, P. E. Raymond, H. B. Whittington, and D. Collins has yielded over 75,000 soft-bodied fossils, most of which are housed at the Smithsonian Institution in Washington, D.C., and the Royal Ontario Museum (ROM) in Toronto.

Although interest in the Burgess Shale fauna has waxed and waned since its discovery, its importance has inspired work on other Lagerstätten and helped galvanize the paleontological community's attention on soft-bodied deposits in general. For example, work on the Burgess Shale has stimulated work on the older Chengjiang fauna (Chapter 3), as well as a number of other Burgess Shale–type localities from around the world (Chapter 5).

In the first descriptions of the Burgess fauna, Walcott placed most of the new taxa (over 110 species) within existing taxonomic groups.

Among other reasons, this was largely because his specimen analysis was based on examination of unprepared single shale slabs. Because many specimens are oriented obliquely to bedding planes, folded over on themselves, and/or contain shale intercalated between carcass segments, subsequent dissection and analysis of part–counterpart slabs allowed more detailed morphological reconstructions and hence more accurate taxonomic assignments (Whittington 1971a). Together with reexcavation of original collecting sites and documentation of new collection sites, re-examination of Burgess Shale specimens using these techniques has yielded a variety of new interpretations about these organisms and provided more accurate phylogenetic, paleoecological, and environmental information about Middle Cambrian marine life.

The fauna of the deposit includes both relatively common skeletonized forms and an abundance of soft-bodied forms seldom preserved in typical Cambrian paleoenvironments. This conservation Lagerstätten is thus unique because it provides the most comprehensive view of a typical benthic paleocommunity during the Cambrian explosion. Early in its history, this deposit was recognized as exceptional (Walcott 1911b), not only because it was the first documented example of a nearly complete Cambrian paleocommunity, but because at that time the Cambrian represented the oldest accepted record of animal life. Faunas from these strata are preserved in fine-grained obrution deposits that were transported via fluidized flows into a poorly oxygenated basin or trough adjacent to a major carbonate escarpment. Rapid burial, low oxygenation, and early diagenetic clay replacement of carcasses allowed the preservation of a variety of soft- *and* hard-bodied organisms representing most major marine phyla, as well as several morphologically distinctive organisms of uncertain taxonomic affinity.

GEOLOGICAL CONTEXT

The Burgess Shale is an informal name for two fossiliferous shale members of the Burgess Shale Formation, which are well exposed near the town of Field, British Columbia (Figure 4.1). Over the past 100 years, the vast majority of the Burgess Shale biota has been collected from two quarries, known as the Walcott and Raymond quarries, both of which are located along the western slope of a ridge between Mount Field and Mount Wapta, in the southern portion of Yoho National Park.

During the Cambrian, three continuous but laterally interfingered lithofacies belts were deposited along the passive margin of the North American Cordilleran margin: the inner detrital belt, the middle carbonate belt, and the outer detrital belt (Palmer 1960; Robison 1960). The Burgess Shale occurs along the margin of the outer two facies belts, and thus the stratigraphic framework for the Burgess Shale includes se-

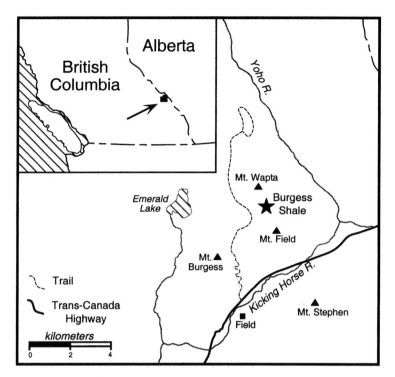

FIGURE 4.1 Part of Yoho National Park, British Columbia, Canada. The star indicates the location of the Walcott and Raymond quarries. (Modified from Whittington 1985)

quences deposited in both settings (Stewart 1989). In the Mount Field region, the margin of the carbonate belt was delineated by a nearly vertical carbonate escarpment composed of a massive, approximately 250 m thick sequence of thin-bedded reef-flat dolomites—these constitute the Cathedral Limestone Formation (Aitken 1971; McIlreath 1974, 1977; Aitken and McIlreath 1984) (Figure 4.2). Beyond the outer edge of this escarpment, deep-water slope limestones and platform-derived carbonate debris accumulated in what is known as the Takakkaw Tongue (Aitken 1997). Above the Takakkaw Tongue is the Burgess Shale Formation consisting of 10 members, most of which are dominated by shales (Fletcher and Collins 1998) (Figure 4.2). Above the four lowest members of the Burgess Shale Formation are the Walcott Quarry Shale Member and the Raymond Quarry Shale Member (Figure 4.2). Although a few soft-bodied fossils occur in the basal shales of the overlying Emerald Lake Oncolite Member, the vast majority of soft-bodied fossils occur in the Walcott Quarry and Raymond Quarry Shale Members. For the purposes of this chapter, the name Burgess Shale is used to denote localities in the Yoho National Park region that occur in the stratigraphic interval encompassing these two members, unless otherwise noted. The Burgess Shale Formation is approximately 350 m thick, and its uppermost member, the Marpole Limestone Member, both caps the underlying nine members in the Burgess Shale Formation and is laterally equivalent with the Cathedral escarpment-capping Waputik Member of the Stephen Formation (Fletcher and Collins 1998) (Figure 4.2). Both the Burgess Shale and Stephen Formations are capped by the laterally extensive limestones of the upper Middle Cambrian Eldon Limestone Formation (Walcott 1908a, 1908b) (Figure 4.2).

The presence of similar assemblages of *Glossopleura* zone trilobites in the Takakkaw Tongue and the Cathedral Limestone Formation suggests coeval deposition (Fritz 1971; Aitken and McIlreath 1984; Fletcher and

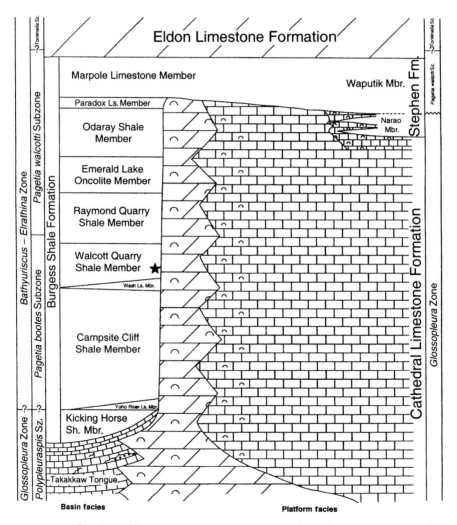

FIGURE 4.2 Stratigraphic context for exposures of Middle Cambrian strata in the Mount Field region. Walcott's original "Phyllopod Bed" is indicated with a star. (Modified from Fletcher and Collins 1998; after original from Fritz 1971 and Aitken and McIlreath 1984)

Collins 1998). Although the basal member of the Burgess Shale Formation contains *Glossapleura* zone trilobites, the remainder of the unit and the adjacent Stephen Formation contain Middle Cambrian *Bathyuriscus-Elrathina* zone trilobites (Walcott 1908a, 1908b, 1917, 1927, 1928; Rasetti 1951), also suggesting coeval deposition of these units. More important, this co-occurrence corroborates lithostratigraphic evidence that significant paleobathymetric disparity existed in this environment during deposition of the Burgess Shale Formation (Fritz 1971; McIlreath 1974, 1977; Aitken and McIlreath 1984; Fletcher and Collins 1998).

Paleoenvironmental Setting

Burgess Shale faunas were deposited in a relatively deep (~200 m below sea level) outer shelf setting, tens of meters oceanward of a massive cliff-like carbonate bank (Aitken 1971; Whittington 1971a; Piper 1972; McIlreath, 1977; Aitken and McIlreath, 1984) (Figure 4.2). During initial deposition of the shales of the Burgess Shale Formation, this carbonate bank extended at least 160 m above the muddy seafloor. Although originally interpreted to reflect an intact reef margin (Aitken and McIlreath 1984), Stewart, Dixon, and Rust (1993) reinterpreted this cliff face as a headwall formed through platform margin collapse. Within the immediate area of the Burgess quarries, strata of the Takakkaw Tongue record post-truncation deposition of associated smaller carbonate debris from this platform into an outer shelf setting (Rasetti 1951; Collins and Stewart 1991). After drowning of the platform, finer-grained siliciclastic sediments built up in the basin, eventually blanketing the surface of the platform.

The fauna of the Burgess Shale includes allochthonous and autochthonous assemblages of pelagic and benthic organisms that were likely transported from near the base of the escarpment into a localized depression or trough by a series of high-density fluidized flows (Whittington 1971a, 1971b; Piper 1972; Allison and Brett 1995). In the Walcott quarry, soft-bodied fossils are preserved in a wide variety of orientations with respect to bedding, and many are compressed at high angles to bedding or exhibit folding of carcasses, suggesting entrainment in a localized flow. The presence of seafloor slumping and graded bedding corroborates this evidence, suggesting transport of faunas (Piper 1972). Some have argued that initiation of these flows was mediated by biological activity (Hecker 1982) or high sediment organic content (Keller 1982). Although the distance of transport of this allochthonous assemblage is debated, Conway Morris (1979a, 1986) estimated maximum transport distances of 0.9 to 1.8 km. Experimental taphonomic studies corroborate the possibility of a far-traveled soft-bodied fauna (Allison 1986; Briggs and Kear 1993), and the presence of photosynthetic algae in the deposit suggests a minimal transport estimate for pelagic forms of 70 m (Briggs, Erwin, and Collier 1994). Individual beds are graded and may possess skeletal remains at the base, including a wide variety of skeletonized and soft-tissue compressions, foldings, and orientations (Walcott 1912a; Whittington 1971a; Piper 1972; Conway Morris 1977d, 1986). The absence of evidence for amalgamation or recycling in the rhythmically layered shales of this deposit further suggests that each layer in the Burgess Shale may represent an individual transport event, and thus paleoecologic inferences (outlined later) drawn from myriad layers in this deposit are constrained by an unknown duration of time-averaging. Although the soft-bodied faunas from the Walcott Quarry Shale Member were almost certainly transported, some fossils from the Raymond Quarry Shale Member were likely deposited *in situ* or suffered minimal transport. For example, trace fossils in the Raymond Member were not transported (Allison and Brett 1995), and some sponges exhibit evidence of burial while their bases were still rooted in the sediment (Collins 1996b).

On a larger scale, strata exposed along the northwest–southeast-trending Mount Wapta–Mount Field ridge closely parallel an important facies change in this region. The schematic diagram in Figure 4.2 illustrates the lateral intersection of the Burgess Shale Formation with the carbonate strata of the Cathedral Limestone Formation. The coarsely crystalline Cathedral Formation carbonates were likely deposited in a shallow, well-oxygenated tropical setting, along the eastern margin of Laurentia (Robison 1976; Scotese et al. 1979; Whittington 1981b). If Cambrian conditions are analogous to modern conditions, then the Burgess Shale may have been accessible to high levels of faunal migration typical of tropical settings, suggesting that the deposit may represent maximum diversity levels for deeper-water tropical Cambrian facies (Conway Morris 1986). Isotopic analyses of bulk organic and carbonate carbon in Burgess Shale fossils seem to corroborate these hypotheses (Butterfield 1990a).

TAPHONOMY

The unique taphonomic and paleoenvironmental conditions of the Burgess Shale are of special importance because they allowed paleontologists to obtain their first view of a typical Cambrian community—in the process, revealing that these communities, much like modern settings, were dominated by soft-bodied forms as well as the more typically preserved mineralized forms. The Burgess Shale fauna includes forms with both mineralized and nonmineralized skeletons, including benthic and pelagic assemblages of faunas and floras that likely inhabited both outer shelf and platform margin settings. These faunas consist of animals living at the site of the deposit, as well as animals and algae that lived nearby and were trapped in a fluidized flow. In addition to living specimens, flows likely entrained a host of discarded or postmortem skeletal elements of trilobites, brachiopods, monoplacophorans, and hyolithids (Whittington 1971a; Piper 1972; Conway Morris 1977d, 1986).

The hard and soft parts of the fauna are largely preserved flat on bedding planes, laid out or squashed in a variety of orientations, although some are oriented at an oblique angle with portions of their appendages intersecting different levels of strata. After quarrying, specimens are usually preserved as part and counterpart on split slabs. Although some of the fossils may have lain on the seafloor for a relatively long time, as evidenced by scattered skeletal fragments, most of these organisms were likely buried rapidly within the fluidized flow. The absence of escape structures, even though the fauna includes effective burrowers such as priapulids (Conway Morris 1977d), and the lack of coiling or osmotic shrinkage in the annelids and arthropods (*sensu* Dean, Rankin, and Hoffman 1964), suggest that Burgess Shale organisms may have been stunned or killed before burial (Conway Morris 1986). Evidence of post-burial decay varies from complete preservation of internal organs to absence of all soft parts. The degree of decay is visible in some specimens by the presence of a squeezed-out intestine and dark stain near the posterior of the organism (Whittington 1971b). This stain is visible as a pyrite patina (Allison and Brett 1995) and is inferred to represent body contents that seeped out into the surrounding sediment during the initial stages of decay.

Although soft-bodied specimens do contain evidence of kerogenized organic carbon films, and some organic-walled fossils have been extracted from the deposit (e.g., *Wiwaxia*), most of the soft tissues have been replaced by hydrous aluminosilicates together with minor pyrite (Conway Morris 1977d; Whittington 1980b; Butterfield 1990a; Orr, Briggs, and Kearns 1998). Soft-bodied fossils are typically preserved at the tops of fining-upward beds, which are commonly laminated and capped with organic detritus (Allison and Brett 1995). Whereas the exact mechanism for this unique soft- and hard-part preservation is still poorly understood, it has been hypothesized to result from the combination of rapid burial and mortality mentioned earlier (Whittington 1981b; Conway Morris 1986), replacement of tissues by clays during early diagenesis (Orr, Briggs, and Kearns 1998; but see Butterfield 1990a; Towe 1996), deposition of this flow in an anoxic environment (Whittington 1985; Allison and Brett 1995), and proximity to structurally resistant geomorphic features (Collins, Briggs, and Conway Morris 1983).

Although the mechanism for replacement of soft tissues by clay minerals is not well constrained, replacement may have been catalyzed by bacteria, by variations in the composition of the decaying tissues, and/or by variations in pore-water chemistry (Orr, Briggs, and Kearns 1998). Another preservation catalyst for this Lagerstätte was the locally fluctuating bottom-water oxygenation. For example, evidence for anoxia in sediments and overlying waters is common in the soft-tissue-rich Walcott Quarry Shale Member and includes rarity of pyrite framboids, even dispersal of pyrite, and absence of trace fossils and shell beds. Where soft-tissue preservation is more sporadic higher in the deposit (e.g., Raymond Quarry Shale Member), higher benthic oxygenation levels are suggested by low-diversity shell beds, complex burrows, and pyritic organic remains (Allison and Brett 1995). Finally, soft-tissue preservation is also mediated by geologic and geomorphic factors. In particular, well-preserved fossils are typically found only in the shales immediately adjacent to the Cathedral Limestone Formation, largely because strata adjacent to this massive carbonate escarpment were shielded from penetrative deformation and development of intense cleavage during subsequent orogenic and metamorphic events (Aitken 1971; Collins, Briggs, and Conway Morris 1983; Aitken and McIlreath 1984).

Mineralized skeletal elements of the Burgess assemblage, such as calcareous skeletal elements, have mostly been replaced or coated by silicates or framboidal pyrite, whereas phosphatic or siliceous skeletal parts such as inarticulate brachiopods or sponge spicules may retain their original composition (Conway Morris 1985). Occurrences of mineralized skeletal elements range from thin, patchy, bedding-plane accumulations to centimeter-thick shell beds; although fragmentation and abrasion are minimal, most fossils are disarticulated and oriented convex up (Allison and Brett 1995).

PALEOBIOLOGY

One of the most notable features of the Burgess Shale fauna is the variety of morphologies and the diversity of taxonomic groups represented. The majority of work on this deposit has focused on describing and interpreting these biota, and it is recommended that readers consult the systematic literature as well as larger review-style atlases (Conway Morris et al. 1982; Whittington 1985; Briggs, Erwin, and Collier 1994) for detailed listings and photographic overviews. A crucial component in understanding the paleobiology and diversity of the faunas is reconciliation of their morphology, a task that has undergone an evolutionary shift. For example, initial work by Walcott and others placed all Burgess faunas within existing clades, whereas later studies suggested that many fossils were of indeterminate origin, leading others to propose that extreme morphologic disparity exists in the Burgess and hence reflects widespread developmental plasticity in the Cambrian (Gould 1989; for a strong opposing view, see Conway Morris 1998, and references therein). Recent work on other related Lagerstätten (Chapters 3 and 5), coupled with new preparation techniques and intensive re-collection of Mount Stephen localities, has caused this interpretive tide to ebb and allowed firmer placement of many of the more problematic faunas into conventional taxonomic groups. Thus, morphologic disparity of the Burgess fauna has been both understated and overstated in the past, with recent consensus that the range in morphologic diversity is comparable to that visible in modern communities (Briggs, Fortey, and Wills 1992). The following is a brief overview of some of the more notable Burgess Shale faunas, including forms that are clearly allied to modern groups and forms whose phylogenetic affinities are more enigmatic.

The most prominent fossils in the Burgess are likely the arthropods, as they encompass a wide variety of forms, including enigmatic taxa such as *Branchiocaris* and *Marella* (Figure 4.3), as well as more easily classified crustaceans. Arthropods also include chelicerates, trilobites, possible ostracods, possible cirripeds, and a number of arachnomorphs of indeterminate grade (Walcott 1908a, 1908b, 1911a, 1912a, 1916, 1918a, 1918b, 1931; Resser 1929, 1938, 1942; Rasetti 1951; Simonetta 1970; Whittington 1971b,

FIGURE 4.3 The arthropod *Marella*. Length of specimen is 2 cm. (From Whittington 1971b: pl. 16, fig. 2. Reproduced with the permission of the Minister of Public Works and Government Services Canada, 2201 and Courtesy of the Geological Survey of Canada)

1974, 1975b, 1980a, 1981b; Hughes 1975; Simonetta and Delle Cave 1975; Briggs 1976, 1977, 1978, 1981, 1992; Bruton 1981; Collins and Rudkin 1981; Conway Morris et al. 1982; Bruton and Whittington 1983; Briggs and Collins 1988). Although there are far too many arthropods to deal with systematically in this chapter, Briggs and Fortey (1989) review the Burgess arthropods' relationships to major arthropod groups, and Briggs, Erwin, and Collier (1994) provide an excellent photographic overview of the more spectacular forms.

Marella splendens (Figure 4.3) is by far the most abundant and well preserved organism in the Burgess Shale, and is preserved in a variety of attitudes relative to bedding planes (Walcott 1912a). This small (~1 to 2 cm long) arthropod has a wedge-shaped head with elongate tapered spines followed by a body with two pairs of jointed appendages and a series of chitinous leg and gill branches, which may have been used to swim or walk on the seafloor (Whittington 1971b; Briggs and Whittington 1985). Unlike many of the other arthropods in the fauna, *Marella* was likely blind—no eyes have been documented from known specimens.

Canadaspis perfecta is the second most common Burgess taxon, and is one of the earliest known crustaceans, perhaps together with various species of *Carnarvonia, Isoxys, Odaraia, Perspicaris, Plenocaris, Tuzoia,* and possibly *Waptia* (Walcott 1908a, 1912a, 1931; Resser 1929; Simonetta and Delle Cave 1975; Briggs 1977, 1978, 1981). *Canadaspis* was a benthic-feeding phyllocarid arthropod that often occurs with exquisitely preserved biramous limbs, abdomen, gut, spiny telson, gill branches, and sometimes antennae visible beyond or under its bivalved carapace (Briggs 1978, 1992). Another common arthropod is *Sidneyia inexpectans,* a large merostome-shaped organism often containing small trilobites, ostracods, and hyoliths in its gut—suggesting that it was also a predator (Bruton 1981). A similarly unusual arthropod is the crustacean-like *Odaraia,* a lobster-like form with large stalked eyes, a cylindrical carapace, and three fin-like projections on its tail (Briggs 1981).

The only described chelicerate from the Burgess is one of the most spectacular forms discovered under the aegis of Desmond Collins's (Collins, Briggs, and Conway Morris 1983) post-Whittington Burgess excavations. *Sanctacaris uncata* constitutes the earliest example of a chelicerate, and is characterized by six pairs of spiny appendages extending from its head (Figure 4.4). This form had a rather flattened telson and wide head shield, and is thought to have been a predator that lived on or near the seafloor (Briggs and Collins 1988). Several taxa with chelicerate affinities are known, and perhaps the most striking example is the blind form *Leanchoilia superlata* (Figure 4.5), which possesses unique whip-like attachments at the end of its frontal appendages and a triangular tail spine extending from its posterior (Bruton and Whittington 1983). Although

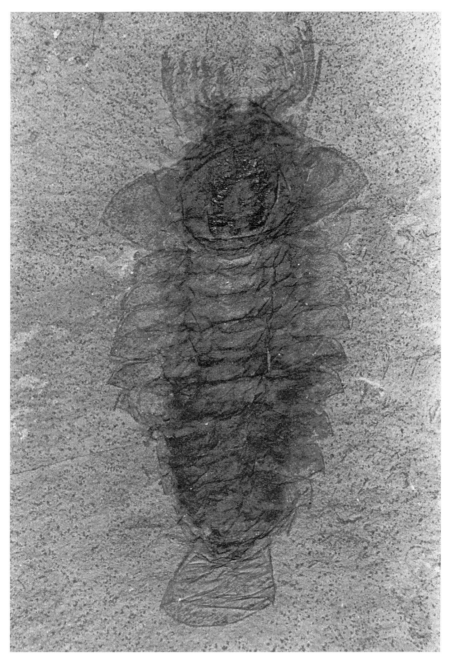

FIGURE 4.4 The chelicerate *Sanctacaris uncata*. Length of specimen is 7 cm. (Photo courtesy of D. Collins, Royal Ontario Museum, Toronto)

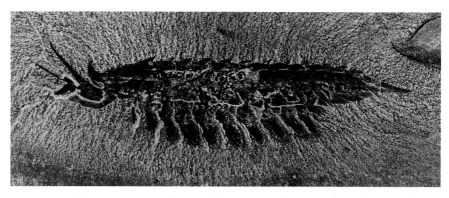

FIGURE 4.5 The arachnomorph *Leancholia superlata.* Length of specimen is 9.5 cm. (Photo courtesy of D. Collins, Royal Ontario Museum, Toronto)

its contents are not distinguishable as phosphatic skeletal debris, the gut trail of this form contains small concentrations of apatite.

Included in the more typical Cambrian arthropods in the Burgess are the 15 genera of trilobites or trilobite-like forms, including agnostids (Walcott 1908b, 1912a, 1916, 1918b, 1931; Resser 1938, 1942; Rasetti 1951; Simonetta and Delle Cave 1975). Among the trilobites, perhaps the most notable is the predatory form *Olenoides,* characterized by a broad cephalon, semicircular pygidium, elongate antennae, biramous head and thoracic limbs, filamentous gill branches, and spinose walking limbs (Whittington 1975b, 1980a). Another well-known trilobite-like arthropod is the burrowing form *Naraoia compacta* (Figure 4.6), an unusual bivalved trilobite that has similar limbs and antennae, but is characterized by a nonmineralized exoskeleton whose dorsal portion consists of two large shields (Whittington 1977). In some examples, traces of the inferred gut and diverticula are visible, although the contents are unknown.

Sponges compose the second most varied group of organisms within the deposit as well as one of the more morphologically ornate components of the fauna. As the dominant sessile epifaunal element of the fauna, this group includes both attached and unattached forms, which are documented in detail in monographs by Walcott (1920) and Rigby (1986). Although hexactinellid and calcareous sponges occur, the vast majority of forms are demosponges. The most common demosponge, *Vauxia gracilenta,* is also possibly one of the most unusual. With tube-like branches and a bush-like appearance, *Vauxia* is notable because its skeleton is composed of a tough organic fibrous network rather than discrete spicules. Other demosponges include forms with delicate radiating spines such as *Choia ridleyi,* the branching heavily spiculate *Pirania muricata,* the conical *Capospongia undulata,* the sac-like *Crumillospongia biporosa,* the elongate cone *Halichondrites elissa,* the double-walled tubular *Leptomitus lineatus,* the wrinkled *Wapkia gransis,* and a variety of forms of

FIGURE 4.6 The trilobite-like *Naraoia compacta.* Length of specimen is 3.6 cm. (Photo courtesy of D. Collins, Royal Ontario Museum, Toronto)

Hazelia and *Vauxia*. Calcareous sponges include forms such as the globe-like *Eiffelia globosa;* hexactinellid sponges include enigmatic forms such as *Protospongia* (thought to be an early lineage of this class) and the sac-like *Diagonialla hindei.*

A variety of polychaetes are also preserved in striking detail, with clear outlines of individual bristle setae, tentacles, trunk segments, appendages, gut, proboscises, and respiratory branchiae (Walcott 1911d; Conway Morris 1979a). *Burgessochaeta setigera* was the most abundant of these forms, a likely infaunal form characterized by long tentacles, biramous appendages, and at least 24 pairs of delicate setae. *Canadia spinosa,* characterized by dorsal and ventral lobes bearing many chitinous bristles, is perhaps one of the most beautiful of the polychaetes, and is also of evolutionary importance because its bristle microstructure is similar to that of *Wiwaxia* (Butterfield 1990b) (Figure 4.7).

Priapulids are relatively common within the Burgess Shale, and include forms like *Ancalagon minor, Fieldia lanceolata, Ottoia prolifica, Lecythioscopa simplex, Louisella pedunculata, Scolecofura rara,* and *Selkirkia columbia* (Walcott 1911d, 1912a; Conway Morris 1977d). Perhaps the best known is *Ottoia,* which may have been a predatory and possibly cannibalistic burrower, characterized by a pronounced proboscis and numerous small hooks around its oral aperture. Several specimens of *Ottoia* contain hyoliths and brachiopods in their guts, and almost all the priapulids exhibit exceptional anatomical details, including setae, gut, proboscis hooks, spines, and papillae.

Cnidarians are also known, including the elongate tubular anthozoan *Mackenzia costalis* and the frondose pennatulacean *Thaumaptilon walcotti*

FIGURE 4.7 The polychaete *Canadia spinosa.* Length of specimen is 3.2 cm. (Photo courtesy of D. Collins, Royal Ontario Museum, Toronto)

(Walcott 1911c; Briggs and Conway Morris 1986; Conway Morris 1993). *Thaumaptilon* is of importance because it may have close affinities to late Neoproterozoic Ediacaran fronds like *Charniodiscus,* and/or may reflect the first documented complex Ediacaran holdover taxon.

Although Middle Cambrian echinoderms are more frequently found articulated in other Burgess Shale–type deposits (Chapter 5), a possible early crinoid, *Echmatocrinus brachiatus,* occurs in the Burgess, as well as examples of more primitive groups of stalked echinoderms, such as the cystoid *Gogia radiata* (Sprinkle 1973; Sprinkle and Collins 1998; for alternative interpretations, see Conway Morris 1993, or Ausich and Babcock 1998). Both of these forms exhibit features thought to be ancestral to those of younger stalked echinoderms, including the presence of an attached holdfast, plated stalk and calyx, and arms radiating from the calyx. However, *Gogia* has rather unusual pores along its plate sutures, as well as brachioles along its ambulacra. All *Echmatocrinus* specimens are attached to hard objects such as the priapulid *Selkirkia* or to hyolithids, suggesting an immobile suspension-feeding life habit (Sprinkle and Collins 1998). Edriaoasteroids such as *Walcottidiscus magister* and *W. typicalis* are also known (Bassler 1935, 1936). *Eldonia ludwigi* is another form that may be allied with the echinoderms, as it possesses tube-feet, a discoidal center with a radial meshwork of fibers, a coiled gut, oral tentacles, and other features suggesting a holothurian affinity (Walcott 1911c; Durham 1974).

Although its affinity remains controversial (Butterfield 1990b), the lancelet-like *Pikaia gracilens* possesses features indicating the presence of a notochord and myotomic muscle tissue, which suggest that it may be one of the earliest known cephalochordates (Walcott 1911d; Conway Morris and Whittington 1979).

Ctenophores also occur, including forms such as the ornate globe-like *Ctenorhabdotus,* the bowl-shaped *Fasciculus vesanus,* and the stringy *Xanioascus canadensis,* all of which have typical features such as cilia, but atypical features such as the presence of many comb-rows and absence of tentacles (Simonetta and Delle Cave 1978; Conway Morris and Collins 1996). Another possible pelagic hydroid or cnidarian is *Scenella amii,* a chondrophorine preserved in dense aggregations of small, flattened cones on bedding planes (Matthew 1902; Babcock and Robison 1988).

In addition to the exceptionally preserved soft-bodied forms, the Burgess Shale contains a number of more traditional skeletonized faunas, including brachiopods, hyolithids, molluscs, and the trilobites mentioned previously. At least six brachiopod genera have been recognized, including relatively ordinary inarticulates such as *Lingulella waptaensis,* as well as more ornate forms like *Micromitra burgessensis,* which commonly exhibit exquisite preservation of elongate delicate setae extending beyond each shell's mantle fringe (Walcott 1912b, 1924; Resser 1938). Articulate brachiopods are also known, including the calcified forms *Diraphora bellicostata* and *Nisusia burgessensis* (Walcott 1924). *Haplophrentis carinatus* is one of the more famous, yet traditional, faunal elements in that it is one of the few examples of an articulated hyolith occurring with the helens, operculum, and conch all preserved and attached to one another (Matthew 1899; Babcock and Robison 1988).

Finally, in addition to metazoa, there are a number of trace fossils and algae in the deposit. Although little systematic work has been done on trace fossils within the more oxygenated layers of the deposit, bed-parallel forms such as *Cruziana* and *Planolites* and vertically inclined forms such as *Arenicolites* and *Monocraterian* occur (Allison and Brett 1995). At least 10 genera of elaborate frondose algae also occur, typically preserved as thin, isolated, shiny films (Walcott 1919, 1931; Walton 1923; Ruedemann 1931; Satterthwait 1976; Collins, Briggs, and Conway Morris 1983; Conway Morris and Robison 1988). Although the flora has not attracted the widespread attention that the fauna has generated, it may include approximately 10 percent of the fossils in the deposit (Conway Morris 1986). Algae include possible green algae, such as the kelp-like *Margaretia dorus* and the stipe-laden *Yuknessia simplex.* A variety of red algae also are found, ranging from simple forms such as the stick-like *Dalya nitens* and *Wahpia insolens,* to more elaborate forms such as the branching *Dalyia racemata,* the finely branching *Waputikia ramosa,* and the shrub-like *Bosworthia simulans.* Possible cyanobacteria such as the tuft-like *Marpolia aequalis,* the filamentous *Marpolia spissa,* and the perforated, sheet-like *Morania confluens* are also found.

PROBLEMATIC FAUNA

Some forms, such as the sponge-like *Chancelloria eros* and the graptolite-like *Chaunograptus scandens,* are of uncertain taxonomic affinity because their morphology is unlike that of modern or other fossil analogues (Walcott 1920; Ruedemann 1931). For example, although relatively large (up to 50 cm tall) specimens of the cone-like *Chancelloria* are known, the nature of their spicular construction is unlike that of modern sponges (Goryanskiy 1973; Bengtson and Missarzhevsky 1981; Bengtson, Collins, and Runnegar 1996; Butterfield and Nicholas 1996), and branching forms from other Burgess Shale–type deposits have not yet been formally described (K. Peterson, personal communication, 1999). Other forms are of enigmatic taxonomic affinity simply because they are known from only a single specimen, such as *Odontogriphus omalus* (Conway Morris 1976a), or because the few examples of these taxa are poorly preserved. Problematic members of the nekton include forms like *Nectocaris pteryx,* a probable chaetognath characterized by large eyes and a dart-shaped finned body (Conway Morris 1976b; Simonetta 1988).

Wiwaxia corrugata is another problematic form characterized by a double row of vertically extrusive and laterally spinose plates extending along its body and sides. The presence of broken and regenerated or replaced plates along the margins of *Wiwaxia* suggest that this organism was involved in predator–prey interactions and that it may have used its plates as defensive armor (Conway Morris 1985, 1992; Butterfield 1990b). Although the taxonomic affinity of *Wiwaxia* is unclear, the presence of shorter chitinous sclerites in interplate body surfaces suggests that *Wiwaxia* may be allied with halkieriids or polychaetes (Conway Morris 1985; Conway Morris and Peel 1995).

Opabinia regalis is one of the more striking problematic forms (Figure 4.8). This segmented organism had a flexible body and a nozzle-snouted proboscis at its anterior characterized by a claw-like apparatus on its end. Like *Anomalocaris*, *Opabinia* has lateral flaps along its trunk, a fan-shaped tail, and lobopodian-type legs, and is of uncertain taxonomic affinity (Whittington 1975a; Bergström 1986, 1987; Briggs and Whittington 1987). *Opabinia* may have swum about the bottom, collecting organic debris. Although the most striking physical aspect of this creature is its five eyes, the combination of the lobopod and arthropod morphologic characters is of greater importance, as they have led to the suggestion that this form reflects an early arthropod stem group (Budd 1996).

In an interesting history of early discovery, pieces of *Anomalocaris canadensis* and *A. nathorsti* were originally interpreted to be several indi-

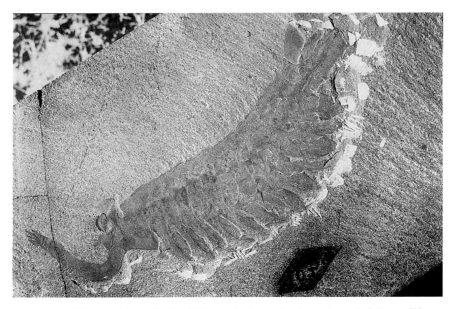

FIGURE 4.8 The problematic *Opabinia regalis*. Length of specimen is 6.4 cm. (Photo courtesy of H. B. Whittington, University of Cambridge)

vidual organisms, including arthropods, crustaceans, and medusoids (Whiteaves 1892; Walcott 1911c, 1912a; Madsen 1957; Conway Morris 1978; Conway Morris and Whittington 1979). Subsequent to these reports, excavation of fossil appendages coupled with discovery of complete specimens revealed that *Anomalocaris* is one of the largest and most bizarre organisms of the Burgess Shale fauna, possessing a large lobed head, large eyes, two spiny shrimp-shaped feeding limbs, a lateral elongate series of flaps, lobopod-like legs, a fan-shaped tail, and a circular or square jaw surrounded by circlets of sharp plates or teeth. At present, *A. canadensis* and its close cousin *Laggania cambria* are among the largest organisms in the Burgess paleocommunity (up to 0.5 m) and are thought to belong to an extinct class of arthropods (Collins 1996a; Hou, Bergström, and Ahlberg 1995; Minicucci 1999). The combination of large lateral eyes, swimming lobes, and unique jaw structure suggests that these taxa may have been formidable mobile predators (Whittington and Briggs 1985).

Dinomischus isolatus, a cup-shaped organism characterized by a long thin stem, supported itself above the substrate by anchoring a bulbous holdfast into the seafloor. *Dinomischus* is also known from the Chengjiang Lagerstätte (Chapter 3), and had a stalk lined with cilia and plates. Although very small, this organism is thought to have been a passive suspension feeder and may represent a type of entoproct (Conway Morris 1977b; Chen and Erdtmann 1991).

Lobopodians are another group that is well constrained at the phylum level, but whose detailed systematic position is still poorly understood. The most well known of these mobile epifaunal organisms are *Asheaia pedunculata* and *Hallucigenia sparsa.* As with *Anomalocaris* and other problematica, the discovery of articulated specimens from other Burgess Shale–type deposits coupled with delicate extraction techniques have allowed us insight into their morphology and paleobiology (Robison 1985; Ramsköld and Hou 1991). *Asheaia* is commonly associated with sponges, and has soft legs with claws at their ends (Walcott 1911d; Whittington 1978; Monge-Najera 1995). *Hallucigenia* was an armored lobopod, originally thought to walk on seven pairs of spines, with seven tentacles on its top (Walcott 1911d; Conway Morris 1977a). Recent discovery of similar organisms in the Chengjiang Lagerstätte has revised this interpretation, indicating that the tentacles are actually walking legs and the spines are on top, possibly for defensive protection (Ramsköld and Hou 1991).

PALEOECOLOGY

Although a field-based paleoecologic study of the Burgess paleocommunity has yet to be published (but see Collins 1996b; Fletcher and Collins 1998), the combination of abundant fossiliferous material, concentration of such samples in a few institutions, widespread soft-tissue preservation, and systematic analysis of the fossils allows us remarkable insight into the distribution, diversity, and niches of the fauna and flora inhabiting the general vicinity of the soft seafloor at the base of the Cathedral Limestone Formation.

Like many modern communities, the Burgess Shale is dominated by arthropods (42 genera), but also contains a host of other organisms, including as many as 20 sponge genera, 18 problematic taxa, 10 algae, seven priapulids, six brachiopods, five polychaetes, four echinoderms, four cnidarians, three ctenophores, two hemichordates, two chordates, two onychophorans, one genus of hyolithid, and several types of trace fossils (syntheses in Conway Morris 1986; Briggs, Erwin, and Collier 1994; Allison and Brett 1995). Although the majority of this information was collected from the classic Walcott and Raymond quarries, rough diversity information compiled from more recent Royal Ontario Museum (ROM) sites in coeval strata exhibits a similar range in diversity, with at least 37 genera representing at least five major phyla (Collins, Briggs, and Conway Morris 1983). Like most fossiliferous deposits, strata in these deposits record burial of animals living at the time of transport, as well as a large number of dead individuals or discarded skeletal elements. Thus, discarded or dead skeletal elements in the Burgess Shale may significantly bias analyses of abundance and diversity in this deposit. Conway Morris (1986), however, was able to minimize these biases in his analyses by compensating for trilobite ecdysis, noting both the presence or absence of opercula and helens in hyolithids and the extended mantle setae in inarticulate brachiopods in his attempts to reconstruct the original relative composition of the Burgess Shale community. After accounting for dead or discarded skeletal elements, his analyses of species diversity yielded results similar to those from analysis of generic diversity—an important correlation because exclusion of dead individuals significantly decreases the relative contribution of shelly components to the total biodiversity.

Abundance estimates were made through systematic analysis of over 65,000 Burgess Shale museum specimens on over 30,000 slabs, with animals comprising up to 87.9 percent; algae, at least 11.3 percent; and indeterminate material, approximately 0.8 percent (Conway Morris 1986; Briggs, Erwin, and Collier 1994). Typical Cambrian shelly marine taxa (trilobites, brachiopods, monoplacophorans, hyolithids, echinoderms, sponges) account for approximately 20 percent of the genera and less than 2 percent of the individuals. In addition, whereas the Burgess is similar to other Cambrian deposits because arthropods dominate its taxonomic composition, it is different because trilobites compose only a small portion (<14 percent of "living" genera and <0.5 percent of "living" individuals) of these assemblages (Conway Morris 1986). Although only a small portion of the information compiled by Conway Morris (1986) is presented here, it reiterates that the vast majority of faunas inhabiting typical Cambrian marine settings would be unknown were it not for Lagerstätte such as the Burgess Shale.

PALEOECOLOGICAL RECONSTRUCTION

Although most of the soft-bodied forms have been transported, Conway Morris (1986) was able to classify Burgess taxa based on their inferred life habits, feeding type, trophic grouping, and position in a trophic web. In doing so, he noted that the Burgess fauna is dominated by infaunal and epifaunal benthic forms that likely lived on and in the muddy substrates adjacent to their locus of final deposition. Although this does not provide *in situ* information about tiering, commensalism, or related habits, much of this information has been gleaned from careful analysis of the paleobiology and taphonomy of individual specimens.

In general, the infauna appears to be dominated by priapulids, including a variety of mobile forms such as *Ottoia* and sessile forms such as *Louisella*. The epifauna is considerably more diverse, including a variety of attached forms such as the sponge *Vauxia* (Collins 1996b) and the enigmatic *Dinomischus* and *Chancelloria*, as well as a variety of mobile forms reflected by the myriad arthropods such as *Marella* and the problematic *Wiwaxia*. Pelagic additions to the fauna include agnostoid and eodiscoid trilobites (Robison 1972) and rare soft-bodied taxa such as the floating holothuroid *Eldonia* and the actively mobile chordate *Pikaia*.

In addition to these guilds, there is strong evidence for ecologic interactions between members of the Burgess fauna. The sponge *Pirania*, for example, has brachiopods growing on it (Whittington 1985), and the sponge *Choia* is often found in clusters (Walcott 1920), suggesting a gregarious habit. In some cases, postmortem links provide clues about the life habits of faunas, such as occurrences of the sponge *Eiffelia* attached to empty tubes of the priapulid *Selkirkia* (Conway Morris et al. 1982).

Feeding preference is also quite variable, with important benthic assemblages dominated by deposit-feeding arthropods, carnivores, or scavengers dominated by forms such as *Sidneyia,* and epifaunal suspension feeders dominated by sponges and brachiopods (Conway Morris 1986). Feeding habits are based largely on analogy with inferred habits of similar extant and fossil taxa, in addition to circumstantial morphologic evidence such as the predator-like mouth apparatus of *Anomalocaris* and *Laggania* (Whittington and Briggs 1985). Evidence for predation is also inferred by identifiable gut contents, including hyolithids and brachiopods in the gut of *Ottoia* (Walcott 1911d), and trilobites, brachiopods, and hyolithids in the gut of *Sidneyia* (Bruton 1981). Possible scavenging is also suggested from occurrences of the lobopod *Hallucigenia* on top of an undescribed worm (Conway Morris 1977a).

Correlation of the data on inferred life habit and feeding type with respect to numbers of individuals and biovolume suggests that epifaunal vagrant deposit feeders are the most dominant Burgess Shale organisms, with secondary importance of infaunal sessile suspension feeders, nektobenthic suspension feeders, epifaunal vagrant carnivores, epifaunal sessile (high-level) suspension feeders, and infaunal vagrant carnivores (Conway Morris 1986).

Analysis of the trophic nucleus suggests that the Burgess Shale community was quite similar to modern communities, in that it was dominated by relatively few species; 10 percent of the benthic species comprise 91 percent of the individuals and 82 percent of the biovolume of the entire community. A hypothetical generalized trophic web constructed by Conway Morris (1986) indicates that primary consumers are an important component of the Burgess community. Suspension feeders are thought to have consumed phytoplankton and suspended detritus, while deposit feeders may have exploited bacterial and microbial forms, benthic algae, and detritus. Carnivores may have played an important part in this trophic system, as they occupied a high level and account for a large amount of the biovolume of the fauna.

One of the most significant contrasts between the Burgess Shale fauna and other Cambrian assemblages is the significant role of predation, as evidenced by the large number of carnivores and "armored" organisms. Although predatorial borings and bite marks are occasional features of deposits from this interval (Conway Morris and Jenkins 1985; Conway Morris and Bengtson 1994), evidence of predation in other Cambrian deposits may be underrepresented due to lack of preservation of mobile predatory organisms. Because most of the carnivorous or scavenging animals from the Burgess Shale are soft-bodied, they would likely never have been preserved under normal taphonomic conditions characteristic of typical Cambrian deposits.

Niche partitioning and possible food selection by epifaunal vagrant deposit feeders may also be inferred from the Burgess fauna. Whereas data from Cambrian paleocommunities where only skeletonized organisms are preserved suggests that Cambrian communities were ecologically generalized, niche partitioning in the Burgess Shale indicates that they may have been relatively complex. For example, epifaunal tiering by sponges, rare eocrinoids, and pseudo-crinoids may have extended to levels as high as 20 cm above the substrate (Conway Morris 1979b).

CONCLUSIONS

The unusual paleoenvironmental and taphonomic regimes of the Burgess Shale allowed the preservation of a rich soft- and hard-bodied Middle Cambrian fauna. Although not unique, it is certainly one of the best-preserved sources of information on Cambrian life. Whereas this fauna currently has analogues in similar Middle and Lower Cambrian deposits, it is important both in a historical context, because it helped focus attention on the Cambrian radiation, and in a paleontological context, because study of the fauna has revealed myriad new and different taxa, including potential new phyla. More important, the Burgess Shale fundamentally changed the way paleontologists thought about the complexity and diversity of Cambrian paleocommunities because it helped redirect shelly fauna-based thinking about the biota and conditions of the Cambrian explosion (Conway Morris 1989; Aronson 1992).

Despite all the accomplishments that have been made through several generations of work on this deposit, it is striking that many of Walcott's original Burgess specimens have yet to be reexamined or described using modern methods of specimen preparation and dissection. Perhaps the discovery of better preserved specimens in the older Cambrian Lagerstätte in Chengjiang (Chapter 3) and additional Middle Cambrian Lagerstätten in sites such as northern Greenland (Chapter 5) has helped alleviate the need for such work. However, correlation of information from the Burgess Shale fauna with that from other Burgess Shale–type faunas is an obvious focus for future work, and may aid in identifying the signature of the metazoan radiation during the Cambrian, as well as evaluating the role of contingency in early animal evolution. Further insights into Burgess Shale paleoecology might also be made through field-based analysis of fossil distribution within the Burgess Shale, perhaps together with taphonomic information and x-radiographic ichnofabric analyses (*sensu* Gaines and Droser 1999).

ACKNOWLEDGMENTS

I would like to thank D. Collins and S. Conway Morris for their constructive comments, which helped improve this contribution.

REFERENCES

Aitken, J. D. 1971. Control of lower Paleozoic sedimentary facies by the Kicking Horse Rim, southern Rocky Mountains, Canada. *Bulletin of Canadian Petroleum Geology* 19:557–569.

Aitken, J. D. 1997. *Stratigraphy of the Middle Cambrian Platformal Succession, Southern Rocky Mountains.* Geological Survey of Canada Bulletin, no. 398. Ottawa: Natural Resources Canada.

Aitken, J. D., and I. A. McIlreath. 1984. The Cathedral Reef escarpment, a Cambrian great wall with humble origins. *Geos* 13:17–19.

Allison, P. A. 1986. Soft-bodied animals in the fossil record: The role of decay in fragmentation during transport. *Geology* 14:979–981.

Allison, P. A., and C. A. Brett. 1995. *In situ* benthos and paleo-oxygenation in the Middle Cambrian Burgess Shale, British Columbia, Canada. *Geology* 23:1079–1082.

Aronson, R. B. 1992. Decline of the Burgess Shale fauna: Ecologic or taphonomic restriction? *Lethaia* 25:225–229.

Ausich, W. I., and L. E. Babcock. 1998. The phylogenetic position of *Echmatocrinus brachiatus,* a probable octocoral from the Burgess Shale. *Palaeontology* 41:193–202.

Ausich, W. I., and D. J. Bottjer. 1982. Tiering in suspension-feeding communities on soft substrata throughout the Phanerozoic. *Science* 216:173–174.

Babcock, L. E., and R. A. Robison. 1988. *Taxonomy and Paleobiology of Some Middle Cambrian Scenella (Cnidaria) and Hyolithids (Mollusca) from Western North America.* University of Kansas Paleontological Contributions, no. 121. Lawrence: University of Kansas Paleontological Institute.

Bassler, R. S. 1935. The classification of Edrioasteroidea. *Smithsonian Miscellaneous Collections* 93:1–11.

Bassler, R. S. 1936. New species of American Edrioasteroidea. *Smithsonian Miscellaneous Collections* 95:1–33.

Bengtson, S., D. H. Collins, and B. Runnegar. 1996. Chancelloriid sclerite formation—Turning the problem inside-out. In J. E. Repetski, ed., *Sixth North American Paleontological Convention: Abstracts of Papers,* p. 29. Paleontological Society Special Publication, no. 8. Washington, D.C.: Paleontological Society.

Bengtson, S., and V. V. Missarzhevsky. 1981. Coeloscleritophora—A major group of enigmatic Cambrian metazoans. *United States Geological Survey Open-file Report* 81–743:19–21.

Bergström, J. 1986. *Opabinia* and *Anomalocaris,* unique Cambrian "arthropods." *Lethaia* 19:241–246.

Bergström, J. 1987. The Cambrian *Opabinia* and *Anomalocaris. Lethaia* 20:187–188.

Briggs, D. E. G. 1976. *The Arthropod* Branchiocaris *n. gen., Middle Cambrian, Burgess Shale, British Columbia.* Geological Survey of Canada Bulletin, no. 264. Ottawa: Geological Survey of Canada.

Briggs, D. E. G. 1977. Bivalved arthropods from the Cambrian Burgess Shale of British Columbia. *Palaeontology* 20:595–621.

Briggs, D. E. G. 1978. The morphology, mode of life, and affinities of *Canadaspis perfecta* (Crustacea: Phyllocarida), Middle Cambrian, Burgess Shale, British Columbia. *Philosophical Transactions of the Royal Society of London, B* 281: 439–487.

Briggs, D. E. G. 1981. The arthropod *Odaraia alata* Walcott, Middle Cambrian, Burgess Shale, British Columbia. *Philosophical Transactions of the Royal Society of London, B* 291:541–585.

Briggs, D. E. G. 1992. Phylogenetic significance of the Burgess Shale crustacean *Canadaspis. Acta Zoologica* 73:293–300.

Briggs, D. E. G., and D. Collins. 1988. A Middle Cambrian chelicerate from Mount Stephen, British Columbia. *Palaeontology* 31:779–798.

Briggs, D. E. G., and S. Conway Morris. 1986. Problematica from the Middle Cambrian Burgess Shale of British Columbia. In A. Hoffman and M. H. Nitecki, eds., *Problematic Fossil Taxa,* pp. 167–183. New York: Oxford University Press.

Briggs, D. E. G., D. H. Erwin, and F. J. Collier. 1994. *The Fossils of the Burgess Shale.* Washington, D.C.: Smithsonian Institution Press.

Briggs, D. E. G., and R. A. Fortey. 1989. The early radiation and relationships of the major arthropod groups. *Science* 246:241–243.

Briggs, D. E. G., R. A. Fortey, and M. A. Wills. 1992. Morphologic disparity in the Cambrian. *Science* 256:1670–1673.

Briggs, D. E. G., and A. J. Kear. 1993. Decay and preservation of polychaetes: Taphonomic thresholds in soft-bodied organisms. *Paleobiology* 19:107–134.

Briggs, D. E. G., and H. B. Whittington. 1985. Modes of life of arthropods from the Burgess Shale, British Columbia. *Transactions of the Royal Society of Edinburgh* 76:149–160.

Briggs, D. E. G., and H. B. Whittington. 1987. The affinities of the Cambrian animals *Anomalocaris* and *Opabinia. Lethaia* 20:185–186.

Bruton, D. L. 1981. The arthropod *Sidneyia inexpectans,* Middle Cambrian, Burgess Shale, British Columbia. *Philosophical Transactions of the Royal Society of London, B* 295:619–656.

Bruton, D. L., and H. B. Whittington. 1983. *Emeraldella* and *Leanchoilia,* two arthropods from the Burgess Shale, British Columbia. *Philosophical Transactions of the Royal Society of London, B* 300:553–585.

Budd, G. E. 1996. The morphology of *Opabinia regalis* and the reconstruction of the arthropod stem-group. *Lethaia* 29:1–14.

Butterfield, N. J. 1990a. Organic preservation of non-mineralizing organisms and the taphonomy of the Burgess Shale. *Paleobiology* 16:272–286.

Butterfield, N. J. 1990b. A reassessment of the enigmatic Burgess Shale fossil *Wiwaxia corrugata* (Matthew) and its relationship to the polychaete *Canadia spinosa* Walcott. *Paleobiology* 16:287–303.

Butterfield, N. J. 1995. Secular distribution of Burgess Shale–type preservation. *Lethaia* 28:1–14.

Butterfield, N. J., and C. J. Nicholas. 1996. Burgess Shale–type preservation of both non-mineralizing and "shelly" Cambrian organisms from the Mackenzie Mountains, northwestern Canada. *Journal of Paleontology* 70:893–899.

Chen, J., and B. D. Erdtmann. 1991. Lower Cambrian fossil Lagerstätte from Chengjiang, Yunnan, China: Insights for reconstructing early metazoan life. In A. M. Simonetta and S. Conway Morris, eds., *The Early Evolution of Metazoa and the Significance of Problematic Taxa,* pp. 57–76. Cambridge: Cambridge University Press.

Collins, D. 1996a. The "evolution" of *Anomalocaris* and its classification in the arthropod class Dinocarida (Nov.) and order Radiodonta (Nov.). *Journal of Paleontology* 70:280–293.

Collins, D. 1996b. The *Leanchoilia-Ottoia* fauna from the Middle Cambrian Burgess Shale of British Columbia. In J. E. Repetski, ed., *Sixth North American Paleontological Convention: Abstracts of Papers,* p. 77. Paleontological Society Special Publication, no. 8. Washington, D.C.: Paleontological Society.

Collins, D., D. E. G. Briggs, and S. Conway Morris. 1983. New Burgess Shale fossil sites reveal Middle Cambrian faunal complex. *Science* 222:163–167.

Collins, D., and D. M. Rudkin. 1981. *Priscansermarinus barnetti,* a probable lepadomorph barnacle from the Middle Cambrian Burgess Shale of British Columbia. *Journal of Paleontology* 55:1006–1015.

Collins, D., and W. D. Stewart. 1991. The Burgess Shale and its environmental setting, Fossil Ridge, Yoho National Park. In P. L. Smith, ed., *Canadian Pale-*

ontology Conference I: A Field Guide to the Paleontology of Southwestern Canada, pp. 104–117. Vancouver: University of British Columbia.

Conway Morris, S. C. 1976a. A new Cambrian lophophorate from the Burgess Shale of British Columbia. *Palaeontology* 19:199–222.

Conway Morris, S. C. 1976b. *Nectocaris pteryx,* a new organism from the Middle Cambrian Burgess Shale of British Columbia. *Neues Jahrbuch für Geologie und Paläontologie, Monatshefte* 12:705–713.

Conway Morris, S. C. 1977a. A new metazoan from the Cambrian Burgess Shale, British Columbia. *Palaeontology* 20:623–640.

Conway Morris, S. C. 1977b. A new entoproct-like organism from the Burgess Shale of British Columbia. *Palaeontology* 20:833–845.

Conway Morris, S. C. 1977c. Aspects of the Burgess Shale fauna, with particular reference to the non-arthropod component. *Journal of Paleontology, Supplement* 51:7–8.

Conway Morris, S. C. 1977d. *Fossil Priapulid Worms.* Special Papers in Palaeontology, no. 20. London: Palaeontological Association.

Conway Morris, S. C. 1978. *Laggania cambria* Walcott: A composite fossil. *Journal of Paleontology* 52:126–131.

Conway Morris, S. C. 1979a. Middle Cambrian polychaetes from the Burgess Shale of British Columbia. *Philosophical Transactions of the Royal Society of London, B* 285:227–274.

Conway Morris, S. C. 1979b. The Burgess Shale (Middle Cambrian) fauna. *Annual Review of Ecology and Systematics* 10:327–349.

Conway Morris, S. C. 1985. Cambrian Lagerstätten: Their distribution and significance. *Philosophical Transactions of the Royal Society of London, B* 311:49–65.

Conway Morris, S. C. 1986. The community structure of the Middle Cambrian phyllopod bed (Burgess Shale). *Palaeontology* 29:423–467.

Conway Morris, S. C. 1989. Burgess Shale faunas and the Cambrian explosion. *Science* 246:339–346.

Conway Morris, S. C. 1992. Burgess Shale–type faunas in the context of the "Cambrian explosion": A review. *Journal of the Geological Society of London* 149:631–636.

Conway Morris, S. C. 1993. The fossil record and the early evolution of the Metazoa. *Nature* 361:219–225.

Conway Morris, S. C. 1995. Enigmatic shells, possibly halkieriid, from the Middle Cambrian Burgess Shale, British Columbia. *Neues Jahrbuch für Geologie und Paläontologie, Abhandlungen* 195:319–331.

Conway Morris, S. C. 1998. *The Crucible of Creation: The Burgess Shale and the Rise of Animals.* Oxford: Oxford University Press.

Conway Morris, S. C., and S. Bengtson. 1994. Cambrian predators: Possible evidence from boreholes. *Journal of Paleontology* 68:1–23.

Conway Morris, S. C., and D. H. Collins. 1996. Middle Cambrian ctenophores from the Stephen Formation, British Columbia, Canada. *Philosophical Transactions of the Royal Society of London, B* 351:279–308.

Conway Morris, S. C., and R. J. F. Jenkins. 1985. Healed injuries in Early Cambrian trilobites from South Australia. *Alcheringa* 9:167–177.

Conway Morris, S. C., and J. S. Peel. 1995. Articulated halkieriids from the Lower Cambrian of north Greenland and their role in early protostome evolution. *Philosophical Transactions of the Royal Society of London, B* 347:305–358.

Conway Morris, S., and R. A. Robison. 1988. *More Soft-Bodied Animals and Algae from the Middle Cambrian of Utah and British Columbia.* University of Kansas

Palaeontological Contributions, no. 122. Lawrence: University of Kansas Paleontological Institute.

Conway Morris, S., and H. B. Whittington. 1979. The animals of the Burgess Shale. *Scientific American* 241:122–133.

Conway Morris, S., H. B. Whittington, D. E. G. Briggs, C. P. Hughes, and D. L. Bruton. 1982. *Atlas of the Burgess Shale.* London: Palaeontological Association.

Dean, D., J. S. Rankin, and E. Hoffman. 1964. A note on the survival of polychaetes and amphipods in stored jars of sediments. *Journal of Paleontology* 38:608–609.

Durham, J. W. 1974. Systematic position of *Eldonia ludwigi* Walcott. *Journal of Paleontology* 48:750–755.

Fletcher, T. P., and D. H. Collins. 1998. The Middle Cambrian Burgess Shale and its relationship to the Stephen Formation in the southern Rocky Mountains. *Canadian Journal of Earth Sciences* 35:413–436.

Fritz, W. H. 1971. Geological setting of the Burgess Shale. In E. L. Yochelson, ed., *Proceedings of the North American Paleontological Convention,* vol. 2, pp. 1155–1170. Lawrence, Kans.: Allen Press.

Gaines, R. R., and M. L. Droser. 1999. Ichnofabric and taphonomy of the Lower Cambrian Latham Shale, San Bernardino County, California: Clues for the decline of soft-bodied preservation. *Geological Society of America Abstracts with Programs* 31:363.

Goryanskiy, V. Y. 1973. O neobkhodimosti isklyucheniya roda *Chancelloria* Walcott iz tipa gubok. In I. T. Zhuravleva, ed., *Problemy paleontologii i biostratigrafii nizhnego kembriya Sibiri i Dal'nego Vostoka,* pp. 39–44. Novosibirsk: Izdatel'stvo Nauka.

Gould, S. J. 1989. *Wonderful Life: The Burgess Shale and the Nature of History.* New York: Norton.

Hecker, B. 1982. Possible benthic fauna and slope instability relationships. In S. Saxon and J. K. Nieuwenhuis, eds., *Marine Slides and Other Mass Movements,* pp. 335–347. New York: Plenum.

Hou, X., J. Bergström, and P. Ahlberg. 1995. *Anomalocaris* and other large animals in the Lower Cambrian Chengjiang fauna of southwest China. *Geologiska Föreningens i Stockholm Förhandlingar* 117:163–183.

Hughes, C. P. 1975. Redescription of *Burgessia bella* from the Middle Cambrian Burgess Shales, British Columbia. *Fossils & Strata* 4:415–436.

Keller, G. H. 1982. Organic matter and the geotechnical properties of submarine sediments. *Geo-Marine Letters* 2:191–198.

Madsen, F. J. 1957. On Walcott's supposed Cambrian holothurians. *Journal of Paleontology* 31:281–282.

Matthew, G. F. 1899. Studies on Cambrian faunas, no. 3: Upper Cambrian fauna of Mt. Stephen, British Columbia. *Transactions of the Royal Society of Canada,* 2nd ser., 5:39–66.

Matthew, G. F. 1902. Notes on Cambrian faunas. *Transactions of the Royal Society of Canada* 8:93–112.

McIlreath, I. A. 1974. Stratigraphic relationships at the western edge of the Middle Cambrian carbonate facies belt, Field, British Columbia. *Geological Survey Papers of Canada* 74–1A:333–334.

McIlreath, I. A. 1977. Accumulation of a Middle Cambrian, deep-water limestone debris apron adjacent to a vertical, submarine carbonate escarpment, southern Rocky Mountains, Canada. In H. E. Cook and P. Enos, eds., *Deep-*

Water Carbonate Environments, pp. 113–124. Special Publication, no. 25. Tulsa, Okla.: Society of Economic Paleontologists and Mineralogists.

Minicucci, J. M. 1999. Forward to the Cambrian—Anomalocarid studies at the end of the millennium. *Palaeontological Association Newsletter* 41:23–32.

Monge-Najera, J. 1995. Phylogeny, biogeography and reproductive trends in the Onychophora. *Zoological Journal of the Linnean Society* 114:21–60.

Orr, P. J., D. E. G. Briggs, and S. L. Kearns. 1998. Cambrian Burgess Shale animals replicated in clay minerals. *Science* 281:1173–1175.

Palmer, A. R. 1960. Some aspects of the early Upper Cambrian stratigraphy of White Pine County, Nevada, and vicinity. In J. W. Boettcher and W. W. Sloan, eds., *Guidebook to the Geology of East Central Nevada*, pp. 53–58. Salt Lake City: Intermountain Association of Petroleum Geologists.

Piper, D. W. 1972. Sediments of the Middle Cambrian Burgess Shale, Canada. *Lethaia* 5:169–175.

Ramsköld, L., and X. Hou. 1991. New Early Cambrian animal and onycophoran affinities of enigmatic metazoans. *Nature* 351:225–227.

Rasetti, F. 1951. Middle Cambrian stratigraphy and faunas of the Canadian Rocky Mountains. *Smithsonian Miscellaneous Collections* 116.

Resser, C. E. 1929. New Lower and Middle Cambrian Crustacea. *Proceedings of the United States National Museum* 76:1–18.

Resser, C. E. 1938. Fourth contribution to nomenclature of Cambrian fossils. *Smithsonian Miscellaneous Collections* 97.

Resser, C. E. 1942. Fifth contribution to nomenclature of Cambrian fossils. *Smithsonian Miscellaneous Collections* 101:1–58.

Rigby, J. K. 1986. *Sponges of the Burgess Shale (Middle Cambrian), British Columbia.* Palaeontographica Canadiana, no. 2. Calgary: Canadian Society of Petroleum Geologists.

Robison, R. A. 1960. Lower and Middle Cambrian stratigraphy of the eastern Great Basin. In J. W. Boettcher and W. W. Sloan, eds., *Guidebook to the Geology of East Central Nevada*, pp. 43–52. Salt Lake City: Intermountain Association of Petroleum Geologists.

Robison, R. A. 1972. Mode of life of agnostoid trilobites. *International Geological Congress* 24:33–40.

Robison, R. A. 1976. Middle Cambrian trilobite biostratigraphy of the Great Basin. *Brigham Young University Geology Studies* 23:93–109.

Robison, R. A. 1985. Affinities of *Aysheaia* (Onychophora) with description of a new Cambrian species. *Journal of Paleontology* 59:226–235.

Ruedemann, R. 1931. Some new Middle Cambrian fossils from British Columbia. *Proceedings of the United States National Museum* 79:1–18.

Satterthwait, D. F. 1976. Paleobiology and paleoecology of Middle Cambrian algae from western North America. Ph.D. diss., University of California, Los Angeles.

Scotese, C. R., R. K. Bambach, C. Barton, R. van der Voo, and A. M. Ziegler. 1979. Paleozoic base maps. *Journal of Geology* 87:217–277.

Simonetta, A. M. 1970. Studies on non-trilobite arthropods of the Burgess Shale (Middle Cambrian). *Palaeontographica Italica* 66:35–45.

Simonetta, A. M. 1988. Is *Nectocaris pteryx* a chordate? *Bolletino di Zoologia* 55:63–68.

Simonetta, A. M., and L. Delle Cave. 1975. The Cambrian non-trilobite arthropods from the Burgess Shale of British Columbia: A study of their comparative morphology, taxonomy and evolutionary significance. *Palaeontographica Italica* 69:1–37.

Simonetta, A. M., and L. Delle Cave. 1978. Notes on new and strange Burgess Shale fossils (Middle Cambrian of British Columbia). *Atti della Società Toscana di Scienze Naturali, A* 85:45–49.

Sprinkle, J. 1973. *Morphology and Evolution of Blastozoan Echinoderms*. Harvard University Museum of Comparative Zoology Special Publication. Cambridge, Mass.: Museum of Comparative Geology, Harvard University.

Sprinkle, J., and D. Collins. 1998. Revision of *Echmatocrinus* from the Middle Cambrian Burgess Shale of British Columbia. *Lethaia* 31:269–282.

Stewart, W. D. 1989. A preliminary report on stratigraphy and sedimentology of the lower and middle Chancellor Formation (Middle and Upper Cambrian) in the zone of facies transition, Rocky Mountain Main ranges, southeastern British Columbia. *Geological Survey of Canada, Current Research* 89–1D:61–68.

Stewart, W. D., O. A. Dixon, and B. R. Rust. 1993. Middle Cambrian carbonate-platform collapse, southeastern Canadian Rocky Mountains. *Geology* 21:687–690.

Towe, K. M. 1996. Fossil preservation in the Burgess Shale. *Lethaia* 29:107.

Walcott, C. D. 1908a. Nomenclature of some Cambrian Cordilleran formations. *Smithsonian Miscellaneous Collections* 53:1–12.

Walcott, C. D. 1908b. Cambrian sections of the Cordilleran area. *Smithsonian Miscellaneous Collections* 53:167–230.

Walcott, C. D. 1911a. Middle Cambrian Merostomata: Cambrian geology and paleontology II. *Smithsonian Miscellaneous Collections* 57:17–40.

Walcott, C. D. 1911b. A geologist's paradise. *National Geographic Magazine* 22:509–521.

Walcott, C. D. 1911c. Middle Cambrian Holothurians and Medusae: Cambrian geology and paleontology II. *Smithsonian Miscellaneous Collections* 57:41–68.

Walcott, C. D. 1911d. Middle Cambrian Annelids: Cambrian geology and paleontology II. *Smithsonian Miscellaneous Collections* 57:09–144.

Walcott, C. D. 1912a. Middle Cambrian Branchiopoda, Malacostraca, Trilobita and Merostomata: Cambrian geology and paleontology II. *Smithsonian Miscellaneous Collections* 57:145–228.

Walcott, C. D. 1912b. *Cambrian Brachiopoda*. United States Geological Survey Monographs, no. 51. 2 vols. Washington, D.C.: Government Printing Office.

Walcott, C. D. 1916. Cambrian trilobites: Cambrian geology and paleontology III. *Smithsonian Miscellaneous Collections* 64:303–456.

Walcott, C. D. 1917. Nomenclature of some Cambrian Cordilleran formations. *Smithsonian Miscellaneous Collections* 67:1–8.

Walcott, C. D. 1918a. Geological explorations in the Canadian Rockies. In Explorations and Fieldwork of the Smithsonian Institution in 1917. *Smithsonian Miscellaneous Collections* 68:4–20.

Walcott, C. D. 1918b. Appendages of trilobites: Cambrian geology and paleontology IV. *Smithsonian Miscellaneous Collections* 67:115–216.

Walcott, C. D. 1919. Middle Cambrian algae: Cambrian geology and paleontology IV. *Smithsonian Miscellaneous Collections* 67:217–260.

Walcott, C. D. 1920. Middle Cambrian Spongiae: Cambrian geology and paleontology IV. *Smithsonian Miscellaneous Collections* 67:261–364.

Walcott, C. D. 1924. Cambrian and Ozarkian Brachiopoda: Cambrian geology and paleontology IV. *Smithsonian Miscellaneous Collections* 67:477–554.

Walcott, C. D. 1927. Pre-Devonian sedimentation in the southern Canadian Rocky Mountains. *Smithsonian Miscellaneous Collections* 75:147–173.

Walcott, C. D. 1928. Pre-Devonian Paleozoic formations of the Cordilleran provinces of Canada. *Smithsonian Miscellaneous Collections* 75:185–367.

Walcott, C. D. 1931. Addenda to descriptions of Burgess Shale fossils. *Smithsonian Miscellaneous Collections* 85:1–46.

Walton, J. 1923. On the structure of a Middle Cambrian alga from British Columbia (*Marpolia spissa* Walcott). *Proceedings of the Cambridge Philosophical Society, Biological Sciences* 1:59–62.

Whiteaves, J. F. 1892. Description of a new genus and species of Phyllocarid Crustacea from the Middle Cambrian of Mount Stephen, B.C. *Canadian Record of Science* 5:205–208.

Whittington, H. B. 1971a. The Burgess Shale: History of research and preservation of fossils. In E. L. Yochelson, ed., *Proceedings of the North American Paleontological Convention,* vol. 2, pp. 1170–1201. Lawrence, Kans.: Allen Press.

Whittington, H. B. 1971b. *Redescription of* Marella splendens *(Trilobitoidea) from the Burgess Shale, Middle Cambrian, British Columbia.* Geological Survey of Canada Bulletin, no. 209. Ottawa: Department of Energy, Mines, and Resources.

Whittington, H. B. 1974. Yohoia *Walcott and* Plenocaris *n. gen., Arthropods from the Burgess Shale, Middle Cambrian, British Columbia.* Geological Survey of Canada Bulletin, no. 231. Ottawa: Department of Energy, Mines, and Resources.

Whittington, H. B. 1975a. The enigmatic animal *Opabinia regalis,* Middle Cambrian Burgess Shale, British Columbia. *Philosophical Transactions of the Royal Society of London, B* 271:1–43.

Whittington, H. B. 1975b. Trilobites with appendages from the Middle Cambrian, Burgess Shale, British Columbia. *Fossils & Strata* 4:97–136.

Whittington, H. B. 1977. The Middle Cambrian trilobite *Naraoia,* Burgess Shale, British Columbia. *Philosophical Transactions of the Royal Society of London, B* 284:165–197.

Whittington, H. B. 1978. The lobopod animal *Aysheaia pedunculata* Walcott, Middle Cambrian, Burgess Shale, British Columbia. *Philosophical Transactions of the Royal Society of London, B* 284:165–197.

Whittington, H. B. 1980a. The significance of the fauna of the Burgess Shale, Middle Cambrian, British Columbia. *Proceedings of the Geologists' Association* 91:127–148.

Whittington, H. B. 1980b. Exoskeleton, moult stage, appendage morphology, and habits of the Middle Cambrian trilobite *Olenoides serratus. Palaeontology* 23:171–204.

Whittington, H. B. 1981a. Cambrian animals: Their ancestors and descendants. *Proceedings of the Linnean Society, New South Wales* 105:79–87.

Whittington, H. B. 1981b. Rare arthropods from the Burgess Shale, Middle Cambrian, British Columbia. *Philosophical Transactions of the Royal Society of London, B* 292:329–357.

Whittington, H. B. 1985. *The Burgess Shale.* New Haven: Yale University Press.

Whittington, H. B., and D. E. G. Briggs. 1985. The largest Cambrian animal, *Anomalocaris,* Burgess Shale, British Columbia. *Philosophical Transactions of the Royal Society of London, B* 309:569–609.

Yochelson, E. L. 1967. Charles Doolittle Walcott, 1850–1927. *Biographical Memoirs, National Academy of Sciences* 39:471–540.

5

Burgess Shale–Type Localities: The Global Picture

James W. Hagadorn

INCREASED AWARENESS OF THE BURGESS SHALE BIOTA HAS led to systematic searching for and accidental discovery of a plethora of soft-bodied faunas throughout Lower and Middle Cambrian strata all across the globe. Many of these soft-bodied forms are known from the Burgess Shale or are closely allied with taxa from the Burgess Shale. Over the past 30 years, these Burgess Shale–type faunas have been documented throughout the Cambrian, as well as from Ordovician and Silurian strata. In some cases, such as Chengjiang (Chapter 3), these faunas have actually begun to approach or eclipse the paleontological impact of the original Burgess quarries.

Furthermore, as this awareness peaked, many obscure, previously undescribed, or isolated soft-bodied specimens from other deposits were recognized as Burgess Shale–type faunas and have been placed into a larger-scale paleobiologic context (e.g., Parker Slate and Kinzers Formation biotas) (Conway Morris 1993; Garcia-Bellido Capdevila and Conway Morris 1999). Although the faunal abundance and preservation in many of these deposits is less than that in the Burgess Shale, many of them have provided additional information about soft-bodied paleocommunities that inhabited a wide range of paleoenvironments during the Cambrian explosion.

Like those in the original locality, Burgess Shale–type faunas are usually preserved in shales, exhibit a range in preservation from exceptional articulated specimens to disarticulated fragments, and were formed by a combination of taphonomic scenarios (e.g., obrution and stagnation). In general, Burgess Shale–type deposits are dominated by nontrilobite arthropods, with a large contingent of priapulid worms, a diverse array of sponges, and a minor assortment of sclerite-bearing lobopodians. Based on these taxa, key observations about the distribution, temporal persistence, mode of life, and identity of soft-bodied faunas have been made, in addition to hypotheses about early animal evolution. Some of these observations are presented in this chapter, as well as a brief review of the key features of the major Burgess Shale–type localities (for a more thorough overview, see Conway Morris 1985, 1989b, and references therein).

GEOLOGICAL CONTEXT

This chapter utilizes a broad definition of a Burgess Shale–type locality, which includes deposits with articulated soft-bodied taxa that are also known from the Burgess Shale, as well as other Cambrian soft-bodied occurrences that may contain different nonmineralized taxa. For taphonomic studies, however, it is probably more prudent to delimit Burgess Shale–type deposits by their mode of preservation (Butterfield 1990b) than by the occurrence of a shared fauna or simply by the presence of soft-tissue preservation, which may have occurred through different taphonomic pathways (Butterfield 1994). Because a strict taphonomic definition eliminates many isolated occurrences of shared taxa, which extend the geographic and temporal range of Burgess Shale taxa, and because these occurrences are utilized in many of the phylogenetic and evolutionary studies focusing on Burgess Shale–type taxa (see "Paleoecology"), these shared taxa occurrences are included in this chapter.

Burgess Shale–type biotas have been documented from at least 40 Cambrian localities distributed across every continent except Antarctica and South America (Conway Morris 1989b, 1990; Allison and Briggs 1991, and references therein) (Figure 5.1). In the Cambrian, most faunas occur in strata ranging in age from the Atdabanian Stage of the Early Cambrian to the top of the Marjuman Stage of the Middle Cambrian (Conway Morris 1989a; Allison and Briggs 1991). A few Burgess Shale–type assemblages are also known from the Upper Cambrian, Ordovician, and Silurian, but most of these younger occurrences consist of isolated specimens (Dawson and Hinde 1889; Dawson 1896; Rigby 1986b; Barskov and Zhuravlev 1988).

Conway Morris (1989b) thoroughly reviewed the geographic distribution of Burgess Shale–type faunas and noted that the majority of these taxa come from a dozen or so well-exposed highly fossiliferous localities, which will be briefly discussed. Although not covered here, isolated occurrences of more common Burgess Shale–type taxa are known from exposures in England (Breadstone and Shineton Shales; Whittard 1953), Sardinia (Vasenapov Suite; Repina and Okuneva 1969), Spain (Murero

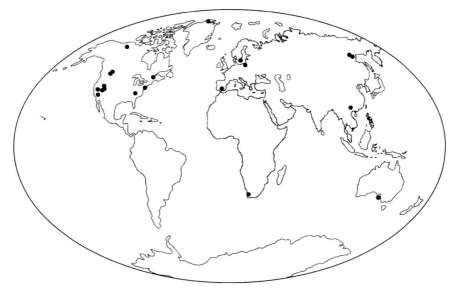

FIGURE 5.1 Distribution of most of the Burgess Shale–type localities (indicated by dots). (Modified from Conway Morris 1989a)

Shale; Conway Morris and Robison 1986), eastern Siberia (Chabdy Suite; Krishtofovich 1953; Barskov and Zhuravlev 1988), north-central Russia (Chopko, Lena, and Sinsk Formations; Voropinov 1957; Goryansky 1977; Rigby 1986a; Barskov and Zhuravlev 1988), Guizhou Province, China (Kaili Formation; Zhao 1994), and Quebec, Canada (Metis Shale; Dawson and Hinde 1889; Dawson 1896), as well as from deposits in California (Carrara Formation; M. Vendrasco, personal communication, 1999), Idaho (Gibson Jack Formation; Robison 1984) (Rennie Shale; Resser 1938), Nevada (unit B7; Stewart and Palmer 1967) (Pioche Shale; McCollum 1994) (Emigrant Formation; Hagadorn 1998), and Utah (Bloomington Formation; Briggs and Robison 1984) (Ute Formation; Briggs and Robison 1984). Many of these localities are relatively non-fossiliferous (e.g., Pioche and Latham Shales), and many are no longer available for paleontological inquiry, due to their subsurface nature (e.g., Zawiszyn Formation), infilling of quarries, or burial by housing developments (e.g., Kinzers Shale, Parker Slate of the eastern United States) (Campbell 1969).

PALEOENVIRONMENTAL SETTINGS

Many Burgess Shale–type localities were characterized by depositional environments broadly similar to those of the original Burgess Shale. Faunas tend to have been deposited in relatively deep settings at or below storm wave base (Rees 1984; Robison 1991), where periodically dysaerobic bottom-waters, anaerobic sediments, and clay–organic interactions coupled with rapid burial may have mediated exceptional soft-bodied preservation (Conway Morris 1989b; Butterfield 1990b; Briggs and Fortey 1992; Allison and Brett 1995). In Laurentia, for example, most localities occur on the seaward side of a large carbonate platform that extended along the Laurentian margin during much of the Cambrian, in settings broadly interpreted to reflect outer shelf environments (Palmer 1960; Robison 1960; overviews in Conway Morris 1986; Conway Morris 1989a, 1989b). Soft-bodied fossils are typically preserved within or at bed interfaces of graded mudstone and siltstone units deposited adjacent to these carbonate banks or in a deep subtidal outer shelf-type setting. Fossils occur in beds with sharp or erosive bases, both in uniformly oriented concentrations on bed surfaces and in quasi-randomly oriented positions within beds. In most cases, Burgess Shale–type deposits reflect deposition outboard of this bank. In a few instances, however, such as the upper Wheeler Formation in Utah and, perhaps, the Mount Cap Formation of northwestern Canada, fossils were deposited landward of the carbonate platform, in a relatively quiescent, muddy subtidal environment (Palmer 1960; Robison 1960, 1984, 1991; Brady and Koepnick 1979; Rees 1984, 1986; Rogers 1984; Butterfield and Nicholas 1996). The Latham Shale faunas of California also may have been deposited in a similar inner detrital belt setting (Briggs and Mount 1982).

Outside North America, there appears to be significantly more variety in paleoenvironments that preserve Burgess Shale–type faunas. An exception to the Laurentian-style deposition that characterizes many of the deeper subtidal Burgess Shale–type localities (e.g., Chengjiang) is the finely laminated siltstones and mudstones of the Soom Shale of South Africa. Because it conformably overlies glacial tillites, the Soom Shale was probably deposited under anoxic conditions in a glacioestuarine to shallow-marine environment, at the initial phase of a glacioeustatic rise in sea level, through turbiditic deposition of outwash silts and muds from retreating glaciers (Theron, Rickards, and Aldridge 1990).

The Venenäs Conglomerate fauna of southeastern Sweden may also represent a paleoenvironmental anomaly, as quasi-Burgess Shale–type faunas occur in sandstones, rather than in finely laminated shales (Pompeckj 1927; Jaeger and Martinsson 1967; Krumbiegel, Deichfuss, and Deichfuss 1980). Unfortunately, soft-bodied fossils are known from only glacial erratics, so little is known about the paleoecology or paleobiology of soft-bodied faunas that may have been living in or adjacent to these coarser clastic Early Cambrian environments. At the one outcrop of this unit, only biomineralized fossils such as *Mobergella* are preserved and occur in calcite-cemented, medium-grained, well-rounded sandstones (Bengtson 1968). Fossils are preserved throughout the conglomerate, suggesting significant reworking, and are sometimes covered by a thin phosphatic film.

TAPHONOMY

Over the years, a variety of mechanisms have been invoked to explain relatively widespread preservation of Burgess Shale–type faunas in the Cambrian, including diagenetic replacement by aluminosilicates (Whittington 1971; Conway Morris 1977, 1986; Orr, Briggs, and Kearns 1998), flattening of specimens with syngenetic formation of euhedral pyrite crystals (Whittington 1974), and preservation as kerogenized organic remains in close association with clay particles (Butterfield 1990a). Others have suggested reduced bioturbation and/or anoxia as key factors mediating Burgess Shale–type preservation (Aronson 1992, 1993; Allison and Briggs 1993a, 1993b, 1994; Pickerill 1994; Allison and Brett 1995).

In general, it appears that most Burgess Shale–type occurrences represent obrution deposits where autochthonous forms were smothered, together with parautochthonous and allochthonous forms that were transported in distal turbidites or high-density fluidized flows. Based on the examples discussed later, together with Chengjiang and the Burgess Shale (Chapters 3 and 4), it seems there existed during this interval a continuum of taphonomic pathways that mediated soft-tissue preservation, most of which were characterized by lack of post-burial disturbance, low oxygen, and clay–organic interactions. However, based on the variety of shelf-like marine environments that exhibit Burgess Shale–type preservation and their widespread geographic distribution, there may have been larger-scale controls acting on primary taphonomic processes during this interval (Allison and Briggs 1991, 1993a). Butterfield (1995), for example, has suggested that long-term changes in marine clay mineralogies, whose anti-enzymatic and/or stabilizing effects on decaying organic molecules could have mediated organic film-type preservation known from many Burgess deposits, may have significantly affected the secular distribution of nonmineralized Burgess-type preservation. However, the Soom Shale of South Africa and the Emu Bay Shale of South Australia provide a contrast to this model (Briggs and Nedin 1997; Nedin 1997; Gabbott 1998).

The fauna from the subtidal Emu Bay Shale exhibits two entirely different modes of soft-tissue preservation, in which the soft tissues of most of the fauna are preserved as red-stained fibrous calcium carbonate. The exception is the enigmatic taxon *Myoscolex*, in which muscle blocks and fibers are preserved as apatite encased in a thin film of fibrous calcite (Briggs and Nedin 1997; Nedin 1997). Unlike most of the other Burgess Shale–type localities, in which chitinous and soft-bodied fossils are preserved as organic residues or aluminosilicate films (Orr, Briggs, and Kearns 1998), the Emu Bay Shale does not show such replacement of soft parts, and little organic or chitinous matter is preserved (McKirdy 1971; Glaessner 1979). The fossiliferous shales of Emu Bay are thought to have formed in a localized euxinic water body ponded below wave base next to a mobile slope (Pocock 1964; Glaessner 1979; Conway Morris and Jenkins 1985).

An additional example is the Soom Shale, where fossils are preserved with high fidelity as illitic and alunite minerals, and appear to reflect autochthonous or parautochthonous forms deposited in a basinal setting characterized by relatively cool anaerobic bottom-waters (summary in Gabbott 1998). Gabbott (1998) and Gabbott, Aldridge, and Theron (1995) suggest that clay mineral replacement of soft tissues likely resulted from clay mineral adsorption onto decaying organic matter. Although similar clay replacement has recently been documented by Orr, Briggs, and Kearns (1998) for the original Burgess Shale, the Soom Shale examples are better constrained. For example, Soom Shale clay mineral adsorption was mediated by overlying waters that were of low pH and rich in cations, coupled with low carbonate, iron, and oxygen concentrations in the sediment.

Another taphonomic factor influencing the distribution of Burgess Shale–type faunas, especially in Laurentia, is their proximity to a major carbonate bank, which in numerous cases helped shield the soft shales from subsequent tectonic and metamorphic overprinting (Collins, Briggs, and Conway Morris 1983; Aitken and McIlreath 1984). For example, in the Kinzers Formation, which outcrops in many places in Pennsylvania, soft-bodied fossils are preserved only adjacent to massive carbonates of the platform margin, where the soft shales have been sheltered from post-Cambrian deformation. Elsewhere, outcrops with exceptionally preserved biomineralized fossils, such as edrioasteroids, occur, but specimens are commonly deformed or vertically cleaved by regional tectonism (Ruedemann 1933; Resser and Howell 1938; Durham 1966; Campbell and Kauffman 1969). This type of post-lithification protection also occurs in Burgess Shale–type deposits in Greenland and Utah (Robison 1991; Conway Morris 1998).

Soft-Bodied Faunas
of the Major Localities

Lowest Cambrian (Upper? Tommotian–Lower Atdabanian) Kalmarsund Sandstone of southeastern Sweden. Although it does not exhibit classic Burgess Shale–type preservation, the Kalmarsund Sandstone is worthy of mention because it contains the earliest known articulated Burgess Shale–type taxon. In particular, two specimens of the oldest onychophoran-like lobopodian, *Xenusion,* have been recovered from glacial erratics derived from this unit (Pompackj 1927; Jaeger and Martinsson 1967; Krumbiegel, Deichfuss, and Deichfuss 1980).

Lower Cambrian (Lower Atdabanian) Zawiszyn Formation of northern Poland. Core cuttings from the Zawiszyn Formation are well known for the cap-shaped fossil *Mobergella holsti* and contain the earliest evidence of Burgess Shale–type soft-tissue preservation, including a number of soft-bodied forms such as the anomalocarid *Cassubia infercambriensis* and the naraoiid *Liwia plana* (Bengtson 1968; Lendzion 1975; Dzik and Lendzion 1988).

Lower Cambrian (Uppermost Atdabanian) Maotianshan Shale near Chengjiang, China. See Chapter 3.

Lower Cambrian (Uppermost Atdabanian) Buen Formation of Peary Land, northern Greenland. Commonly known as the Sirius Passet fauna, Burgess Shale–type faunas from the Buen Formation are dominated by a diverse suite of nontrilobite arthropods, such as *Isoxys volucris;* the soft-bodied *Tegopelte;* the more common naraoiid *Buenaspis forteyi;* the lobopod *Kerygmachela kierkegaardi;* a variety of undescribed polychaete, priapulid, and palaeoscolecid worms; and the demosponges *Choia hindei* and *Vauxia? gracilenta* (Rigby 1986b; Conway Morris et al. 1987; Conway Morris and Peel 1990; Budd 1993, 1997, 1999; Williams, Siveter, and Peel 1996; Budd and Peel 1998). The most notable fossils extracted from these localities are fully articulated halkieriids, which are conspicuous scleritized sluglike metazoans previously known only from individual plates (Conway Morris and Peel 1990, 1995).

Lower Cambrian (Toyonian) Kinzers Formation of Pennsylvania. The Kinzers Formation contains a variety of soft-bodied taxa, including *Anomalocaris?* cf. *pennsylvanica;* the bivalved arthropods *Hymenocaris dubia, Serracaris lineata,* and *Tuzoia getzi;* sponges such as *Hazelia walcotti;* priapulid worms such as *Selkirkia? columbia* and *Tubulella flagellum;* several worms of uncertain phylogenetic affinity, such as *Atalotaenia adela* and *Kinzeria crinita;* the alga *Margaretia dorus;* and a trilobite with preserved antennae (Stose and Jonas 1922; Dunbar 1925; Resser 1929; Resser and Howell 1938; Campbell 1969; Campbell and Kauffman 1969; Conway Morris 1977, 1985; Briggs 1978, 1979; Rigby 1987; Conway Morris and Robison 1988; Garcia-Bellido and Conway Morris 1999).

Lower Cambrian (Toyonian) Parker Slate of northern Vermont. The poorly known Parker Slate contains a number of soft-bodied forms, including the demosponge *Leptomitus zitteli* and bivalved arthropods such as *Protocaris marshi* and *Tuzoia vermontensis* (Resser and Howell 1938; Briggs 1976, 1979; Rigby 1987). Potentially the most important fossil from this quarry is the frondlike *Emmonsaspis cambriensis,* represented by three specimens, which is thought to represent an Ediacaran holdover taxon, perhaps related to the pennatulaceans (Conway Morris 1993).

Lower Cambrian (Upper Toyonian) Latham Shale of southern California. Although poorly preserved due to regional metamorphic overprinting, the Latham Shale is characterized by classic Burgess Shale–type preservation of isolated appendages of *Anomalocaris* (Briggs and Mount 1982), worms (Mount 1980; Conway Morris 1985), possible annelids (Conway Morris 1977), and abundant frondescent and globose algae such as *Morania* (Waggoner and Hagadorn 1999). Also preserved intact are easily disarticulated biomineralized taxa, such as the eocrinoid *Gogia ojenai* (Durham 1978).

Lower Cambrian (Botomian) Emu Bay Shale of Kangaroo Island, South Australia. The Emu Bay Shale possesses several Burgess Shale–type taxa, notably the pelagic bivalved arthropods *Isoxys communis* and *Tuzoia australis,* the sponge-like *Chancelloria,* the priapulid worm *Palaeoscolex antiquus, Anomalocaris briggsi,* the *Opabinia*-like *Myoscolex ateles,* the arachnomorph *Xandarella,* and the trilobite-like *Naraoia* (Glaessner 1979; Conway Morris and Jenkins 1985; Bengtson et al. 1990; McHenry and Yates 1993; Nedin 1995a, 1995b, 1997; Briggs and Nedin 1997). Other than trilobites with preserved antennae (McHenry and Yates 1993; Nedin 1995b), the most common soft-bodied taxon in the deposit is the unusually phosphatized *Myoscolex,* represented by over 2,800 specimens, some with preserved appendages (Briggs and Nedin 1997).

Lower Cambrian (Upper Toyonian) Mount Cap Formation of Northwest Territories, Canada. The Mount Cap Formation soft-bodied fauna includes anomalocarid appendages, several types of bivalved nontrilobite arthropods such as *Isoxys,* articulated hyolithids, articulated chancelloriids, a palaeoscolecid worm, *Wiwaxia* sclerites, bipectinate setae attributed to filter-feeding crustaceans, and filamentous algae/cyanobacteria (Butterfield 1993, 1994; Butterfield and Nicholas 1996).

Middle Cambrian (Upper Delamaran–Marjuman) Spence Shale and Wheeler and Marjum Formations of Utah. Exposures of the Wheeler and Marjum Formations and the Spence Shale in the Wellsville Mountains, Drum Mountains, and House Range are well known for their trilobites (such as *Elrathia kingii,* which adorns many a refrigerator magnet), having been quarried by commercial and avocational paleontologists for decades. Although Burgess Shale–type faunas are rare within these units, extensive collecting efforts and generous donations of specimens by avocational paleontologists have provided the most well documented assemblage of Burgess Shale–type faunas outside the Chengjiang Lagerstätte.

The fauna includes numerous nontrilobite arthropods, including *Alalcomenaeus, Anomalocaris nathorsti, Branchiocaris pretiosa, Cambropodus gracilis* (Figure 5.2), and *Naraoia compacta;* a number of bivalved forms, such as *Canadaspis* and *Tuzoia guntheri;* a variety of algae, including *Margaretia dorus, Marpolia spissa* (Figure 5.3), and *Yuknessia simplex;* the lobopodian *Aysheaia prolata;* the annelid worm *Palaeoscolex ratcliffei* (Figure 5.4); the priapulids *Ottoia prolifica* and *Selkirkia willoughbyi;* and the medusoid *Eldonia ludwigi* (Willoughby and Robison 1979; Gunther and Gunther 1981; Robison and Richards 1981; Conway Morris and Robison 1982, 1988; Robison 1984, 1985, 1991; Briggs and Robison 1984). The House Range deposits are also well known for their sponges, including demosponges such as *Choia carteri, Hamptonia bowerbanki, Hazelia palmata,* and *Leptomitus metta,* and hexactinellid sponges such as *Diagoniella cyathiformis* (Figure 5.5), *Hintzespongia bilamina, Testiispongia venula, Protospongia hicksi* (Figure 5.6), and *Valospongia gigantis* (Rigby 1966, 1969, 1978, 1983; Rigby and Gutschick 1976; Rigby, Gunther, and Gunther 1997). These units are also well known for exquisite preservation of easily disarticulated biomineralized taxa, such as articulated hyolithids, carpoids, eocrinoids (Figure 5.7), and several edrioasteroids (Sprinkle 1985; Ubaghs and Robison 1985, 1988; Babcock and Robison 1988).

Middle Cambrian (Lowermost Marjuman) Burgess Shale near the Walcott Quarry, Canadian Rockies. Burgess Shale–type localities located within 30 km of the Walcott quarry contain several new forms not known from

FIGURE 5.2 The uniramous arthropod *Cambropodus gracilis* from the Wheeler Formation in the Drum Mountains, Utah. Body length of specimen is 1 cm. (Photo courtesy of R. Robison, University of Kansas)

FIGURE 5.3 The alga *Marpolia spissa* from the upper Spence Shale in the Wellsville Mountains, Utah. Specimen is 3.1 cm across. (Photo courtesy of R. Robison, University of Kansas)

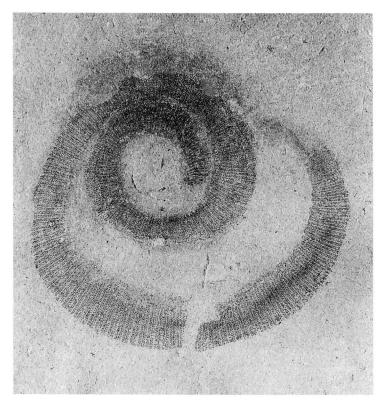

FIGURE 5.4 The worm *Palaeoscolex ratcliffei* from the Spence Shale in the Wellsville Mountains, Utah. Length of specimen is 9 cm. (Photo courtesy of R. Robison, University of Kansas)

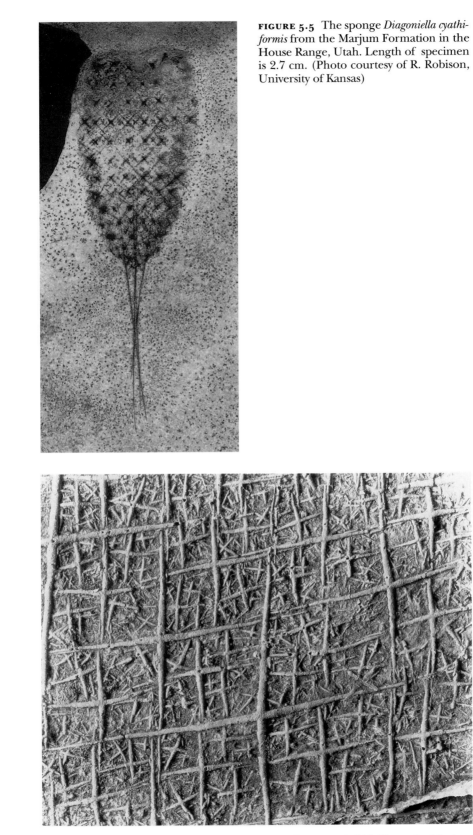

FIGURE 5.5 The sponge *Diagoniella cyathiformis* from the Marjum Formation in the House Range, Utah. Length of specimen is 2.7 cm. (Photo courtesy of R. Robison, University of Kansas)

FIGURE 5.6 Spicule fabric in wall of the sponge *Protospongia hicksi* from the Marjum Formation in the House Range, Utah. Field of view is 2.1 cm. (Photo courtesy of J. K. Rigby Sr., Brigham Young University)

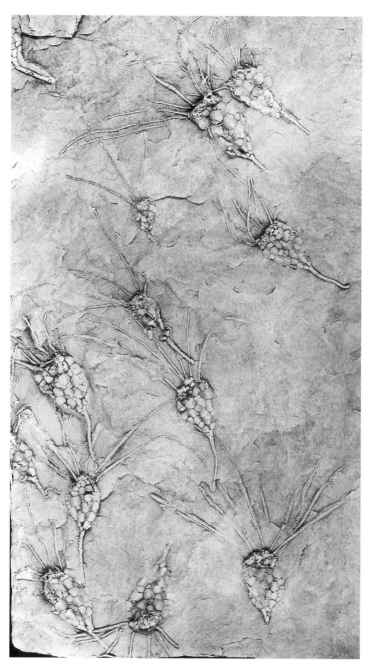

FIGURE 5·7 Multiple specimens of the eocrinoid *Gogia* on a bedding plane from the Spence Shale, Wellsville Mountains, Utah. Field of view is 10.7 cm wide. (Photo courtesy of R. Robison, University of Kansas)

the original quarries, such as the spectacular chelicerate *Sanctacaris un-cata* (Collins, Briggs, and Conway Morris 1983; Briggs and Collins 1988). Most of the localities, however, contain similar faunas, but in very different proportions (Collins, Briggs, and Conway Morris 1983; Collins 1985). In total, at least five distinct faunal assemblages are recognized and occur at four stratigraphic levels spanning two trilobite zones. Some of these localities are dominated by the large predator *Anomalocaris.* In contrast, other horizons do not contain *Marella* (the most common animal in the Burgess Shale), but include abundant *Alalcomenaeus* and *Branchiocaris,* which are extremely rare in the Walcott quarry (Collins, Briggs, and Conway Morris 1983; Briggs and Collins 1988).

Upper Ordovician (Ashgill) Soom Shale Member of the Cedarberg Formation of Cape Province, South Africa. Although isolated occurrences of Burgess Shale–type taxa, such as *Palaeoscolex* and *Diagoniella,* are known from Silurian deposits (Rigby 1986b; Conway Morris 1977, 1989b), the youngest occurrence of an entire Burgess Shale–type fauna occurs in the Soom Shale, where gut traces, fibrous musculature, and other soft tissues are preserved. In addition to algae, two localities of the Soom Shale have revealed giant conodont apparatuses (for which it is famous), the naraoiid arthropod *Soomaspis splendida,* the eurypterid *Onychopterella augusti,* and a scolecodont apparatus (Theron, Rickards, and Aldridge 1990; Chesselet 1992; Aldridge, Gabbott, and Theron 1993; Aldridge and Theron 1993; Braddy, Aldridge, and Theron 1995; Fortey and Theron 1995; Gabbott, Aldridge, and Theron 1995, 1997).

PALEOBIOLOGY AND PALEOECOLOGY

The discovery of additional Burgess Shale–type biotas has strongly influenced our understanding of the diversity, abundance, and taphonomy of Burgess Shale–type faunas. Although analysis of morphologic change within individual clades awaits further study (Conway Morris and Robison 1986), we now know much more about the longevity of individual soft-bodied taxa, such as the sponge *Diagoniella* and the palaeoscolecid worms, which appear to span the Early Cambrian to Silurian epochs (Rigby 1986a, 1986b; Conway Morris 1989b). Another important impact of Burgess Shale–type faunas is their widespread geographic occurrence in Lower Cambrian Burgess Shale–type deposits, which provided early evidence suggesting rapid dispersal of Burgess Shale–type faunas shortly after the terminal Proterozoic–Cambrian boundary and has hinted at a possible earlier phase of metazoan development (Conway Morris 1989b),

more recently confirmed by metazoan-grade faunas in terminal Neoproterozoic deposits (Chapter 2).

Although a variety of soft-bodied and problematic taxa occur within
the Burgess Shale–type localities, many of the localities are little different from other shelly Cambrian localities, in that they are dominated by
trilobites, brachiopods, hyoliths, and monoplacophorans (Conway Morris 1986, 1989a). Most Lower Cambrian Burgess Shale–type localities
contain only a few members of the original Burgess quarry soft-bodied
fauna, with the exception of the Kinzers Formation of Pennsylvania
(Conway Morris 1985), the Buen Formation of Greenland (Conway Morris et al. 1987), and the Mount Cap Formation of the Northwest Territories (Butterfield 1994). However, fossiliferous Middle Cambrian deposits are much more similar to one another. As an example, approximately 75 percent of the genera found in the Wheeler and Marjum
Formations of Utah also occur in the Burgess Shale (Conway Morris and
Robison 1986).

Perhaps the most significant impact of newly discovered Burgess
Shale–type faunas has been on our understanding of problematic taxa,
early animal evolution, and ecologic interactions. A few brief examples
of these are now considered.

Problematic Taxa

Like *Microdictyon* from Chengjiang, exceptionally preserved sclerite-
bearing faunas from newer Burgess Shale–type deposits have provided a
phylogenetic and paleobiologic link for many types of isolated sclerites
and plates known from other Cambrian deposits. For example, the Sirius Passet fauna in Greenland contains the only known articulated
halkieriid, *Halkiera evangelista,* which was previously known only from isolated sclerites. Similarly, although conodont teeth are known worldwide,
the Soom Shale provides one of only three articulated examples of these
enigmatic soft-bodied agnathans, and exceptionally preserved eye muscles, trunk musculature, and feeding apparatuses have provided insight
on conodont animal anatomy previously unavailable elsewhere (Briggs,
Clarkson, and Aldridge 1983; Aldridge and Theron 1993; Gabbott,
Aldridge, and Theron 1995). Even in cases where articulated specimens
are not known from a given deposit, the preservation of disarticulated elements can be so exceptional as to allow larger-scale evolutionary inferences to be made. For example, in the Mount Cap Formation, the exceptional preservation of surficial textures and internal structure of
organic-walled chancelloriid sclerites has allowed identification of features that suggest a link between chancelloriids and modern horny
sponges, hinting at a poriferan affinity and contradicting the previous

interpretation that these forms were closely allied with the cnidarians (Butterfield and Nicholas 1996).

Early Animal Evolution

Many of the soft-bodied taxa in Burgess Shale–type deposits, especially sclerite-bearing and arthropod forms, appear to be morphologically and phylogenetically intermediate between established groups (Wills, Briggs, and Fortey 1994; Conway Morris and Peel 1995; Hou and Bergström 1995). In the Soom Shale, for example, exquisitely preserved lamellate book-gills in the eurypterid *Onychopterella augusti* provide early evidence for a respiratory system adapted to aquatic habitats, together with features such as Kiemenplatten, which possibly could have been used for aerial respiration. Considered together, these features provide firm evidence for a close relationship between eurypterids and scorpions (Braddy et al. 1999).

Several taxa from Burgess Shale–type deposits are also important because they have stimulated discussions about possible evolutionary links between major clades. For example, the large lobopod *Xenusion* is morphologically similar to *Aysheaia* and was originally hypothesized to reflect an ancestral link between the Onychophora and the Tardigrada (Dzik and Krumbiegel 1989). In this interpretation, *Xenusion* would have been an ancestral walking form that evolved from a priapulid-like crawling ancestor (Dzik and Krumbiegel 1989). Subsequent work on other better-preserved lobopodians, however, suggests that the Onychophora compose their own monophyletic group and do not share any synapamorphies with the Tardigrada (Ramsköld and Chen 1998). In other examples, lobopods such as the gilled *Kerygmachela* have been allied with the biramous arthropods and have been used as evidence for a biphyletic origin of the arthropod body plan (Budd 1993).

Another example comes from the Sirius Passet (Buen Formation) fauna, where *Halkiera evangelista* provides a strong evolutionary link to the similarly sclerite-covered *Wiwaxia*, to the primitive polychaete annelid *Canadia spinosa*, and possibly to an ancestor of the annelid crown group. Conway Morris and Peel (1995) have also proposed that the presence of two shell-like plates at the posterior and anterior of *Halkiera* provide a link between annelid and brachiopod stem lineages, which would suggest placement of lophophorates within the protostomes (Halanych et al. 1995).

Perhaps the most intense evolutionary debate has focused on the Arthropoda. Widespread preservation of nontrilobite arthropods in most of the Burgess Shale–type deposits has allowed examination of the early radiation and relationships of the major arthropod groups (Briggs and Fortey 1989; Bergström 1992; Briggs, Fortey, and Wills 1992; Wills,

Briggs, and Fortey 1994; Wills et al. 1998) and prompted numerous debates on the levels of body plan diversity in the Cambrian. For example, some have suggested that Cambrian arthropods exhibit significant morphologic novelty, whereas others argue that they show no more morphologic complexity than do modern groups. This debate is primarily due to overestimation of morphologic complexity and taxonomic separation of problematic taxa and to inadequate modern-centric taxonomies (Briggs and Conway Morris 1986; Briggs and Fortey 1989; Gould 1989; Briggs, Fortey, and Wills 1992). For a recent review of salient issues in this hotly debated area of inquiry, see Wills et al. (1998).

Ecologic Interactions

Along with phylogenetic questions, Burgess Shale–type localities have enhanced our understanding of the variety of life habits present in Cambrian seas (Conway Morris 1985), utility of filter feeding in early Paleozoic trophic systems (Butterfield 1994), and roles of predation (Conway Morris 1977, 1979, 1981, 1985). Cumulatively, Burgess Shale–type faunas reflect a variety of life habits, ranging from actively swimming pelagic forms to infaunal benthic forms. The variety of life habits indicated in Burgess Shale–type deposits does not differ significantly from those found in the Burgess Shale or Chengjiang and are thus not reviewed here. However, several Burgess Shale–type deposits provide striking new evidence on early trophic webs and feeding strategies, beyond the relatively common occurrences of coprolites filled with skeletal debris and preservation of faunas with their gut contents intact. For example, in the Mount Cap Formation, elaborate crustacean filter-feeding structures with a modern aspect are preserved, together with phosphatized hyolithid guts and abundant acritarch-laden fecal strings and fecal pellets (Butterfield 1996). Together, these provide evidence for the early record of nonbioturbating zooplantonic/nectonic filter-feeding activities and suggest the development and presence of micrograzing by the Early Cambrian (Butterfield 1994). More direct behavioral information is recorded in deposits like the Emu Bay Shale, which contains information on early predation. In the Emu Bay, evidence of bite marks on both mineralized and nonmineralized trilobitomorphs as well as trilobite-rich coprolites suggest the first link between the activities of the presumed predator *Anomalocaris* and other faunas inhabiting Cambrian seas (Nedin 1999).

CONCLUSIONS

In general, Burgess Shale–type faunas are characterized by a distinct temporal, geographic, and paleoenvironmental distribution in the early Paleozoic. Although they span nearly the entire globe, from Greenland to South Australia, most are concentrated in the uppermost Lower Cambrian and Middle Cambrian. From this interval, these deposits are characterized by a taxonomically and paleoecologically diverse fauna that has provided profound insights into the relationship between enigmatic and well-known clades, has helped resolve the identity of previously problematic scleritized taxa, and has offered clues about early animal ecology. Clearly there still are rewards available for those who continue to examine and collect soft-bodied fossils from these early deposits. For example, we still know very little about the microfauna and microflora of many deposits (*sensu* Mankiewicz 1992; Butterfield and Nicholas 1996) or the taphonomic characteristics of some of the more prominent Burgess Shale–type deposits (e.g., House Range, Sirius Passet). Furthermore, new occurrences of Ediacaran-style soft-bodied preservation in Cambrian strata (Jensen, Gehling, and Droser 1998; Hagadorn, Fedo, and Waggoner 2000) suggest that we have much more to learn about the nature of the possibly nonoverlapping taphonomic windows characterizing Burgess Shale–type and Ediacaran-type preservation during the late Proterozoic–Phanerozoic transition.

ACKNOWLEDGMENTS

I would like to thank R. Robison for reviewing this chapter, and D. Briggs, S. Conway Morris, C. Nedin, A. Palmer, and B. Waggoner for their insightful discussions and aid in locating literature.

REFERENCES

Aitken, J. D., and I. A. McIlreath. 1984. The Cathedral Reef escarpment, a Cambrian great wall with humble origins. *Geos* 13:17–19.

Aitken, J. D., and I. A. McIlreath. 1990. Comment on "The Burgess Shale: Not in the shadow of the Cathedral Escarpment." *Geoscience Canada* 17:111–115.

Aldridge, R. J., S. E. Gabbott, and J. N. Theron. 1993. The Soom Shale: A unique Ordovician Konservat-Lagerstätte from South Africa. *Palaeontology Newsletter* 20:9.

Aldridge, R. J., and J. N. Theron. 1993. Conodonts with preserved soft-tissue from a new Ordovician *Konservat-Lagerstätte. Journal of Micropaleontology* 12:113–117.

Allison, P. A. 1988. *Konservat-Lagerstatten*: Cause and classification. *Paleobiology* 14:331–344.

Allison, P. A. 1993. Paleo-oxygenation and *in-situ* benthos within the Burgess Shale: The fossils you never heard about. *Geological Society of America Abstacts with Programs* 25:458.

Allison, P. A., and C. Brett. 1995. *In situ* benthos and paleo-oxygenation in the Middle Cambrian Burgess Shale, British Columbia, Canada. *Geology* 23:1079–1082.

Allison, P. A., and D. E. G. Briggs. 1991. Taphonomy of nonmineralized tissues. In P. A. Allison and D. E. G. Briggs, eds., *Taphonomy: Releasing the Data Locked in the Fossil Record*, pp. 26–71. New York: Plenum.

Allison, P. A., and D. E. G. Briggs. 1993a. Exceptional fossil record: Distribution of soft-tissue preservation through the Phanerozoic. *Geology* 21:527–530.

Allison, P. A., and D. E. G. Briggs. 1993b. Burgess Shale biotas; burrowed away? Discussion. *Lethaia* 26:184–185.

Allison, P. A., and D. E. G. Briggs. 1994. Exceptional fossil record: Distribution of soft-tissue preservation through the Phanerozoic. Reply. *Geology* 22:184.

Aronson, R. B. 1992. Decline of the Burgess Shale fauna: Ecologic or taphonomic restriction? *Lethaia* 25:225–229.

Aronson, R. B. 1993. Burgess Shale–type biotas were not just burrowed away: Reply. *Lethaia* 26:185.

Ausich, W. I., and D. J. Bottjer. 1982. Tiering in suspension-feeding communities on soft substrata throughout the Phanerozoic. *Science* 216:173–174.

Babcock, L. E., and R. A. Robison. 1988. *Taxonomy and Paleobiology of Some Middle Cambrian Scenella (Cnidaria) and Hyolithids (Mollusca) from Western North America.* University of Kansas Paleontological Contributions, no. 121. Lawrence: University of Kansas Paleontological Institute.

Barskov, I. S., and A. Y. Zhuravlev. 1988. Soft-bodied organisms of the Siberian Platform. *Paleontological Journal* 22:1–7.

Bengtson, S. 1968. The problematical genus *Mobergella* from the Lower Cambrian of the Baltic area. *Lethaia* 1:325–351.

Bengtson, S., and S. Conway Morris. 1984. A comparative study of Lower Cambrian *Halkieria* and Middle Cambrian *Wiwaxia. Lethaia* 17:307–329.

Bengtson, S., S. Conway Morris, B. J. Cooper, P. A. Jell, and B. N. Runnegar. 1990. Early Cambrian fossils from South Australia. *Memoir of the Association of Australasian Palaeontologists* 9:1–364.

Bergström, J. 1992. The oldest arthropods and the origin of the Crustacea. *Acta Zoologica* 73:287–291.

Braddy, S. J., R. J. Aldridge, S. E. Gabbott, and J. N. Theron. 1999. Lamellate book-gills in a Late Ordovician eurypterid from the Soom Shale, South Africa: Support for a eurypterid–scorpion clade. *Lethaia* 32:72–74.

Braddy, S. J., R. J. Aldridge, and J. N. Theron. 1995. A new eurypterid from the Late Ordovician Table Mountain Group, South Africa. *Palaeontology* 38:563–582.

Brady, M. J., and R. B. Keopnick. 1979. A Middle Cambrian platform-to-basin transition, House Range, west-central Utah. *Brigham Young University Geology Studies* 26:1–7.

Briggs, D. E. G. 1976. *The Arthropod Branchiocaris n. gen., Middle Cambrian, Burgess Shale, British Columbia.* Bulletin of the Geological Survey of Canada, no. 264. Ottawa: Geological Survey of Canada.

Briggs, D. E. G. 1978. A new trilobite-like arthropod from the Lower Cambrian Kinzers Formation, Pennsylvania. *Journal of Paleontolgy* 52:132–140.

Briggs, D. E. G. 1979. *Anomalocaris,* the largest known Cambrian arthropod. *Palaeontology* 22:631–664.

Briggs, D. E. G., E. N. K. Clarkson, and R. J. Aldridge. 1983. The conodont animal. *Lethaia* 16:1–14.

Briggs, D. E. G., and D. Collins. 1988. A Middle Cambrian chelicerate from Mount Stephen, British Columbia. *Palaeontology* 31:779–798.

Briggs, D. E. G., and S. Conway Morris. 1986. Problematica from the Middle Cambrian Burgess Shale of British Columbia. In A. Hoffman and M. H.

Nitecki, eds., *Problematic Fossil Taxa*, pp. 167–183. New York: Oxford University Press.

Briggs, D. E. G., and R. A. Fortey. 1989. The early radiation and relationships of the major arthropod groups. *Science* 246:241–243.

Briggs, D. E. G., and R. A. Fortey. 1992. The Early Cambrian radiation of arthropods. In J. H. Lipps and P. W. Signor, eds., *Origin and Early Evolution of the Metazoa*, pp. 335–373. New York: Plenum.

Briggs, D. E. G., R. A. Fortey, and M. A. Wills. 1992. Morphological disparity in the Cambrian. *Science* 256:1670–1673.

Briggs, D. E. G., and J. D. Mount. 1982. The occurrence of the giant arthropod *Anomalocaris* in the Lower Cambrian of southern California, and the overall distribution of the genus. *Journal of Paleontology* 56:1112–1118.

Briggs, D. E. G., and C. Nedin. 1997. The taphonomy and affinities of the problematic fossil *Myoscolex* from the Lower Cambrian Emu Bay Shale of South Australia. *Journal of Paleontology* 71:22–32.

Briggs, D. E. G., and R. A. Robison. 1984. *Exceptionally Preserved Non-trilobite Arthropods and* Anomalocaris *from the Middle Cambrian of Utah.* University of Kansas Paleontological Contributions, no. 111. Lawrence: University of Kansas Paleontological Institute.

Budd, G. E. 1993. A Cambrian gilled lobopod from Greenland. *Nature* 364:709–711.

Budd, G. E. 1997. Stem group arthropods from the Lower Cambrian Sirius Passet fauna of North Greenland. In R. A. Fortey and R. H. Thomas, eds., *Arthropod Relationships*, pp. 125–138. London: Chapman and Hall.

Budd, G. E. 1999. A nektaspid arthropod from the Early Cambrian Sirius Passet fauna, with a description of retrodeformation based on functional morphology. *Palaeontology* 42:99–122.

Budd, G. E., and J. S. Peel. 1998. A *Xenusion*-like lobopod from the Lower Cambrian Sirius Passet fauna. *Palaeontology* 41:1201–1213.

Butterfield, N. J. 1990a. A reassessment of the enigmatic Burgess Shale fossil *Wiwaxia corrugata* (Matthew) and its relationship to the polychaete *Canadia spinosa* Walcott. *Paleobiology* 16:287–303.

Butterfield, N. J. 1990b. Organic preservation of non-mineralizing organisms and the taphonomy of the Burgess Shale. *Paleobiology* 16:272–286.

Butterfield, N. J. 1993. A new Burgess Shale–type fossil assemblage from the Northwest Territories, Canada. *Palaeontology Newsletter* 20:10.

Butterfield, N. J. 1994. Burgess Shale–type fossils from a Lower Cambrian shallow-shelf sequence in northwestern Canada. *Nature* 369:477–479.

Butterfield, N. J. 1995. Secular distribution of Burgess Shale–type preservation. *Lethaia* 28:1–14.

Butterfield, N. J. 1996. Guts and guano: A record of Cambrian alimentation. *Palaeontology Newsletter* 32:12.

Butterfield, N. J., and C. J. Nicholas. 1996. Burgess Shale–type preservation of both non-mineralizing and "shelly" Cambrian organisms from the Mackenzie Mountains, northwestern Canada. *Journal of Paleontology* 70:893–899.

Campbell, L. D. 1969. Stratigraphy and paleontology of the Kinzers Formation, southeastern Pennsylvania. Master's thesis, Franklin and Marshall College.

Campbell, L. D., and M. E. Kauffman. 1969. *Olenellus* fauna of the Kinzers Formation, southeastern Pennsylvania. *Proceedings of the Pennsylvania Academy of Science* 43:172–176.

Chesselet, P. 1992. Disarticulated remains of an Ordovician metazoan from the Cedarberg Formation, South Africa: A re-interpretation of *Eohostimella parva* Kovács-Endrödy. *Palaeontology of South Africa* 29:11–20.

Cisne, J. L. 1973. Beecher's Trilobite Bed revisited: Ecology of an Ordovician deep water fauna. *Postilla* 160:1–25.

Collins, D. 1985. A new Burgess Shale–type fauna in the Middle Cambrian Stephen Formation on Mt. Stephen, British Columbia. *Geological Society of America Abstracts with Programs* 17:550.

Collins, D. 1986. Paradise revisited. *Rotunda* 19:30–39.

Collins, D., D. E. G. Briggs, and S. Conway Morris. 1983. New Burgess Shale fossil sites reveal Middle Cambrian faunal complex. *Science* 222:163–167.

Conway Morris, S. 1977. *Fossil Priapulid Worms.* Special Papers in Palaeontology, no. 20. London: Palaeontological Association.

Conway Morris, S. 1979. The Burgess Shale (Middle Cambrian) fauna. *Annual Reviews of Ecology and Systematics* 10:327–349.

Conway Morris, S. 1981. The Burgess Shale fauna as a mid-Cambrian community. *United States Geological Survey Open-File Report* 81–743:47–49.

Conway Morris, S. 1985. Cambrian Lagerstätten: Their distribution and significance. *Philosophical Transactions of the Royal Society of London, B* 311:49–65.

Conway Morris, S. 1986. The community structure of the Middle Cambrian phyllopod bed (Burgess Shale). *Palaeontology* 29:423–467.

Conway Morris, S. 1989a. Burgess Shale faunas and the Cambrian explosion. *Science* 246:339–346.

Conway Morris, S. 1989b. The persistence of Burgess Shale–type faunas: Implications for the evolution of deeper-water faunas. *Transactions of the Royal Society of Edinburgh: Earth Sciences* 80:271–283.

Conway Morris, S. 1990. Late Precambrian and Cambrian soft-bodied faunas. *Annual Reviews of Earth and Planetary Sciences* 18:101–122.

Conway Morris, S. 1993. Ediacaran-like fossils in Cambrian Burgess Shale–type faunas of North America. *Palaeontology* 36:593–635.

Conway Morris, S. C. 1998. *The Crucible of Creation: The Burgess Shale and the Rise of Animals.* Oxford: Oxford University Press.

Conway Morris, S., and R. J. F. Jenkins. 1985. Healed injuries in Early Cambrian trilobites from South Australia. *Alcheringa* 9:167–177.

Conway Morris, S., and J. S. Peel. 1990. Articulated halkieriids from the Lower Cambrian of north Greenland. *Nature* 345:802–805.

Conway Morris, S., and J. S. Peel. 1995. Articulated halkieriids from the Lower Cambrian of North Greenland and their role in early protostome evolution. *Philosophical Transactions of the Royal Society of London, B* 347:305–358.

Conway Morris, S., J. S. Peel, A. K. Higgins, N. J. Soper, and N. C. Davis. 1987. A Burgess Shale–like fauna from the Lower Cambrian of north Greenland. *Nature* 326:181–183.

Conway Morris, S., and R. A. Robison. 1982. The enigmatic medusoid *Peytoia* and a comparison of some Cambrian biotas. *Journal of Paleontology* 56:116–122.

Conway Morris, S., and R. A. Robison. 1986. *Middle Cambrian Priapulids and Other Soft-bodied Fossils from Utah and Spain.* University of Kansas Paleontological

Contributions, no. 117. Lawrence: University of Kansas Paleontological Institute.

Conway Morris, S., and R. A. Robison. 1988. *More Soft-bodied Animals and Algae from the Middle Cambrian of Utah and British Columbia.* University of Kansas Paleontological Contributions, no. 122. Lawrence: University of Kansas Paleontological Institute.

Conway Morris, S., and H. B. Whittington. 1979. The animals of the Burgess Shale. *Scientific American* 241:122–133.

Daily, B., A. R. Milnes, C. R. Twidale, and J. A. Bourne. 1979. Geology and geomorphology. In M. J. Tyler, J. K. Ling, and C. R. Twidale, eds., *Natural History of Kangaroo Island*, pp. 1–38. Adelaide: Royal Society of South Australia.

Daily, B., P. S. Moore, and B. R. Rust. 1980. Terrestrial–marine transition in the Cambrian rocks of Kangaroo Island, South Australia. *Sedimentology* 27:379–399.

Dawson, J. W. 1896. Additional notes on fossil sponges and other organic remains from the Quebed Group of Little Metis on the lower St. Lawrence; with notes on some of the specimens by Dr. G. H. Hinde. *Transactions of the Royal Society of Canada* 2:91–121.

Dawson, J. W., and G. I. Hinde. 1889. On new species of fossil sponges from the Siluro-Cambrian at Little Metis on the lower St. Lawrence. *Transactions of the Royal Society of Canada* 7:31–55.

Dunbar, C. O. 1925. Antennae in *Olenellus getzi*, n. sp. *American Journal of Science* 9:303–308.

Durham, J. W. 1966. *Camptostroma*, an Early Cambrian supposed scyphozoan, referable to Echinodermata. *Journal of Paleontology* 40:1216–1220.

Durham, J. W. 1978. A Lower Cambrian eocrinoid. *Journal of Paleontology* 52:195–199.

Dzik, J., and G. Krumbiegel. 1989. The oldest "onychophoran" *Xenusion*: A link connecting phyla? *Lethaia* 22:169–181.

Dzik, J., and K. Lendzion. 1988. The oldest arthropods of the East European platform. *Lethaia* 21:29–38.

Fletcher, T. P., and D. H. Collins. 1998. The Middle Cambrian Burgess Shale and its relationship to the Stephen Formation in the southern Rocky Mountains. *Canadian Journal of Earth Sciences* 35:413–436.

Fortey, R. A., and J. N. Theron. 1995. A new Ordovician arthropod, *Soomaspis*, and the agnostid problem. *Palaeontology* 37:841–861.

Fritz, W. H. 1990. In defense of the escarpment near the Burgess Shale fossil locality. *Geoscience Canada* 17:106–110.

Gabbott, S. E. 1998. The taphonomy of the Ordovician Soom Shale *Lagerstätte*: An example of soft tissue preservation in clay minerals. *Palaeontology* 41:631–667.

Gabbott, S. E., R. J. Aldridge, and J. N. Theron. 1995. A giant conodont with preserved muscle tissue from the Upper Ordovician of South Africa. *Nature* 374:800–803.

Gabbott, S. E., R. J. Aldridge, and J. N. Theron. 1997. A new enigmatic fossil from the Ordovician Soom Shale Lagerstätte. *Palaeontology Newsletter* 36: 22.

Garcia-Bellido Capdevila, D., and S. Conway Morris. 1999. New fossil worms from the Lower Cambrian of the Kinzers Formation, Pennsylvania, with some comments on Burgess Shale–type preservation. *Journal of Paleontology* 73:394–402.

Glaessner, M. F. 1979. Lower Cambrian Crustacea and annelid worms from Kangaroo Island, South Australia. *Alcheringa* 3:21–31.

Gorjansky, V. Y. 1977. First find of sponge remains in the Lower Cambrian of eastern Siberia. *Ezhegodnik Vsesoëiûznogo Paleontologicheskogo Obshchestva* 20:274–276.

Gould, S. J. 1989. *Wonderful Life: The Burgess Shale and the Nature of History.* New York: Norton.

Gunther, L. F., and V. G. Gunther. 1981. Some Middle Cambrian fossils of Utah. *Brigham Young University Geology Studies* 28:1–81.

Hagadorn, J. W. 1998. Restriction of a late Neoproterozoic biotope: Ediacaran faunas, microbial structures, and trace fossils from the Proterozoic–Phanerozoic transition, Great Basin, USA. Ph.D. diss., University of Southern California.

Hagadorn, J. W., C. W. Fedo, and B. M. Waggoner. 2000. Lower Cambrian Ediacara fossils from the Great Basin, USA. *Journal of Paleontology* 74:731–740.

Halanych, K. M., J. D. Bacheller, A. M. A. Aguinaldo, S. M. Liva, D. M. Hillis, and J. A. Lake. 1995. Evidence from 18S Ribosomal DNA that the lophophorates are protostome animals. *Science* 267:1641–1643.

Hamman, W., R. Laske, and G. L. Pillola. 1990. *Tariccoia arusensis* n.g. n.sp., an unusual trilobite-like arthropod: Rediscovery of the "phylllocarid" beds of Taricco (1922) in the Ordovician "Puddinga" sequence of Sardinia. *Bollettino della Società Paleontologica Italiana* 29:163–178.

Hou, X. 1987a. Two new arthropods from Lower Cambrian, Chengjiang, eastern Yunnan. *Acta Palaeontologica Sinica* 26:236–256.

Hou, X. 1987b. Early Cambrian large bivalved arthropods from Chengjiang, eastern Yunnan. *Acta Palaeontologica Sinica* 26:286–297.

Hou, X., and J. Bergström. 1995. Cambrian lobopodians—Ancestors of extant onychophorans? *Zoological Journal of the Linnean Society* 114:3–19.

Hou, X., J. Chen, and H. Lu. 1989. Early Cambrian new arthropods from Chengjiang, Yunnan. *Acta Palaeontologica Sinica* 28:42–57.

Hou, X., L. Ramsköld, and J. Bergström. 1991. Composition and preservation of the Chengjiang fauna—A Lower Cambrian soft-bodied fauna. *Zoological Scripta* 20:395–411.

Hou, X., and W. Sun. 1988. Discovery of Chengjiang fauna at Meishucun, Jinning, Yunnan. *Acta Palaeontologica Sinica* 28:32–41.

Ivantsov, A. Y. 1990. The first Lower Cambrian phyllocarids from Yakutia. *Paleontological Journal* 24:134–136.

Jaeger, H., and A. Martinsson. 1967. Remarks on the problematic fossil *Xenusion auerswaldae. Geologiska Föreningens i Stockholm Förhandlingar* 88:435–452.

Jensen, S., J. G. Gehling, and M. L. Droser. 1998. Ediacara-type fossils in Cambrian sediments. *Nature* 393:567–569.

Kristofovich, A. N. 1953. Discovery of clubmoss plant in the Cambrian deposits of eastern Siberia. *Doklady Akademii nauk SSSR* 91:1377–1379.

Krumbiegel, G., H. Deichfuss, and H. Deichfuss. 1980. Ein neuer fund von *Xenusion. Hallesches Jahrbuch für Geowissenschaften* 5:97–99.

Lendzion, K. 1975. Fauna of the *Mobergella* zone in the Polish Lower Cambrian. *Kwartalnik Geologiczny* 19:237–242.

Ludvigsen, R. 1989. The Burgess Shale: Not in the shadow of the Cathedral Escarpment. *Geoscience Canada* 16:51–59.

Ludvigsen, R. 1990. Reply to comments by Fritz and Aitken and McIlreath. *Geoscience Canada* 17:116–118.

Mankiewicz, C. 1992. *Obruchevella* and other microfossils in the Burgess Shale: Preservation and affinity. *Journal of Paleontology* 66:717–729.

McCollum, L. B. 1994. Faunal extinction and replacement at the Lower–Middle Cambrian boundary, southern Great Basin. *Geological Society of America Abstracts with Programs* 26:54.

McHenry, B., and A. Yates. 1993. First report of the enigmatic metazoan *Anomalocaris* from the Southern Hemisphere and a trilobite with preserved appendages from the Early Cambrian of Kangaroo Island, South Australia. *Records of the South Australian Museum* 26:77–86.

McIlreath, I. A. 1977. Accumulation of a Middle Cambrian, deep-water limestone debris apron adjacent to a vertical, submarine carbonate escarpment, southern Rocky Mountains, Canada. In H. E. Cook and P. Enos, eds., *Deep-Water Carbonate Environments*, pp. 113–124. Special Publication, no. 25. Tulsa, Okla.: Society of Economic Paleontologists and Mineralogists.

McKirdy, D. M. 1971. An organic geochemical study of the Australian Cambrian and Precambrian. Master's thesis, University of Adelaide.

Mikulic, D. G., D. E. G. Briggs, and J. Kluessendorf. 1985a. A new exceptionally preserved biota from the lower Silurian of Wisconsin, U.S.A. *Philosophical Transactions of the Royal Society of London, B* 311:75–85.

Mikulic, D. G., D. E. G. Briggs, and J. Kluessendorf. 1985b. A Silurian soft-bodied biota. *Science* 228:715–717.

Mount, J. D. 1980. Characteristics of Early Cambrian faunas from eastern San Bernadino County, California. *Special Publication of the Southern California Paleontological Society* 2:19–29.

Nedin, C. 1995a. The Emu Bay Shale, a Lower Cambrian fossil Lagerstätten, Kangaroo Island, South Australia. *Memoir of the Association of Australasian Palaeontologists* 18:31–40.

Nedin, C. 1995b. The palaeontology and palaeoenvironment of the Early Cambrian Emu Bay Shale, Kangaroo Island, South Australia. Ph.D. diss., University of Adelaide.

Nedin, C. 1997. Taphonomy of the Early Cambrian Emu Bay Shale Lagerstätte, Kangaroo Island, South Australia. *Bulletin of the National Museum of Natural Science* 10:133–141.

Nedin, C. 1999. *Anomalocaris* predation on nonmineralized and mineralized trilobites. *Geology* 27:987–990.

Orr, P. J., D. E. G. Briggs, and S. L. Kearns. 1998. Cambrian Burgess Shale animals replicated in clay minerals. *Science* 281:1173–1175.

Palmer, A. R. 1960. Some aspects of the early Upper Cambrian stratigraphy of White Pine County, Nevada and vicinity. In J. W. Boettcher and W. W. Sloan, eds., *Guidebook to the Geology of East Central Nevada*, pp. 53–58. Salt Lake City: Intermountain Association of Petroleum Geologists.

Pickerill, R. K. 1994. Exceptional fossil record: Distribution of soft-tissue preservation through the Phanerozoic: Discussion. *Geology* 22: 183–184.

Plotnick, R. E. 1986. Taphonomy of a modern shrimp: Implications for the arthropod fossil record. *Palaios* 1:286–293.

Pocock, K. J. 1964. *Estaingia*, a new trilobite genus from the Lower Cambrian of South Australia. *Palaeontology* 7:458–471.

Pompeckj, J. F. 1927. Ein neues Zeugnis uralten Lebens. *Paläontologische Zeitschrift* 9:287–313.

Ramsköld, L., and J. Chen. 1998. Cambrian lobopodians: Morphology and phylogeny. In G. D. Edgecombe, ed., *Arthropod Fossils and Phylogeny*, pp. 107–150. New York: Columbia University Press.

Ramsköld, L., and X. Hou. 1991. New Early Cambrian animal and onyctophoran affinities of enigmatic metazoans. *Nature* 351:225–227.

Rees, M. N. 1984. A fault-controlled trough through a carbonate platform, Middle Cambrian House Range embayment, Utah and Nevada. Ph.D. diss., University of Kansas.

Rees, M. N. 1986. A fault-controlled trough through a carbonate platform: The Middle Cambrian House Range embayment. *Geological Society of America Bulletin* 97:1054–1069.

Repina, L. N., and O. P. Okuneva. 1969. Cambrian arthropods of the Maritime Territory. *Palaeontologischeskii Zhurnal* 1:95–103.

Resser, C. E. 1929. New Lower and Middle Cambrian Crustacea. *Proceedings of the United States National Museum* 76:1–18.

Resser, C. E. 1938. Middle Cambrian fossils from Pend Oreille Lake, Idaho. *Smithsonian Miscellaneous Collections* 97:1–12.

Resser, C. E., and B. F. Howell. 1938. Lower Cambrian *Olenellus* zone of the Appalachians. *Geological Society of America Bulletin* 49:195–248.

Rigby, J. K. 1966. *Protospongia hicksi* Hinde from the Middle Cambrian of western Utah. *Journal of Paleontology* 40:549–554.

Rigby, J. K. 1969. A new Middle Cambrian hexactinellid sponge from western Utah. *Journal of Paleontology* 43:125–128.

Rigby, J. K. 1978. Porifera of the Middle Cambrian Wheeler Shale, from the Wheeler Amphitheater, House Range, in western Utah. *Journal of Paleontology* 52:1325–1345.

Rigby, J. K. 1980. The new Middle Cambrian sponge *Vauxia magna* from the Spence Shale of northern Utah and taxonomic position of the Vauxiidae. *Journal of Paleontology* 54:234–240.

Rigby, J. K. 1983. Sponges of the Middle Cambrian Marjum Limestone from the House Range and Drum Mountains of western Millard County, Utah. *Journal of Paleontology* 57:240–270.

Rigby, J. K. 1986a. *Sponges of the Burgess Shale (Middle Cambrian) British Columbia.* Palaeontolographica Canadiana, no. 2. Calgary: Canadian Society of Petroleum Geologists.

Rigby, J. K. 1986b. Cambrian and Silurian sponges from north Greenland. *Rapport Grønlands gologiske Undersøgelse* 132:51–63.

Rigby, J. K. 1987. Early Cambrian sponges from Vermont and Pennsylvania, the only ones described from North America. *Journal of Paleontology* 61:451–461.

Rigby, J. K., L. F. Gunther, and F. Gunther. 1997. The first occurrence of the Burgess Shale demosponge *Hazelia palmata* Walcott, 1920, in the Cambrian of Utah. *Journal of Paleontology* 71:994–997.

Rigby, J. K., and R. C. Gutschick. 1976. Two new lower Paleozoic hexactinellid sponges from Utah and Oklahoma. *Journal of Paleontology* 50:79–85.

Robison, R. A. 1960. Lower and Middle Cambrian stratigraphy of the eastern Great Basin. In J. W. Boettcher and W. W. Sloan, eds., *Guidebook to the Geol-*

ogy of East Central Nevada, pp. 43–52. Salt Lake City: Intermountain Association of Petroleum Geologists.

Robison, R. A. 1984. *New Occurrences of the Unusual Trilobite* Naraoia *from the Cambrian of Idaho and Utah*. University of Kansas Paleontological Contributions, no. 112. Lawrence: University of Kansas Paleontological Institute.

Robison, R. A. 1985. Affinities of *Aysheaia* (Onychophora), with descriptions of a new Cambrian species. *Journal of Paleontology* 59:226–235.

Robison, R. A. 1991. Middle Cambrian biotic diversity: Examples from four Utah Lagerstätten. In A. M. Simonetta and S. Conway Morris, eds., *The Early Evolution of Metazoa and the Significance of Problematic Taxa*, pp. 77–98. Cambridge: Cambridge University Press.

Robison, R. A., and B. C. Richards. 1981. *Larger Bivalve Arthropods from the Middle Cambrian of Utah*. University of Kansas Paleontological Contributions, no. 106. Lawrence: University of Kansas Paleontological Institute.

Rogers, J. C. 1984. Depositional environments and paleoecology of two quarry sites in the Middle Cambrian Marjum and Wheeler Formations, House Range, Utah. *Brigham Young University Geology Studies* 31:97–115.

Ruedemann, R. 1933. *Camptostroma*, a Lower Cambrian floating hydrozoan. *United States National Museum, Proceedings* 82:1–8.

Seilacher, A. 1970. Begriff und Bedeutung der Fossil-Lagerstätten. *Neues Jahrbuch für Geologie und Paläontologie, Abhandlungen* 1970:34–39.

Seilacher, A., W. E. Reif, and F. Westphal. 1985. Sedimentological, ecological, and temporal patterns of fossil Lagerstätten. *Philosophical Transactions of the Royal Society of London, B* 311:5–23.

Sprinkle, J. 1985. *New Edrioasteroids from the Middle Cambrian of Western Utah*. University of Kansas Paleontological Contributions, no. 116. Lawrence: University of Kansas Paleontological Institute.

Stewart, J. H., and A. R. Palmer. 1967. Callaghan Window, a newly discovered part of the Roberts thrust, Toiyabe range, Lander County, Nevada. *United States Geological Survey Professional Paper* 575D:56–63.

Stewart, W. D., O. A. Dixon, and B. R. Rust. 1993. Middle Cambrian carbonate-platform collapse, southeastern Canadian Rocky Mountains. *Geology* 21:687–690.

Stose, G. W., and A. I. Jonas. 1922. The lower Paleozoic section of southeastern Pennsylvania. *Journal of the Washington Academy of Science* 12:358–366.

Sun, W., and X. Hou. 1987a. Early Cambrian medusae from Chengjiang, Yunnan, China. *Acta Palaeontologica Sinica* 26:257–271.

Sun, W., and X. Hou. 1987b. Early Cambrian worms from Chengjiang, Yunnan, China: *Maotianshania* gen. nov. *Acta Palaeontologica Sinica* 26:299–305.

Theron, J. N., R. B. Rickards, and R. J. Aldridge. 1990. Bedding plane assemblages of *Promissum pulchrum*, a new giant Ashgill conodont from the Table Mountain Group, South Africa. *Palaeontology* 33:577–594.

Ubaghs, G., and R. A. Robison. 1985. *A New Homoiostelean and a New Eocrinoid from the Middle Cambrian of Utah*. University of Kansas Paleontological Contributions, no. 115. Lawrence: University of Kansas Paleontological Institute.

Ubaghs, G., and R. A. Robison. 1988. *Homalozoan Echinoderms of the Wheeler Formation (Middle Cambrian) of Western Utah*. University of Kansas Paleontological Contributions, no. 120. Lawrence: University of Kansas Paleontological Institute.

Vermeij, G. J. 1987. *Evolution and Escalation: An Ecological History of Life.* Princeton, N.J.: Princeton University Press.

Voropinov, V. S. 1957. The first finds of fauna in the upper Lena Formation. *Doklady Akademii nauk SSSR* 114:1291–1293.

Waggoner, B. M., and J. W. Hagadorn. 1999. Unusual Lower Cambrian fossils from the Latham Shale: What, if anything, is a Lagerstätten? *Geological Society of America Abstracts with Programs, Cordilleran Section* 31:105.

Whittard, W. F. 1953. *Palaeoscolex piscatorum* gen. et sp. nov., a worm from the Tremadocian of Shropshire. *Quarterly Journal of the Geological Society of London* 109:125–136.

Whittington, H. B. 1971. The Burgess Shale: History of research and preservation of fossils. In E. L. Yochelson, ed., *Proceedings of the North American Paleontological Convention* vol. 2, pp. 1170–1201. Lawrence, Kans.: Allen Press.

Whittington, H. B. 1974. Yohoia *Walcott and* Plenocaris *n. gen., Arthropods from the Burgess Shale, Middle Cambrian, British Columbia.* Geological Survey of Canada Bulletin, no. 231. Ottawa: Department of Energy, Mines, and Resources.

Williams, M., D. J. Siveter, and J. S. Peel. 1996. *Isoxys* (Arthropoda) from the Early Cambrian Sirius Passet Lagerstätte, north Greenland. *Journal of Paleontology* 70:947–954.

Willoughby, R. H., and R. A. Robison. 1979. Medusoids from the Middle Cambrian of Utah. *Journal of Paleontology* 53:494–500.

Wills, M. A., D. E. G. Briggs, and R. A. Fortey. 1994. Disparity as an evolutionary index: A comparison of Cambrian and Recent arthropods. *Paleobiology* 20:93–130.

Wills, M. A., D. E. G. Briggs, R. A. Fortey, M. Wilkinson, and P. H. A. Sneath. 1998. An arthropod phylogeny based on fossil and recent taxa. In G. D. Edgecombe, ed., *Arthropod Fossils and Phylogeny*, pp. 33–106. New York: Columbia University Press.

Zhang, W., and X. Hou. 1985. Preliminary notes on the occurrence of the unusual trilobite *Naraoia* in Asia. *Acta Palaeontologica Sinica* 24:591–595.

Zhao, Y. L. 1994. Middle Cambrian Kaili fauna in Taijiang, Guizhou. *Acta Palaeontologica Sinica* 33:263–375.

6
Orsten Deposits from Sweden: Miniature Late Cambrian Arthropods

Carol M. Tang

SINCE THE EIGHTEENTH CENTURY, MARINE INVERTEBRATE fossils have been collected from limestone nodules called *orsten* from the Alum Shale of southern Sweden. However, it was not until more than 200 years later that the unique preservation of very fine morphological details and soft integuments was recognized. In 1975, while searching for conodonts under the microscope, Klaus J. Müller discovered, by chance, a rich fauna of tiny phosphatized arthropods—including agnostids, trilobites, and early crustaceans—preserved in fine detail in three dimensions. Because the unique phosphatic preservation has occurred only on fossil specimens less than 2 mm in length, it is not surprising that this unique conservation Lagerstätte had not been recognized earlier.

Since then, Müller and colleagues have developed special techniques to search for, extract, and study these exceptionally well preserved, phosphatized fossils. Over the years, they have examined over 1,200 kg of limestone, collected about 100,000 specimens, and described dozens of genera and species, almost all of them new.

The deposit provides extraordinary insight into the paleoecological structure of a flocculent-layer community and is the oldest well-documented example of meiofauna in the fossil record. The flocculent layer is the fine-grained, unconsolidated layer above the seafloor inhabited by very small organisms called meiofauna. Orsten have yielded the oldest fossil members of the Pentastomida, worm-like internal parasites that inhabit most modern terrestrial vertebrates. Although trilobites are abundant in orsten deposits and surrounding shales, the additional preserved faunal elements indicate that other arthropod taxa played a larger role in Late Cambrian marine communities than previously realized (Müller 1983; Budd 2001). Studies have shown that the conodonts and arthropods are surprisingly diverse, complex, and advanced for this time interval.

Because of the exceptional preservation of even very small features, the orsten material has allowed for the identification, description, and interpretation of eyes, hairs, spines, muscle scars, joints, pores, and, very rarely, soft body parts. Many arthropod larval stages have also been found, thus documenting ontogenetic series for several species for the first time. The morphological details have led to the recognition of intraspecific variation, interpretations of functional morphology, reconstructions of life habitats, and reexaminations of the evolutionary relationships between modern and fossil arthropod groups. Despite the significance of this Lagerstätte, it has not received as much attention as other deposits, possibly due to the small size of the fossils and the publication of many of the results as systematic and phylogenetic studies appealing mainly to arthropod specialists.

Further research has demonstrated the presence of orsten-type fossils in other regions from strata of Early and Middle Cambrian age (Walossek 1999). With these newly discovered faunas, orsten fossils overlap in age with those found in the Burgess Shale and Burgess Shale–type localities worldwide (Chapters 3, 4, 5). Comparative study of these Lagerstätten should provide exceptional insight into the unique faunal assemblages of the Cambrian.

GEOLOGICAL CONTEXT

Orsten are anthraconite limestone nodules embedded in the dark bituminous shales of the Alum Shale Formation of northern Europe. The Alum Shale is a very condensed series deposited during the Middle Cambrian to Early Ordovician (Bergstrom and Gee 1985). The orsten limestones are most common in the Upper Cambrian to Lower Ordovician and are present either as concretions ranging from 10 cm to 2 m in diameter or as flat lenses.

The orsten fauna has been collected from over 20 localities in southern Sweden, as well as from northern Germany and Poland (Walossek and Szaniawski 1991). The Swedish outcrops are often found in quarries of Lower Cambrian strata mined in the nineteenth century. Fossiliferous samples have been collected over a wide area in Sweden (Figure 6.1), especially in Västergötland and the Isle of Öland. In northern Germany, orsten deposits are found as drift boulders. In Poland, Upper Cambrian deposits and orsten-type fossils similar to those found in Sweden have been identified from boreholes (Walossek and Szaniawski 1991).

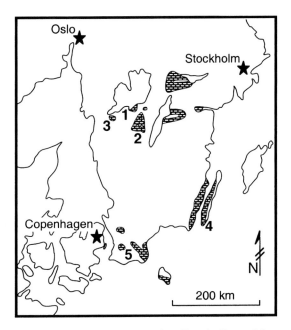

FIGURE 6.1 Southern Sweden with orsten localites indicated by numbered patterned areas: (1) Kinnekulle, Västergötland; (2) Falbygden-Billingen, Västergötland; (3) Hunneberg; (4) Öland; and (5) Skåne. (Modified from Walossek 1993)

Six types of limestone lithology have been recognized in the region (Müller and Hinz 1991): (1) black to gray micritic to sparry thin-layered limestones with orsten and generally sorted, slightly compressed megafossils; (2) beige sparry limestones with fossil preservation similar to that just described; (3) black fine-grained limestones with high organic content and lower fossil concentration; (4) light gray micritic to sparry limestones composed of coarse fossil hash; (5) black micritic thin-layered fossil hash limestones; and (6) whitish to gray weathered limestone with fine fossil hash. These lithologies are not restricted in time and thus occur throughout the sequence.

Because of the large number of trilobites within the shale and limestones, the biostratigraphy has been well established. The most productive orsten deposits are from the Upper Cambrian (Merioneth Epoch) and span six biostratigraphic zones. Soft-body preservation is concentrated to a greater degree in the first and last biostratigraphic zones: *Agnostus pisiformis* zone (sometimes ranging to the *Olenus gibbosus* with *Homagnostus obesus* zone) and *Peltura* sp. zone.

PALEOENVIRONMENTAL SETTING

The orsten deposits are believed to have been restricted to shallower parts of the Alum Shale Sea—a large, shallow, circumscribed basin.

Müller and Walossek (1985a) hypothesize that the seafloor was soft and muddy. Petrologic and sedimentologic analyses of the shales suggest that they were deposited slowly offshore under anoxic conditions, while the carbonates formed during times of increased productivity and oxygenation (Dvoratzek 1987). The distribution of pyrite within the limestones and its high content of organic detrital matter also suggest that there were reduced levels of oxygen within seafloor sediments during deposition. However, the water column itself and the surface of the seafloor at times did contain enough oxygen to support a diverse macrofaunal and meiofaunal assemblage (Müller and Walossek 1991; Walossek 1993).

The presence of complete conodont clusters and the general absence of mechanical abrasion indicate that quiet water conditions prevailed during deposition (Müller and Walossek 1985a). However, some strata with randomly oriented sclerites and fossil hash (with no soft-part preservation) suggest that occasional intervals of higher energy were present as well (Müller and Walossek 1985a).

TAPHONOMY

Fossils with phosphatized integument are rare overall, but when found can be very abundant. Even within short distances, fossils vary widely in abundance and preservation (Müller 1990). This local variability and concentration can make it difficult to find fossiliferous nodules in the field. To study the phosphatic fossils, Müller has collected 1,200 kg of limestone for laboratory processing. The material was etched in 15 percent acetic acid on plastic screens. All residues are examined under the microscope at 80 to 100 times magnification at least once to ensure that all critical fossils are collected.

Because of the delicate nature of the fossils, they cannot be processed with typical sieving techniques, magnetic separation, or separation by heavy liquids. After the fossils are hand picked, they must be mounted, coated with gold platinum, and examined under a scanning electron microscope (SEM). Thus, studying the fossils from this type of Lagerstätte is very time consuming and labor intensive. However, the excellent preservation of even minute structures, such as hairs, makes this line of research very productive.

The phosphatic fossils form two distinct groups: (1) organisms that originally had primarily phosphatic hard parts (such as the phosphatocopines, inarticulate brachiopods, and conodonts), and (2) arthropods possessing cuticles that were replaced with phosphate or secondarily coated with fine to fairly coarse-grained phosphatic matter. Only small fossil fragments (<2 mm) have been secondarily phosphatized, and thus this type of preservation is limited to small growth stages and isolated appendages belonging to larger organisms (Müller 1985). The best preservation is found in organisms about 100 to 500 µm in size (Walossek and Müller 1997).

Preservation is variable, possibly due to differences in organic content and/or physical weathering (Müller and Hinz 1991). Phosphatization is not always complete or even. Both phosphatic coating and phosphatic replacement can be found within the same sample and occasionally even within the same specimen. Replacement generally produces better preservation of details than does the coating of cuticle (Müller and Walossek 1985b). Generally, preservation is better in the anterior portions and worse in the distal appendages (Müller 1979), but preservation is not restricted to strongly sclerotized portions of the body (Müller and Walossek 1985a). Typically, orsten arthropods are preserved as hollow molds, but some (e.g., pentastomid larvae) are solid (Walossek and Müller 1994).

It appears that the arthropod cuticle is somehow a suitable catalyst for the precipitation of phosphates (Müller 1985). However, there are different patterns of phosphatization between taxa. For example, although calcareous trilobite remains are common, only one trilobite fragment (excluding agnostids) has ever been found phosphatized (Walossek 1993).

The source of the phosphate is unknown. It is assumed that the biomass was insufficient to provide enough phosphate to account for the number of preserved fossils. One hypothesis is that the phosphatic material came from weathering of granitic rocks on the mainland near the Alum Shale Sea (Müller 1985). Another hypothesis is that the sediment deposited below the flocculent layer provided significant amounts of phosphate needed for preservation (Walossek 1993).

It has been suggested that studying the varying degrees of decomposition in different organisms and different larval stages may shed light on the tiering within the flocculent-layer community (Walossek 1993). The rationale is that the closer to the anoxic zone below the flocculent layer the organism lived, the less distance the organisms had to travel in order to become preserved, and thus would be better preserved than organisms that lived at higher tiers above the anoxic subsurface.

There do not appear to be large differences in the taxonomic composition of different layers within orsten nodules. Whereas the fossils are commonly flattened and distorted in the shales, those in the limestones are not. The nodules may have been formed by replacement of pore waters with calcium carbonate, which prevented great compaction of the fossils (Müller and Walossek 1985b).

PALEOBIOLOGY AND PALEOECOLOGY

Like other Cambrian Lagerstätten, the orsten fauna is dominated by arthropods (Figures 6.2–6.4). However, this fauna appears to differ from

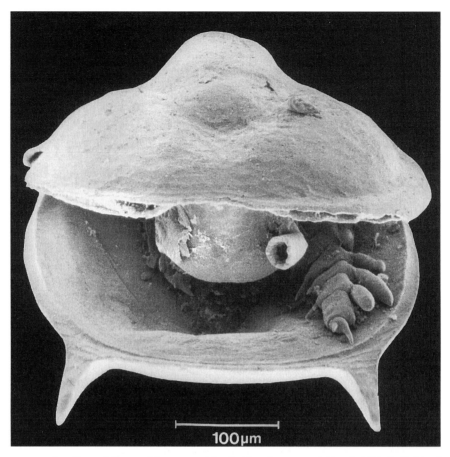

FIGURE 6.2 Larval form of the agnostid trilobite *Agnostus pisiformis*. This individual is in its interpreted, enrolled life position. Specimen is 0.6 mm across. (Photo courtesy of K. Müller, Rheinische Friedrich-Wilhelms-Universität, Bonn)

other Middle Cambrian and Early Ordovician paleocommunities, which had a rich benthic macrofauna. Even in comparison with other Cambrian Lagerstätten, there is apparently no morphological or taxonomic overlap between the fossils found in orsten and in Burgess Shale and Burgess Shale–type localities around the world (Walossek and Müller 1992). Thus, the fossils from orsten provide systematic information not available elsewhere and, therefore, have been the basis for extensive hypotheses regarding the evolutionary history of arthropods and crustaceans.

In addition, orsten is the first meiofaunal assemblage yet recognized in the fossil record. Although meiofauna of the flocculent layer are not often studied by paleontologists, this community plays an important role in marine ecosystems and nutrient-cycling processes. Close study of this fauna can offer insights not only into the systematics and evolutionary

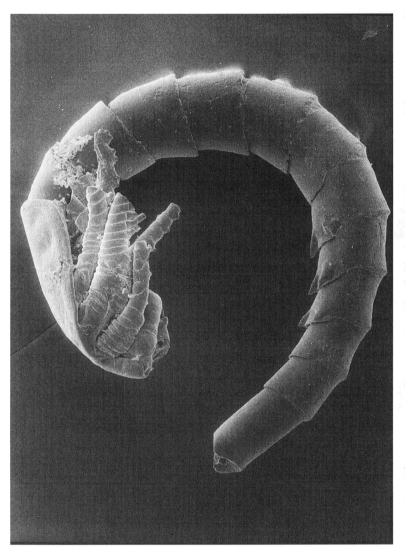

FIGURE 6.3 The crustacean *Skara anulata*. The field of view is 0.9 mm across. (Photo courtesy of K. Müller, Rheinische Friedrich-Wilhelms-Universität, Bonn)

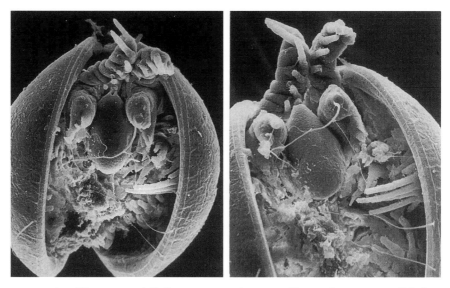

FIGURE 6.4 The ostracod *Falites,* representing a small growth stage, possibly just hatched. Test length is 0.24 mm. (Photo courtesy of K. Müller, Rheinische Friedrich-Wilhelms-Universität, Bonn)

history of these organisms, but also into the development of ecological niches within this important ecosystem (Budd 2001).

Paleoecology

Due to reduced oxygenation levels in parts of the Alum Shale Sea, the fauna is dominated by small arthropods, while large trilobites and brachiopods are virtually restricted to the light-colored layers of the Alum Shale deposited under times of higher oxygenation (Müller and Hinz 1991). Other fossils include sponge spicules, horny and calcareous brachiopods, tubular hyolithid-like cones, and some enigmatic microfossils (Müller and Walossek 1985a).

The phosphatized fauna of orsten consists of tiny arthropods, most of which had swimming adaptations (Müller and Walossek 1985a). However, they are not interpreted as pelagic organisms, but as epibenthic organisms that coexisted in the flocculent detrital layer above the seafloor (Müller and Walossek 1985a). When these organisms died, they slid to the anoxic seafloor bottom and were preserved.

The assemblage is quite diverse and abundant despite these environmental conditions. The diversity of locomotion and feeding adaptations suggests that the organisms inhabited a number of niches within this unconsolidated, high-nutrient layer and exploited different food resources (Budd 2001). Although undisputed predators have not yet been identified, the presence of specimens with lost appendages and fecal pellets containing larvae and setae indicates that predation was present

(Walossek 1993). Some parasitic organisms have also been identified, suggesting that this life habitat evolved early in Earth history (Walossek and Müller 1994).

Arthropods

Arthropods are by far the most dominant organisms within the orsten and Alum Shale. Trilobites dominate the faunal assemblages of the Alum Shale Formation to the point that trilobite remains, mostly exuviae, can be rock forming at times. However, excluding agnostids, only one trilobite fragment has been found phosphatized within an orsten nodule (Walossek 1993).

Agnostids have not been well understood. Before the orsten-based studies, very little was known about their soft body parts, ventral morphology, ontogenetic sequence, or even life habits (Müller and Walossek 1987). *Agnostus* has one pair of antennae, three pairs of head limbs, and five pairs of trunk limbs, as well as large head and tail shields (Figure 6.2). Müller and Walossek (1987) suggest that *Agnostus* lived most of its life at least partially enrolled, with head and tail shields forming a protective "shell" (Figure 6.2). In fact, they propose that *Agnostus* had locked joints that prevented it from stretching out the body completely. Agnostids may have clamped their "shells" completely closed in order to escape predation and/or adverse bottom-water conditions. If so, some of the specimens may have died and been preserved in this life position (Müller and Walossek 1987).

After Müller and Walossek (1987) described *Agnostus pisiformis* in detail, they proposed that Agnostina represents a peculiar side branch in the development of Trilobita and that it exhibits many traits unlike those of other trilobites. There are, however, some morphological similarities between stem-lineage crustaceans and *Agnostus pisiformis* that may shed light on the nature of the relationship between trilobites and crustaceans (Walossek and Müller 1990).

Using cladistic analysis, Walossek (1993) and Walossek and Müller (1990, 1992, 1997, 1998) have categorized other orsten arthopods as representatives either of extinct, ancestral crustacean-like arthropods (referred to as the crustacean stem-lineage group) (Walossek and Müller 1990, 1998) or of living crustacean groups (referred to as crown-group crustaceans). These categorizations and the morphological characters used to make them "involve a fascinating redefinition of the crustaceans" (Schram and Hof 1998:241). Based on orsten fossils and interpretations based on them, it appears that the major radiations among Crustacea occurred by Late Cambrian time (Walossek 1993). If these findings and interpretations become generally accepted (Lauterbach 1988; Schram and Hof 1998), revisions and new phylogenetic interpretations of both Crustacea and Arthropoda would become necessary. The use of orsten arthropods has been cited as evidence to challenge evolutionary hypotheses based on molecular phylogenetic work (Walossek and Müller 1994, 1998).

Martinssonia, Goticaris, and Cambrypachycope are among the taxa interpreted as representatives of stem-group crustaceans by Walossek and Müller (1990, 1992, 1997, 1998; Walossek 1993). Both adult and juvenile forms of *Martinssonia elongata* have been found, and they appear to have been bottom dwellers that fed off particles they stirred up from the seafloor. Although *Martinssonia* resembles crustaceans in some ways, it also possesses many significantly different traits, some of which appear to be similar to those of *Agnostus* (Walossek and Müller 1990). This genus, therefore, was first considered to be a "pre-crustacean" (Müller and Walossek 1986) but is now thought to be a representative of the Crustacea stem group (Walossek and Müller 1990).

Other orsten crustacean taxa, such as *Skara* (Figure 6.3), *Bredocaris, Dala,* and *Rehbachiella,* are hypothesized to be direct ancestors of modern crustacean lineages (crown group) (Walossek and Müller 1990, 1992, 1997, 1998; Walossek 1993). The morphology of the different taxa indicates that they were adapted to different niches and exploited different food sources within and above the flocculent layer. The excellent preservation of morphological details has allowed workers to recognize eyes ranging from simple bumps to stalked eyes to compound eyes, and to identify hairs used for cleaning and swimming. *Skara*—possessing morphologies similar to those of copepods—are interpreted to have been swimming sweep-net feeders with soft head shields. Possible male and female morphotypes have been identified (Müller and Walossek 1985b). *Dala* had specialized bristles; some were used as sensory organs, and others were used for efficient swimming, which it probably did upside down, as do many living crustaceans (Müller and Walossek 1985a). *Bredocaris,* known from larval to adult stages, had fairly large paired eyes and well-developed thoracic swimming appendages. It probably swam in or above the flocculent layer (Müller and Walossek 1988). The genus *Rehbachiella* is now known from 30 larval stages. Due to the similarity of the filter-feeding structure of *Rehbachiella kinnekullensis* with that of the Branchiopoda, Walossek (1993) recognizes this genus as an ancestral marine representative of this taxon.

The Phosphatocopina are bivalved arthropods and are the dominant faunal element in the Swedish orsten (Müller 1979, 1982). They are found most commonly as isolated head shields and exuvia, although some soft-part morphology has also been preserved (Figure 6.4). The phosphatocopines, including the genera *Vestrogothia, Falites* (Figure 6.4), and *Hesslandona,* are interpreted to be nectobenthic, filter-feeding organisms adapted to a number of different niches within the flocculent-layer environment. Phosphatocopina generally exhibit characters and limb morphology that places them below the crown group of Crustacea (Walossek and Müller 1992). Müller (1982) considers the Phosphato-copida as a side branch of ostracod evolution.

The orsten have also yielded the larval form of the first pentastomid in the fossil record (Walossek and Müller 1994). Not much is known about the physiology and life cycles of the living Pentastomida—except that they infest almost all terrestrial tetrapods—and even less is known about their evolution and origin. The fossil specimens range in size from 200 to 700 μm, and because the morphologies are so dissimilar, the larvae have been recognized as representatives of distinct taxa (Walossek and Müller 1994). Two types of larvae—hammer-headed and round-headed—have been divided into three genera and seven species. These forms show evidence for a parasitic lifestyle, including the presence of claw limbs adapted for grasping and anchoring onto host tissue. The difference in limb morphology among taxa suggests possible dissimilarities in life habitats, such as host attachment points.

The presence of pentastomids with parasitic adaptations in this marine Upper Cambrian deposit demonstrates that the group did not originate from free-living, nonparasitic terrestrial or freshwater arthropods. Their original hosts are still unidentified, but must have been marine and may be early representatives of Chordata, possibly conodont animals. Walossek and Müller (1994) suggest that the larvae may have been able to crawl around at the sediment-water interface and once they found a suitable host, became internal ectoparasites living in cavities such as gill chambers. Most of the major morphological characters of Pentastomida were established by the Upper Cambrian, and Walossek and Müller (1994) propose that the morphological stasis exhibited by pentastomids may be due to a conservatism in the coevolution of host-parasite relationships. In other words, it is possible that pentastomids were already highly adapted as internal parasites on chordate hosts by Cambrian times and thus did not have to change much morphologically, even when their hosts invaded new environments, including terrestrial habitats.

Vertebrates

Conodonts are a very common and diverse faunal element in orsten. In fact, conodonts were what Müller was originally searching for before he discovered the soft-part preservation of arthropods. Surprisingly, in this deposit, proto- and para-conodonts are associated with already highly developed euconodonts. Previously, the earliest euconodonts were described mainly from the latest Cambrian close to the Cambrian–Ordovician boundary. However, the orsten have yielded *Acodus cambricus* and a highly developed conodont genus, *Cambropustula,* from the low-ermost Upper Cambrian. These are the first known occurrences of the euconodont Conodontophorida (Müller and Hinz 1991). The excellent preservation of uncompressed fossils within the orsten allows for the reconstruction of the individual conodont sclerites as circular clusters and not crescent-like, as had been proposed (Müller and Hinz 1991).

Conclusions

The Swedish orsten deposit is unlike most other Lagerstätten in that its exceptional preservation is invisible without extensive preparation and high-powered microscopy. Although the faunal elements are limited by preservational size bias (only those less than 2 mm in size), the three-dimensional nature of the fossils allows for much more detailed morphological reconstructions than is usually expected from the fossil record. This exquisite preservation allows for detailed morphological, functional, paleoecological, and systematic studies of meiofauna, an ecological group not often examined by paleontologists.

More systematic work, which may shed light on the systematic relationships between fossil and modern arthropod taxa, is still being conducted on the orsten fauna. The deposits themselves, because of the localized concentration of fossils and difficulty in sampling, may still yield new faunal elements.

Geochemical and sedimentological work is still needed in order to understand the phosphatization process and the large range of preservational quality present in the assemblage. In addition, a better understanding of the paleoenvironmental and paleoceanographic conditions of the Alum Shale Sea could provide information relevant to the paleoecological interpretations of the life habits of these organisms.

Several other Lagerstätten containing orsten-type phosphatic preservation of cuticle and soft tissue have been identified, including the Lower Cretaceous Santana Formation of Brazil, the Triassic of Spitsbergen, the Upper Devonian of the Carnic Alps of Italy, the Middle Cambrian of Australia, and the Lower Cambrian of Comley, United Kingdom (Walossek 1999). Now that Müller, Walossek, and colleagues have perfected a method of searching for and preparing fossils from this type of Lagerstätten, one hopes that even more examples of such exceptional, phosphatic preservation throughout the Phanerozoic can be found and investigated in as much detail as the Cambrian orsten deposits of Sweden.

Acknowledgments

K. Müller and D. Walossek were very helpful in providing reviews, photos, and reprints for this chapter. Special thanks go to D. Walossek for his careful critique of an earlier draft of this manuscript.

REFERENCES

Bergstrom, J., and D. G. Gee. 1985. The Cambrian in Scandinavia. In D. G. Gee, ed., *The Caledonide Orogen: Scandinavia and Related Areas,* vol. 1, pp. 247–271. Chichester: Wiley.

Budd, G. E. 2001. Ecology of nontrilobite arthropods and lobopods in the Cambrian. In A. Y. Zhuravlev and R. Riding, eds., *The Ecology of the Cambrian Radiation,* pp. 404–427. New York: Columbia University Press.

Clarkson, E. N. K. 1993. *Invertebrate Palaeontology and Evolution.* London: Chapman and Hall.

Dvoratzek, M. 1987. Sedimentology and petrology of carbonate intercalations in the Upper Cambrian Olenid shale facies of southern Sweden. *Sveriges Geologiska Undersökning, C* 81:1–73.

Lauterbach, K. E. 1988. Zur position angeblicher Crustacea aus dem Ober-Kambrium im phylogenetischen system der Mandibulata (Arthropoda). *Abhandlungen des naturwissenschaflichen vereins in Hamburg,* n.s., 30: 409–467.

Müller, K. J. 1979. Phosphatocopine ostracodes with preserved appendages from the Upper Cambrian of Sweden. *Lethaia* 12:1–27.

Müller, K. J. 1982. *Hesslandona unisulcata* sp. nov. with phosphatised appendages from Upper Cambrian "Orsten" of Sweden. In R. H. Bate, E. Robinson, and L. M. Sheppard, eds., *Fossil and Recent Ostracods,* pp. 276–304. Chichester: Ellis Horwood.

Müller, K. J. 1983. Crustacea with preserved soft parts from the Upper Cambrian of Sweden. *Lethaia* 16:93–109.

Müller, K. J. 1985. Exceptional preservation in calcareous nodules. *Philosophical Transactions of the Royal Society of London, B* 311:647–673.

Müller, K. J. 1990. Upper Cambrian "orsten." In D. E. G. Briggs and P. R. Crowther, eds., *Palaeobiology: A Synthesis,* pp. 274–277. Oxford: Blackwell Science.

Müller, K. J., and I. Hinz. 1991. *Upper Cambrian Conodonts from Sweden.* Fossils & Strata, no. 28. Oslo: Universitetsforlag.

Müller, K. J., and D. Walossek. 1985a. A remarkable arthropod fauna from the Upper Cambrian orsten of Sweden. *Transactions of the Royal Society of Edinburgh: Earth Sciences* 76:161–172.

Müller, K. J., and D. Walossek. 1985b. *Skaracarida, a New Order of Crustacea from the Upper Cambrian of Västergötland, Sweden.* Fossils & Strata, no. 17. Oslo: Universitetsforlag.

Müller, K. J., and D. Walossek. 1986. *Martinssonia elongata* gen. et sp. n., a crustacean-like euarthropod from the Upper Cambrian of Sweden. *Zoologica Scripta* 15:73–92.

Müller, K. J., and D. Walossek. 1987. *Morphology, Ontogeny, and Life-habit of the* Agnostus pisiformis *from the Upper Cambrian of Sweden.* Fossils & Strata, no. 19. Oslo: Universitetsforlag.

Müller, K. J., and D. Walossek. 1988. *External Morphology and Larval Development of the Upper Cambrian Maxillopod* Bredocaris admirabilis. Fossils & Strata, no. 23. Oslo: Universitetsforlag.

Müller, K. J., and D. Walossek. 1991. Ein Blick durch das "Orsten"-Fenster in die Arthropodenwelt vor 500 Millionen Jahren. *Verhandlungen der deutschen zoologischen Gesellschaft* 84:281–294.

Schram, F. R., and C. H. J. Hof. 1998. Fossils and the interrelationships of major crustacean groups. In G. D. Edgecombe, ed., *Arthropod Fossils and Phylogeny*, pp. 233–302. New York: Columbia University Press.

Walossek, D. 1993. *The Upper Cambrian* Rehbachiella *and the Phylogeny of Branchiopoda and Crustacea*. Fossils & Strata, no. 32. Oslo: Universitetsforlag.

Walossek, D. 1999. On the Cambrian diversity of Crustacea. In F. R. Schram and J. C. von Vaupel Klein, eds., *Crustaceans and the Biodiversity Crisis*, pp. 3–27. Proceedings of the Fourth International Crustacean Congress, Amsterdam. Leiden: Brill.

Walossek, D., and K. J. Müller. 1990. Upper Cambrian stem-lineage crustaceans and their bearing upon the monophyletic origin of Crustacea and the position of *Agnostus*. *Lethaia* 23:409–427.

Walossek, D., and K. J. Müller. 1992. The "Alum Shale Window"—Contribution of "orsten" arthropods to the phylogeny of Crustacea. *Acta Zoologica*, 73:305–312.

Walossek, D., and K. J. Müller. 1994. Pentastomid parasites from the lower Paleozoic of Sweden. *Transactions of the Royal Society of Edinburgh: Earth Sciences* 85:1–37.

Walossek, D., and K. J. Müller. 1997. Cambrian "orsten"-type arthropods and the phylogeny of Crustacea. In R. A. Fortey and R. Thomas, eds., *Arthropod Relationships*, pp. 139–153. London: Chapman and Hall.

Walossek, D., and K. J. Müller. 1998. Early arthropod phylogeny in light of the Cambrian "orsten" fossils. In G. D. Edgecombe, ed., *Arthropod Fossils and Phylogeny*, pp. 185–232. New York: Columbia University Press.

Walossek, D., and H. Szaniawski. 1991. *Cambrocaris baltica* n. gen. n. sp., a possible stem-lineage crustacean from the Upper Cambrian of Poland. *Lethaia* 24:363–378.

7

Beecher's Trilobite Bed: Ordovician Pyritization for the Other Half of the Trilobite

Walter Etter

THE ORDOVICIAN UTICA AND LORRAINE FORMATIONS OF New York State are composed of shales, siltstones, and limestones that have yielded a considerable amount of remarkable fossils, including large articulated eurypterids. Parts of these formations might be considered as Lagerstätten by themselves, but it is a very thin layer within the Frankfort Shales of the Utica Formation that has attracted much attention and has become world famous: the so-called Beecher's Trilobite Bed. This layer has become especially well known for the preservation of pyritized trilobites, which show not only perfectly articulated exoskeletons, but preservation of soft parts, including legs and antennae, muscles, and parts of the digestive tract. Sedimentological evidence indicates that Beecher's Trilobite Bed formed as a consequence of a sudden influx of sediment, carrying bottom-dwelling animals for some distance before burying them under a cover of fine siltstone. Beecher's Trilobite Bed is thus a classic example of an obrution deposit.

The first trilobites with preserved appendages were discovered in 1884 by the amateur collector William S. Valiant in loose blocks along Six Mile Creek near Rome, in upstate New York (Whiteley 1998, 2000). It was not until 1892, however, that he was able to locate the trilobite bed in outcrop. This exciting discovery, first published in 1893 (Matthew 1893), quickly drew the attention of Charles Emerson Beecher, a paleontologist at Yale University. Beecher started excavating the locality in 1893, but by 1895 the trilobite bed was thought to be mined out, and quarrying ceased (Briggs and Edgecombe 1992, 1993). Beecher's sampling efforts yielded several tons of fossiliferous rock, and he subsequently published many papers on the anatomy and development of the trilobites from this bed (Beecher 1893, 1894, 1895a, 1895b, 1896; for a more complete list of Beecher's publications, see Cisne 1973a, 1981).

Beecher died in 1904 while drawing a trilobite from the fossiliferous bed. After his death, most of the unprepared material he had collected was lost or destroyed, and the exact location of the bed was forgotten (Cisne 1973a; Briggs and Edgecombe 1992, 1993). The trilobite bed was named in honor of Beecher, and the investigation of the fossils from it, especially the trilobites, has continued to the present (Cisne 1975, 1981; Whittington and Almond 1987). The paleoecology and taphonomy of Beecher's Trilobite Bed has remained somewhat obscure. It was not until the 1970s that the first detailed paper on paleoecologic aspects of the trilobite bed, based on material collected earlier in the century, was published (Cisne 1973a). By this time, the locality was thought to be exhausted. In 1984 Beecher's Trilobite Bed was rediscovered by amateur fossil collectors Dan Cooper and Thomas Whiteley, who, recognizing the importance of their finding, reported the discovery to several museums (Briggs and Edgecombe 1993; Whiteley 1998). In 1989 a major excavation was started in cooperation with the National Museum of Natural History in Washington, D.C., and the American Museum of Natural History in New York. This excavation not only enabled the recovery of new material, but provided the first opportunity to investigate the taphonomy of the bed (Briggs, Bottrell, and Raiswell 1991).

Geological Context

Beecher's Trilobite Bed is a very localized occurrence, found only near the city of Rome in northern New York State (Figure 7.1). The exact location is Cleveland's Glen, or Six Mile Creek, near Rome (Cisne 1973a; Briggs and Edgecombe 1993). The richly fossiliferous bed is a thin microturbidite layer within a uniform sequence of deep-water turbidites in the Upper Ordovician (Caradocian) Frankfort Shale of the Utica Formation (Cisne 1973a; Briggs, Bottrell, and Raiswell 1991) that were shed into the Taconian foredeep basin. The Frankfort Shale consists of an alternation of siltstones and fine sandstones with intercalated clayey graptolitic layers, the last representing condensed accumulations during periods of low sedimentation (Briggs, Bottrell, and Raiswell 1991).

The trilobite bed is a 40 mm thick dark gray, graded siltstone "characterized by the abundant occurrence of exceptionally preserved trilobites" (Cisne 1973a) (Figure 7.2). It rests on a very thin (3 mm) layer of black mudstone that contains mainly the flattened rhabdosomes of the graptoloid *Climacograptus* sp. (Figure 7.2). Other common fossils in this mudstone include the brachiopods *Schizocrania* and *Camarotoecia*, the

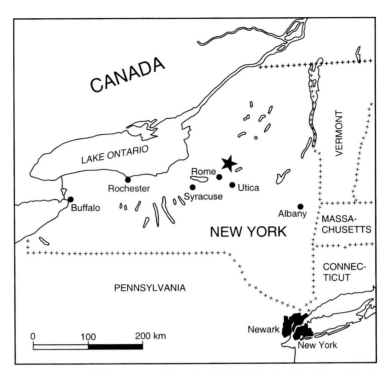

FIGURE 7.1 Geographic location of Beecher's Trilobite Bed, indicated by the star, in upper New York State.

trilobites *Triarthrus* and *Cryptolithus,* and the dendroid graptolite *Inocaulis.* The mudstone is bioturbated and contains mainly horizontal, silt-filled traces (Cisne 1973a), probably belonging to the ichnogenus *Planolites.*

The sedimentological features of the trilobite bed include graded bedding, sole marks, and internal directional features. It is obvious that the 40 mm thick bed represents a single depositional event and that this layer can be called a microturbidite (Cisne 1973a). The turbidity current that deposited the bed was flowing over a muddy bottom, now represented by the underlying mudstone. During its initial phases, the current eroded away material from the muddy bottom and produced ripple marks. With decreasing current velocity, the overlying layers were deposited successively. All the sedimentological features indicate that deposition of the trilobite bed was very rapid (Cisne 1973a).

The microstratigraphy of Beecher's Trilobite Bed has been discussed in detail by Cisne (1973a). The basal layer of the trilobite bed is a less than 1 mm thick layer of coarse, gray siltstone, resting with a sharp erosive contact on the underlying mudstone. Except for some graptolites, the basal layer is barren of fossils. It grades upward into laminated, medium-grained siltstone about 4 mm thick with some ripple marks at

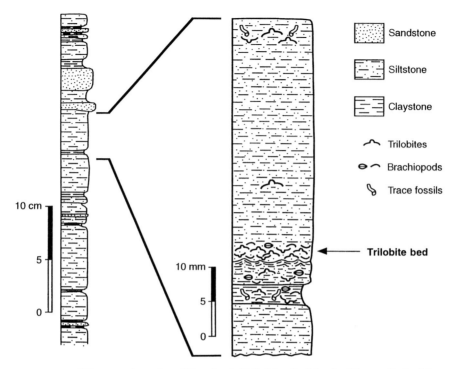

FIGURE 7.2 Microstratigraphy of Beecher's Trilobite Bed in the Upper Ordovician (Caradocian) Frankfort Shale. (Modified from Cisne 1973a and Briggs et al. 1991)

the top. The maximum concentration of pyritized trilobites occurs immediately above the laminated part of the microturbidite, in a laterally continuous, 2 mm thick layer lying 4 to 6 mm above the contact with the underlying mudstone (Cisne 1973a) (Figure 7.2). The famous trilobites and other macrofossils are concentrated in this thin layer, which has been termed the *trilobite layer* of Beecher's Trilobite Bed. The upper 30 mm of the siltstone contain almost no macrofossils, and only in the top layer of the microturbidite are fossils and burrows common again.

PALEOENVIRONMENTAL SETTING

The depositional environment of Beecher's Trilobite Bed was oxygenated, as is obvious from the presence of trace fossils (Cisne 1973a; Briggs, Bottrell, and Raiswell 1991; Briggs and Edgecombe 1993). The perfect preservation of these animals strongly suggests that they were alive just before the catastrophic burial event. The concentration of completely preserved trilobites and other macrofossils in the basal part of Beecher's Trilobite Bed was therefore the result of a mass killing (Beecher 1894; Cisne 1973a). The killing may have been simply the consequence of sediment blanketing or, as has been speculated by Cisne

(1973a), the result of a temperature shock. The temperature of the turbidity current was probably somewhat higher than the ambient temperature to which the animals were acclimatized (Cisne 1973a). Quite strong unidirectional current activity during deposition is evident from the orientation of the trilobites and graptolites in the trilobite layer. The trilobites show, however, no preferred longitudinal orientation, and about equal numbers are embedded with their dorsal as with their ventral surfaces upward (Cisne 1973a).

The animals preserved in the lower part of the trilobite bed were probably not transported very far because the same taxa are present in the underlying mudstone (Cisne 1973a). It was thus certainly a within-habitat dislocation of animals. The very short distance calculated by Cisne (20 m) might, however, be too low an estimate, because it is well known that living or freshly killed arthropods can survive long-distance transport in quite strong currents without serious damage (Allison 1986). After the deposition of the microturbidite, very much the same conditions as before were established, except for variations in grain size of the substratum, and bioturbation in the top of the turbidite layer (Cisne 1973a).

TAPHONOMY

The splendid preservation of the trilobites near the base of Beecher's Trilobite Bed, with exquisitely pyritized appendages, muscles, and parts of digestive tracts (Cisne 1975, 1981), requires very special taphonomic circumstances. The trilobite bed, apart from the topmost few millimeters, has no bioturbation. The turbiditic layer obviously protected the buried carcasses from scavengers (Cisne 1973a; Briggs, Bottrell, and Raiswell 1991). Most of the trilobites and some other soft-bodied animals, including specimens of the polychaete worm *Protoscolex* sp., were subsequently pyritized. The excellent three-dimensional preservation of delicate, originally nonmineralized structures indicates that pyritization started very early during diagenesis, before considerable decay of the soft parts and compaction of the sediment occurred.

To determine the exact timing of pyritization, Briggs, Bottrell, and Raiswell (1991) analyzed the sulfur isotopes of the pyrite, both in the trilobites and in the surrounding sediment. During microbial sulfate reduction, the lighter ^{32}S isotope is preferentially reduced by microbes. In the pore water, dissolved sulfide will thus become enriched in the lighter isotope, and pyrite of very early diagenetic origin will show this lighter isotope signature (Briggs, Bottrell, and Raiswell 1991). When the system becomes closed to the overlying seawater as a result of a thick sediment cover, the pore water sulfide will progressively become enriched in the heavier ^{34}S isotope, and pyrite, which forms later during diagenesis, will become isotopically heavier (positive δ^{34}S values).

In Beecher's Trilobite Bed, especially in the trilobite layer, a massive isotope excursion of very heavy pyrite sulfur values can be observed, with the peak value attaining $+30.7‰$. Isotopically light ($-15‰$ to $-20‰$ $\delta^{34}S_{CDT}$) framboidal pyrite is found scattered in the sediment, indicating formation when sulfate from the overlying seawater could still diffuse into the sediment (Briggs, Bottrell, and Raiswell 1991). The pyrite that lines and replaces the exoskeleton of the trilobites has $\delta^{34}S_{CDT}$ values of $-3‰$ to $+21‰$. For every single trilobite, the isotopic values for the limbs ($\delta^{34}S_{CDT}$ $+10$ to $+27‰$) were significantly heavier than for the exoskeleton. Obviously, the pyritization of the dorsal exoskeleton started even earlier than that of the limbs, probably because the dissolution of the carbonate skeleton buffered the pH locally at a higher value, thus enhancing the supersaturation of iron monosulfides and promoting localized precipitation of pyrite (Briggs, Bottrell, and Raiswell 1991; Briggs et al. 1996). The heavier sulfur isotope composition of the pyritized limbs indicates that pyritization of soft parts continued after the pore-water system was already partially closed to the overlying seawater (Briggs, Bottrell, and Raiswell 1991).

The geochemical analysis conducted by Briggs, Bottrell, and Raiswell (1991) produced some additional results. The concentration of organic carbon is low throughout the whole sequence, including the trilobite layer (0.02 to 0.2 percent; however, these values are somewhat depressed as a result of increased burial temperature) (Briggs, Bottrell, and Raiswell 1991). Pyrite sulfur concentrations are low as well (<2 percent), but reactive iron, which can form iron monosulfides in the presence of sulfide, is still present in the sediment (1.4 to 2.7 percent, measured as acid-soluble Fe). The degree of pyritization is therefore low for the sediment, indicating deposition under well-oxygenated bottom-waters (Raiswell et al. 1988).

In summary, geochemical analysis shows that pyritization of soft parts in Beecher's Trilobite Bed was limited neither by the amount of reactive iron (which is still abundant) nor by the concentration of sulfate (the system was at least initially open to the overlying seawater). The limiting factor was clearly the amount of readily metabolizable organic matter (Briggs, Bottrell, and Raiswell 1991; Briggs et al. 1996), which was concentrated only in the buried carcasses. Therefore, the process of pyritization was more or less confined to the trilobites, but all the specimens in the trilobite layer became pyritized.

PALEOBIOLOGY AND PALEOECOLOGY

In Beecher's Trilobite Bed two distinct fossil assemblages have been recognized: one termed the *mud assemblage,* preserved in the lower few millimeters of the bed, including the trilobite layer and the underlying mudstone, and one occurring in the top layer of the microturbidite, the *silt assemblage* (Cisne 1973a).

In the mud assemblage, which is of primary interest, 28 species have been recorded, of which five are considered planktonic or pseudoplanktonic (Cisne 1973a). These include three inarticulate brachiopod species, a cirripedian, and the graptoloid *Climacograptus* sp. All the species except for the favositid coral and the two problematic taxa have been described (Ruedemann 1925, 1926). The benthic assemblage is dominated by a few species, mainly (in descending order of abundance) the trilobite *Triarthrus eatoni* (Figures 7.3 and 7.4), the dendroid graptolite *Inocaulis arborescens*, the brachiopod *Schizocrania filosa*, and the trilobite *Cryptolithus bellulus*. Echinoderms are absent in the mud assemblage but present in the overlying silt assemblage (Cisne 1973a). The benthic fauna comprises mainly epibenthic and shallow endobenthic species, and its composition, except for the presence of preserved soft-bodied annelids, is by no means extraordinary for Late Ordovician deep-water muddy bottoms of the northern Appalachians.

The state of preservation makes the fauna of Beecher's Trilobite Bed exceptional. The most common species, the trilobite *Triarthrus eatoni* (Figures 7.3 and 7.4), shows such a remarkable state of preservation that it has become the textbook example of a trilobite (Whittington 1992; Levi-Setti 1993). The three-dimensional preservation of pyritized appendages, muscles, and parts of the digestive tract allows a very detailed morphological and functional analysis of this trilobite. The special

FIGURE 7.3 Complete specimen of *Triarthrus eatoni* shown in cross-sectional view. Note the preservation of completely articulated legs and antennae. Body length is 3.5 cm. (Photo by T. Whiteley [YPM 204] courtesy of the Peabody Museum of Natural History, Yale University, New Haven, Conn.)

FIGURE 7.4 Complete specimen of *Triarthrus eatoni* shown in plan view. Note the preservation of completely articulated legs and antennae. Body length in 4.0 cm. (Photo by T. Whiteley [YPM 228] courtesy of the Peabody Museum of Natural History, Yale University, New Haven, Conn.)

preservation style allows not only the investigation of the external features of carefully prepared specimens, but also the visualization of internal organs and the examination of specimens still embedded in the rock by means of x-radiography. Using stereopairs of x-radiographs, Cisne (1975, 1981) was able to reconstruct in a very detailed manner the external morphology, including the appendages, as well as internal structures of *Triarthrus*, although some of his conclusions (insertion of leg branches, presence of a post-pygidial abdomen) were challenged (Whittington and Almond 1987).

Adult specimens of *Triarthrus eatoni*, which attain a maximum length of only 40 mm, show one pair of pre-oral appendages, the long uniramous antennae (Figures 7.3 and 7.4). The three post-oral cephalic legs and the thoracic legs are very similar (Figures 7.3 and 7.4), all bearing an exopodite with many filaments arranged in a blade-like manner, which probably functioned as gills (Whittington and Almond 1987). The most proximal part of the legs shows a marked and bristle-bearing inward projection, thus enclosing a median groove. This narrow ventral space is, in analogy with some modern crustaceans, interpreted as a food groove. The mouth of *Triarthrus* is small and posteriorly directed. It thus seems quite probable that *Triarthrus* collected finely particulate matter with its legs from the seafloor (Whittington and Almond 1987) and transported the particles in the median food groove forward to the mouth (Cisne 1981; Whittington and Almond 1987). Such a trunk–limb feeding mechanism is still found in modern cephalocarid, branchiopod, and phyllocarid crustaceans (Cisne 1981). Specialized masticatory mouthparts were not present, and the muscular esophagus seems to have been developed for suctorial ingestion of food. The anatomy of the digestive system, which includes a J-shaped foregut and extensive glands in the head region, supports the view that this trilobite fed only on small particles (Cisne 1981).

The exceptional preservation of the trilobites allows detailed reconstruction of adult morphology and inferences about their life habits. Since larval and juvenile stages have been preserved as well (mostly as exuviae), a reconstruction of the ontogeny seems possible. Using material collected earlier in the century, Cisne (1973b) concluded that the very small larval (protaspis) stages (0.6 to 0.75 mm) of *Triarthrus eatoni* were most probably planktonic, whereas the juvenile (meraspis) and subadult and adult (holaspis) stages were benthic (Cisne 1973b). The planktonic stage would have lasted about one month, and maximum longevity was at least four years (Cisne 1973b). The growth curve for overall length is approximately linear, whereas the survivorship curve decreases exponentially, indicating a high larval and early juvenile mortality (Cisne 1973b). This pattern is typical for species that produce a large number of small eggs and spend the first ontogenetic stages in the plankton. However, Cisne (1973b) mistook the protaspid of another trilobite (*Stenoblepharum beecheri*) for that of *Triarthrus eatoni* (Briggs and Edgecombe 1992, 1993). Protaspis stages of *Triarthrus* are actually missing in Beecher's Trilobite Bed (Briggs and Edgecombe 1992, 1993). Ironically, it is the absence of these protaspids which suggests that the larval stage of *Triarthrus* did, indeed, spend its life elsewhere, perhaps in the plankton (Briggs and Edgecombe 1992, 1993).

CONCLUSIONS

Apart from the Hunsrück Slate (Chapter 8), Beecher's Trilobite Bed is probably the best example of a fauna showing localized soft-part preservation as a consequence of early diagenetic pyritization. Furthermore, it is one of only a few Ordovician soft-bodied faunas known to date. The specimens of the trilobite *Triarthrus eatoni* show such splendid three-dimensional preservation of the antennae, limbs, parts of the muscular system, and digestive tract that this trilobite is now probably the best-known member of the class (Whittington 1992; Levi-Setti 1993). A third example of soft-part pyritization is the Jurassic of La Voulte-sur-Rhône (Chapter 16).

The accumulation of fossils in the trilobite bed is the result of sudden burial under a turbidite layer. This sediment cover prevented the destruction of the embedded animals by scavengers and provided ideal conditions for early diagenetic mineralization. Since the rediscovery of Beecher's Trilobite Bed in 1984, its taphonomy has been thoroughly studied. The geochemical investigation of these pyritized trilobites and the surrounding sediment suggests that pyritization started very early during diagenesis, and continued even after the system became semi-enclosed and the supply of sulfate from the seawater ceased. Pyritization remained a very localized phenomenon, however, and was basically confined to trilobite soft parts. The low organic content of surrounding sediment prevented disseminated pyritization and rapid depletion of pore-water sulfide. Although detailed study of Beecher's Trilobite Bed has contributed much to our understanding of early diagenetic pyritization of soft parts, it is still puzzling that this mode of soft-part preservation is so rare in the fossil record.

ACKNOWLEDGMENTS

Thanks to D. Briggs, who carefully read the manuscript, offered many helpful comments, and provided references.

REFERENCES

Allison, P. A. 1986. Soft-bodied animals in the fossil record: The role of decay in fragmentation during transport. *Geology* 14:979–981.

Allison, P. A., and D. E. G. Briggs. 1991. Taphonomy of nonmineralized tissues. In P. A. Allison and D. E. G. Briggs, eds., *Taphonomy: Releasing the Data Locked in the Fossil Record,* pp. 25–70. New York: Plenum.

Bartels, C., and G. Brassel. 1990. *Fossilien im Hunsrückschiefer: Dokumente des Meereslebens im Devon.* Idar-Oberstein: Museum Idar-Oberstein.

Beecher, C. E. 1893. On the thoracic legs of *Triarthrus. American Journal of Science,* 3rd ser., 46:467–470.

Beecher, C. E. 1894. On the mode of occurrence, structure and development of *Triarthrus becki. American Geologist* 13:38–43.

Beecher, C. E. 1895a. Structure and appendages of *Trinucleus. American Journal of Science,* 3rd ser., 49:307–311.

Beecher, C. E. 1895b. The larval stages of trilobites. *American Geologist* 16:166–197.

Beecher, C. E. 1896. The morphology of *Triarthrus. American Journal of Science,* 4th ser., 1:250–256.

Briggs, D. E. G., S. H. Bottrell, and R. Raiswell. 1991. Pyritization of soft-bodied fossils: Beecher's Trilobite Bed, Upper Ordovician, New York State. *Geology* 19:1221–1224.

Briggs, D. E. G., and G. D. Edgecombe. 1992. Gold bugs. *Natural History* 101:36–43.

Briggs, D. E. G., and G. D. Edgecombe. 1993. Beecher's Trilobite Bed. *Geology Today* 9:97–102.

Briggs, D. E. G., R. Raiswell, S. H. Bottrell, D. Hatfield, and C. Bartels. 1996. Controls on the pyritization of exceptionally preserved fossils: An analysis of the

Lower Devonian Hunsrück Slate of Germany. *American Journal of Science* 296:633–663.

Cisne, J. L. 1973a. Beecher's Trilobite Bed revisted: Ecology of an Ordovician deepwater fauna. *Postilla* 160:1–25.

Cisne, J. L. 1973b. Life history of an Ordovician trilobite, *Triarthrus eatoni*. *Ecology* 54:135–142.

Cisne, J. L. 1975. Anatomy of *Triarthrus* and the relationship of the Trilobita. *Fossils & Strata* 4:45–63.

Cisne, J. L. 1981. *Triarthrus eatoni* (Trilobita): Anatomy of its exoskeletal, skeletomuscular, and digestive systems. *Palaeontographica Americana* 9:99–142.

Levi-Setti, R. 1993. *Trilobites*. 2nd ed. Chicago: University of Chicago Press.

Matthew, W. D. 1893. On the antennae and other appendages of *Triarthrus becki*. *American Journal of Science*, 3rd ser., 46:121–125.

Raiswell, R., F. Buckley, R. A. Berner, and T. F. Anderson. 1988. Degree of pyritization of iron as a paleoenvironmental indicator of bottom water oxygenation. *Journal of Sedimentary Petrology* 58:812–819.

Ruedemann, R. 1925. The Utica and Lorraine Formations of New York. Part II, systematic paleontology, number 1. *New York State Museum Bulletin* 262:1–172.

Ruedemann, R. 1926. The Utica and Lorraine Formations of New York. Part III, systematic paleontology, number 2. *New York State Museum Bulletin* 272:1–227.

Whiteley, T. E. 1998. Fossil Lagerstätten of New York, part 1: Beecher's Trilobite Bed. *American Paleontologist* 6:2–4.

Whiteley, T. E. 2000. Beecher's Trilobite Bed: A historical overview. *Geological Society of America, Abstracts with Programs* 32:67.

Whittington, H. B. 1992. *Trilobites*. Woodbridge, Eng.: Boydell Press.

Whittington, H. B., and J. E. Almond. 1987. Appendages and habits of the Upper Ordovician trilobite *Triarthrus eatoni*. *Philosophical Transactions of the Royal Society of London, B* 317:1–46.

8

Hunsrück Slate: Widespread Pyritization of a Devonian Fauna

Walter Etter

THE LOWER DEVONIAN HUNSRÜCK SLATE (*HUNSRÜCK-schiefer*) of Germany is a conservation Lagerstätte that has become famous for the occurrence of pyritized echinoderms, arthropods, and other taxa. The pyritization is associated not only with skeletons, but also with nonmineralized soft parts. This situation is reminiscent of Beecher's Trilobite Bed (Chapter 7). But while pyritization in the latter locality is confined to trilobites, the Hunsrück Slate has yielded preserved soft parts in a variety of taxa, including cnidarians, ctenophores, cephalopods, tentaculites, annelids, arthropods, and echinoderms (Bartels and Brassel 1990; Bergström 1990; Bartels 1994; Bartels, Briggs, and Brassel 1998; Raiswell, Bartels, and Briggs 2001). These fossils were mainly benthic and nektobenthic animals that were embedded by rapid episodic burial (Bartels and Brassel 1990; Bergström 1990; Brett and Seilacher 1991). Thus, as with Beecher's Trilobite Bed, the Hunsrück Slate is classified as an obrution deposit (Seilacher, Reif, and Westphal 1985; Brett and Seilacher 1991).

It is not known exactly when quarrying in the Hunsrück near the German village of Bundenbach started, but by the end of the Middle Ages the production of roofing slates was already well established (Bartels 1995). During the eighteenth and nineteenth centuries, quarrying of the slates had become a major branch of industry in this mainly agricultural region (Bartels 1995; Bartels, Briggs, and Brassel 1998). Quarrying techniques have changed little over the centuries, and during roof-slate production, fossils were undoubtedly found a few hundred years ago (Bartels 1995; Bartels, Briggs, and Brassel 1998). It was not before

1862, however, that the first fossils from the Hunsrück Slate were figured and described (Roemer 1862–1864; Kutscher 1969a). From the turn of the twentieth century and continuing into the 1930s, voluminous monographs about various groups of Hunsrück Slate fossils were published (review in Bartels 1995). The first comprehensive review of the Hunsrück Slate and its fossils appeared in 1932 (Opitz 1932). It was also at that time that the Hunsrück Slate gained widespread attention outside Germany and that many fossil specimens were acquired by museums around the world.

Most of the earlier work was devoted to the description of the fauna and the stratigraphic position of the Hunsrück Slate. However, major debates about the depositional environment were taking place as early as the 1930s (Koenigswald 1930a, 1930b; Kutscher 1931; Richter 1931), concentrating on water depth and bottom-water oxygenation. Some of these controversies are still unresolved and are discussed later in this chapter.

A major step in the history of research was the introduction of radiographic techniques to analyze pyritized specimens of the Hunsrück Slate. First applied by Lehmann (1938; Stürmer 1980), this procedure was later considerably refined. Stürmer and co-workers were primarily responsible for detection of the spectacular soft-part preservation in a variety of fossils (Stürmer 1970, 1980; Stürmer and Bergström 1973, 1976, 1978; Blind and Stürmer 1977; Bergström, Stürmer, and Winter 1980; Stanley and Stürmer 1983, 1987; Yochelson, Stürmer, and Stanley 1983; Bergström et al. 1987; review in Bartels and Wuttke 1994a). These findings made the Hunsrück Slate one of the most renowned Lagerstätten in the world. Recent accounts on many aspects of the Hunsrück Slate, including descriptions of fossils, can be found in Bartels and Brassel (1990), Bartels (1994, 1995), and Bartels, Briggs, and Brassel (1998).

GEOLOGICAL CONTEXT

Located just west of the Rhine River and south of the Mosel River, the gentle hills of the Hunsrück rise to about 800 m above sea level (Bartels and Brassel 1990; Bartels, Briggs, and Brassel 1998) (Figure 8.1). Geologically, the Hunsrück is part of the Rhenish Massif in western Germany, which extends farther to the east into the Taunus region and to the north into the Eifel region (Solle 1950; Mittmeyer 1980a; Bartels and Brassel 1990; Bartels 1995). To the south, the Hunsrück is bordered by a major fault with strongly metamorphosed rocks (Bartels and Brassel 1990). In the Hunsrück, the argillaceous slates of Early Devonian age may exceed 3,000 m in thickness (Mittmeyer 1980a). The silty and clayey sediments were deformed by the Variscan orogeny, and the resulting slate shows a marked cleavage (Bartels and Brassel 1990). Although the

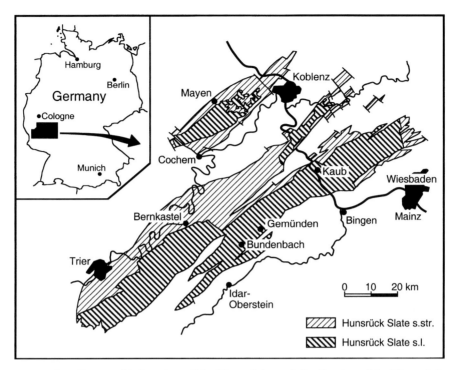

FIGURE 8.1 Geographic location of the Hunsrück, and distribution of the Hunsrück Slate. (Modified from Mittmeyer 1980a)

sediments were not severely metamorphosed, fossils can be found only where the cleavage is roughly parallel to the original bedding planes (Bartels and Brassel 1990). The best localities are around the villages of Bundenbach and Gemünden in the southern part of the Hunsrück, but good exposures also exist around Mayen in the southern Eifel region as well as in the western Taunus region (Bartels and Brassel 1990).

The Hunsrück Slate is a fine-grained metamorphic slate composed of illite and muscovite with varying amounts of chlorites and kaolinite, with minor amounts of smectite and mixed layer minerals (Mosebach 1952; Zimmerle 1992). Accessory minerals include rutile, apatite, pyrite, and sometimes plagioclase feldspar, traces of glauconite, and chamosite (Mosebach 1952; Zimmerle 1992). The dark gray to black color of the slate is caused by disseminated pyrite (0.2 to 0.6 percent) and organic material (0.5 to 0.8 percent) (Mosebach 1952). Intercalated with the homogeneous fine-grained slates are sandy and silty layers that are mostly less than 1 mm in thickness, but sometimes more than a few centimeters thick (Kutscher 1931; Seilacher and Hemleben 1966). The silty layers can appear either as thin and laterally continuous laminae or as more than 1 mm thick lenticular beds (Seilacher and Hemleben 1966). Sedimentary structures in the Hunsrück Slate are not very well known, but do occur in the silty and sandy layers. Cross-bedding in small-scale ripple marks is most common, originally noted by Richter (1931). More obvious are various tool marks and flute casts described from the soles of thin silty layers (Seilacher 1960; Seilacher and Hemleben 1966). A few layers show convolute bedding (Bartels and Brassel 1990).

Although many fossil species are known from the Hunsrück Slate, the exact stratigraphic position of this formation has been disputed for a long time. Many taxa are exclusively known from the Hunsrück Slate, which makes stratigraphic correlation with contemporaneous units difficult (Solle 1950; Mittmeyer 1980a). However, analysis of the brachiopod, goniatite, and trilobite faunas shows that the Hunsrück Slate is of early Emsian age (Solle 1950; Chlupáč 1976; Mittmeyer 1980a). Mittmeyer (1980a) made the distinction between a Hunsrück Slate *sensu lato,* which comprised all the argillaceous slates in the region from late Siegenian (= late Pragian) to early Emsian age, and a Hunsrück Slate *sensu stricto,* which included only the predominately clayey slates of earliest Emsian age (Figure 8.1). This distinction was based on the occurrence (Hunsrück Slate *s.l.*) or absence (Hunsrück Slate *s.str.*) of intercalated volcanic tuff layers (Mittmeyer 1980a). Subsequently, however, tuff layers were found in slates formerly regarded as typical Hunsrück Slates *s.str.* (Bartels and Brassel 1990; Wollanke and Zimmerle 1990), so perhaps such a distinction cannot be made.

During Early Devonian times, the region of the Hunsrück was situated in the Rhenish Basin, which was part of the Variscian geosynclinal system known as the Rhenohercynian Basin (Ziegler 1990). Extensional tectonics led to subsidence and even to crustal separation farther to the east (ophiolitic complexes in the eastern part of the Rhenish Massif) (Ziegler 1990). Volcanism was locally intense. During this time, the Rhenish Basin developed into a back-arc basin, bordered to the north by the Old Red continent, and to the south by the Mid-German High (the Hunsrück Island of Solle 1970; Ziegler 1990). Detrital sediment was shed mainly from the north, but to a smaller extent also from the south (Bartels and Brassel 1990). Differences in subsidence led to a highly structured depositional system, with several basins separated by sills (Mittmeyer 1980a). The thickness of the Hunsrück Slate is quite variable and ranges between 200 and 3,000 m (Mittmeyer 1980a). For the central region around Bundenbach, the sedimentation rate has been calculated as 2 mm of uncompacted sediment per year (Stürmer and Bergström 1973), but there are many indications that sediment accumulation was a highly episodic process. Compaction of the sediment was probably around 1:10 (Bartels 1995). In different regions, the slates developed in somewhat different facies and are known by different local names (Solle 1950; Mittmeyer 1980a). Laterally, the Hunsrück Slate interfingers with more sandy sediments, and both the underlying and the overlying formations are sandstones and siltstones or quartzites (Solle 1950; Mittmeyer 1980a; Bartels and Brassel 1990).

PALEOENVIRONMENTAL SETTING

Much of the debate about the origin of the Hunsrück Slate has been centered around two major themes:

1. Was the seafloor permanently populated by a normal fauna, or was the benthic environment, at least temporarily, hostile to benthic life?
2. Was the sedimentary environment located in shallow or deep water?

Early notions postulating a catastrophic origin of the fossil assemblage include the work of Koenigswald (1930a), who suggested that the seafloor was normally populated by a sparse benthic fauna. Episodic storms stirred up the hydrogen sulfide-enriched bottom sediment, which killed all benthic life. The currents generated by such storms preferentially transported cephalopods, arthropods, and fish, leaving behind an assemblage relatively enriched in echinoderms. This death assemblage was subsequently embedded in the settling sediment (Koenigswald 1930a).

Similarly, Gürich (1931) envisioned poisoning of the benthic fauna through hydrogen sulfide, although he proposed a different mechanism. According to Gürich (1931), the bottom-water was oxygen depleted. The absence of currents led to increasing stagnation and, after a threshold of oxygen deficiency was reached, caused the mass mortality of the benthic fauna through hydrogen sulfide poisoning (Gürich 1931). Recolonization of the environment would have occurred only after episodic currents, which also deposited silty and sandy layers, replenished the bottom-water oxygenation. A quite different, although still catastrophic, view of the genesis of this Lagerstätte was proposed by Lehmann (1957), who depicted the Hunsrück Slate assemblage as a *submarine Pompeii,* where the fossils were the result of repeated sudden burial events by volcanic ashes.

Both Gürich (1931) and Lehmann (1957) interpreted the observed orientation of the fossils (mainly in asteroids and ophiuroids) as "frozen" life positions. This view is still held by some authors (Stemvers-van Bemmel 1989). It is easy to demonstrate, however, that the orientation of the arms in asteroids and ophiuroids found in the Hunsrück Slate was the result of currents (Koenigswald 1930a, 1930b; Seilacher 1960; Bartels and Brassel 1990) (Figure 8.2). Sedimentary structures indicative of current activity (ripple marks, tool marks, flute casts) occur throughout the sections (Seilacher and Hemleben 1966), as do trace fossils (Richter

1931, 1941; Seilacher and Hemleben 1966; Sutcliffe, Briggs, and Bartels 1999).

Additional evidence against a hostile benthic environment comes from paleoecology. The fossil assemblage is diverse and includes both juvenile and adult specimens (Kutscher 1931; Richter 1931; Bartels and Brassel 1990; Bartels 1995). Most workers therefore assume that the benthic environment during deposition of the Hunsrück Slate was normally oxygenated or only slightly oxygen depleted at most (Kutscher 1931; Richter 1931; Seilacher 1960; Seilacher and Hemleben 1966; Stürmer and Bergström 1973; Bartels and Brassel 1990; Bartels 1995; Bartels, Briggs, and Brassel 1998; Sutcliffe, Briggs, and Bartels 1999).

The controversy concerning the water depth of the Hunsrück Slate depositional environment is not yet resolved. Early proponents of a deep-water environment noted the absence of thick-shelled bivalves and the rarity of brachiopods compared with their occurrence in the contemporaneous sandy sediments in the surrounding regions (Frech 1889). It was also assumed that the orientation observed in asteroids and ophiuroids (Figure 8.2) was the result of tectonic deformation alone (Jaeckel 1895). With the detection of current orientation in fossils, a quiet-water environment was no longer acceptable, and a shallow-water origin of the Hunsrück Slate was postulated (Koenigswald 1930a, 1930b). In particular, Kutscher (1931) and Richter (1931, 1935, 1954) observed many features they thought to be indicative of a very shallow to even intertidal marine setting. These features include bimodal orientation of fossils as the result of wave action and sedimentary structures believed to originate only in intermittent subaerially exposed environments—for example, flow marks (*Gefliessmarken*) and foam impressions (*Schaumblasenmarken*) (Richter 1935, 1954). According to these findings, the Hunsrück Slate was deposited in a tidal flat environment (Kutscher 1931; Richter 1931, 1935, 1954; Solle 1950). For a long time, the tidal flat model was (and perhaps still is) the most popular depositional model (Solle 1970; Bartels and Brassel 1990). New evidence cited in favor of this model includes shell beds (Herrgesell 1978, cited in Bartels and Brassel 1990) and raindrop impressions (Bartels and Brassel 1990).

An alternative view to the paleoenvironmental setting was proposed by Seilacher (1960). A reevaluation of the published work showed that there was no unambiguous bimodal orientation in asteroids and ophiuroids (Seilacher 1960). The observed patterns were, rather, the result of two different modes of azimuthal orientation (central parts of the body anchored and arms oriented by currents, versus an umbrella position with anchored arms and elevated central body part) (Seilacher 1960). Furthermore, most of the sedimentary structures observed in the Hunsrück Slate have analogues in flysch deposits (Seilacher and Hemleben 1966). These structures do not indicate a tidal flat environment. In particular,

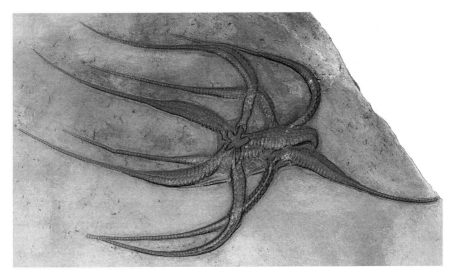

FIGURE 8.2 Two specimens of the ophiuroid *Furcaster zitteli* showing current orientation. Diameter of specimens is 15 cm. (Photo courtesy of C. Bartels, Deutsches Bergbau-Museum, Bochum)

typical structures of intertidal deposits (e.g., channels) are absent. Therefore, Seilacher and Hemleben (1966) concluded in their investigation that the depositional environment was situated in bathyal deposits.

Seilacher's (1960) and Seilacher and Hemleben's (1966) conclusions were not corroborated by Stürmer and Bergström (1973). Based on the development of trilobite eyes, they concluded that water depth was no more than 200 m during deposition of the Hunsrück Slate (there are neither blind trilobites nor species with excessively large eyes) (Stürmer and Bergström 1973). In addition, the presence of light-dependent receptaculites indicates relatively shallow water (Bartels and Blind 1995).

Today, the depositional environment is known in much more detail than it was a few decades ago, and perhaps the ongoing discussion about shallow or deep water is merely the result of the complex paleoenvironment. Mud accumulated mainly in depressions between submarine fans (Bartels, Briggs, and Brassel 1998). These basins were below storm wave base, but most likely not deeper than 200 m (Raiswell, Bartels, and Briggs 2001). The surrounding sills extended locally into the intertidal environment, as is indicated by foam and raindrop impressions (Richter 1954; Bartels and Brassel 1990; Bartels, Briggs, and Brassel 1998).

TAPHONOMY

Fossils from the Hunsrück Slate are aesthetically very appealing because they are pyritized and often fully articulated (Figures 8.2–8.5). Unfortunately, the preparation of such specimens is time consuming and can be

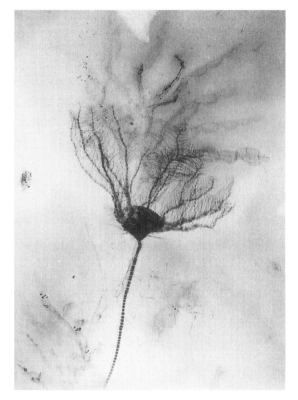

FIGURE 8.3 Radiograph showing the crinoid *Imitatocrinus gracilior* and the pycnogonid *Palaeoisopus problematicus.* Preserved arm length of *Imitatocrinus* is 6 cm. (Photo courtesy of J. Bergström, Swedish Museum of Natural History, Stockholm)

accomplished only by using very fine tools (Bartels and Brassel 1990). Pyritization is not the only preservational mode in the Hunsrück Slate. Especially in the region around Mayen, brachiopods, bivalves, and trilobites can be silicified (Bartels and Brassel 1990). In the central region around Bundenbach, the calcareous shells of brachiopods and molluscs are most commonly dissolved, leaving only a shadow-like impression in the sediment unless the periostracum is pyritized (Bartels and Brassel 1990). There are also differences in the degree of pyritization among different beds and different regions. Nearly complete pyritization of skeletons and even soft parts is confined mainly to many thin horizons around Bundenbach (Bartels and Brassel 1990; Bartels 1995). Most other beds yield predominantly fragmentary and only incompletely pyritized remains, and it must be noted that fossils are rare in the Hunsrück Slate. Bartels and Brassel (1990) therefore reject the term "conservation Lagerstätte" for the Hunsrück Slate as a whole, and prefer to speak of conservation horizons within this formation.

Soft-part preservation in the Hunsrück Slate was first detected in crinoids (Schmidt 1934) that are commonly found fully articulated, with preserved arms, stems, and root structures (Figure 8.3). Surprisingly to Schmidt (1934), the obviously membranaceous tentacles on the arms of the crinoid genus Bathericrinus (and some specimens of *Imitatocrinus* [Figure 8.3] and *Macarocrinus*) are also preserved (Bartels and Wuttke 1994a). Other soft parts found in echinoderms from the Hunsrück Slate include mainly cuticular structures observed in asteroids (Lehmann 1957; Bartels and Brassel 1990; Bartels and Wuttke 1994a) and the digestive tract in the holothurian *Palaeocucumaria hunsrueckiana* (Lehmann 1958; Bartels and Wuttke 1994a; Bartels and Blind 1995).

Various examples of soft-part preservation are found among the arthropods of the Hunsrück Slate. In trilobites of the genera *Chotecops* and *Asteropyge,* the fine structure of the appendages, the structure and innervation of the compound eyes, and the digestive tract are especially well known (Stürmer 1970; Stürmer and Bergström 1973; Bergström and Brassel 1984) (Figure 8.4). The alimentary system can be observed in the sea spider *Palaeoisopus problematicus* (Bergström, Stürmer, and Winter 1980) and the crustacean *Nahecaris stuertzi* (Bergström et al. 1987), and details of the leg musculature are also known in these two arthropods (Bergström, Stürmer, and Winter 1980; Bergström et al. 1987).

The interpretation of soft-part structures in cephalopods is rather uncertain. First detected in goniatites and orthoceratids of the Middle Devonian Wissenbach Slate (Zeiss 1969), traces of the mantle, digestive tract, and tentacles were also described from ectocochleate (Stürmer 1969; Zeiss 1969) and endocochleate cephalopods of the Hunsrück Slate (Bandel, Reitner, and Stürmer 1983; Stürmer 1985). In addition, soft parts have been discovered in tentaculites (Brassel, Kutschner, and Stürmer 1971), gastropods (Houbrick, Stürmer, and Yochelson 1988), and brachiopods (Südkamp 1997).

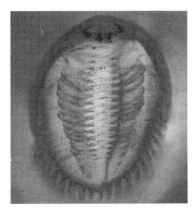

FIGURE 8.4 Radiograph of the trilobite *Asteropyge* sp. showing preserved legs. Length of specimen is 5 cm. (Photo courtesy of J. Bergström, Swedish Museum of Natural History, Stockholm)

Most remarkable are splendidly preserved, pyritized organisms that usually do not survive in the fossil record. Only a few specimens of polychaete annelids from the Hunsrück Slate have been figured or described (Bartels and Brassel 1990; Bergström 1990; Bartels 1994; Bartels and Blind 1995). Even more notable, however, are the findings of medusae in the Hunsrück Slate. They include the velellid hydromedusa *Plectodiscus discoideus,* of which several specimens with preserved tentacles and the float are known (Rauff 1939; Yochelson, Stürmer, and Stanley 1983; Bartels and Brassel 1990; Bartels and Blind 1995) (Figure 8.5). Even more exceptional is the preservation of ctenophores. Two species, *Palaeoctenophora brasseli* and *Archaeocydippida hunsrueckiana,* were discovered during radiographic examination of slabs, and both species are known from single specimens only (Stanley and Stürmer 1983, 1987). Fossil ctenophores were hitherto reported in only some specimens from the Cambrian Burgess Shale of Canada (Briggs, Erwin, and Collier 1994; Conway Morris and Collins 1996).

Although more than 35 international specialists have observed soft-part preservation in fossils of the Hunsrück Slate, Otto (1994) concluded that there is no soft-part preservation at all in the Hunsrück Slate. He accepted only the preservation of delicate chitinous structures in arthropods (e.g., the gills of trilobites) and of spongin fibers in sponges, but concluded that these organic materials are really functional hard parts

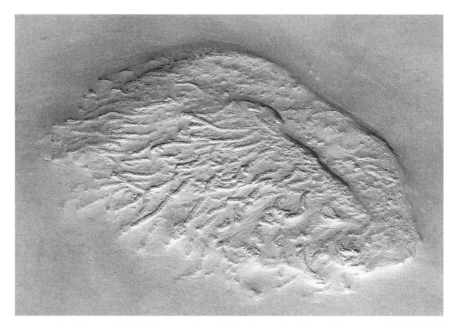

FIGURE 8.5 The velellid hydromedusa *Plectodiscus discoideus.* Diameter of disk is 6.5 cm. (Photo courtesy of C. Bartels, Deutsches Bergbau-Museum, Bochum)

and therefore must be excluded from the phenomenon of soft-part preservation. Thus, in Otto's (1994) opinion, all the published examples of soft-part preservation in the Hunsrück Slate would be based on misinterpretations of radiographs. However, as was pointed out in a reply (Bartels and Wuttke 1994b), identical soft-part preservation has been documented in various specimens of the same species, such as in trilobites, and the medusa *Plectodiscus discoideus* (Bartels and Brassel 1990; Bartels and Blind 1995). Yet there is the danger of misinterpreting radiographs. Several specimens initially interpreted as tunicates (Fauchald, Stürmer, and Yochelson 1988) were later correctly described as head shields of phacopide trilobites (Fauchald and Yochelson 1990).

Two mechanisms were obviously involved in the spectacular preservation of the Hunsrück Slate fossils: rapid burial and early diagenetic pyritization (Kott and Wuttke 1987; Bartels and Brassel 1990; Bartels 1994, 1995; Bartels and Wuttke 1994a). Rapid burial was necessary for the preservation of articulated crinoids, asteroids, and ophiuroids in the Hunsrück Slate (Bartels and Brassel 1990). Unless they are embedded in sediment, the multielement skeletons of these organisms decay within hours to days (Schäfer 1972). The beds enclosing the Hunsrück fossils may be a few centimeters thick. In this case the fossils are often oriented at an angle within the beds (Bartels and Brassel 1990). More commonly, however, the blanketing sediment layers are only a few millimeters thick (Bartels and Brassel 1990).

The burial events were caused by mudflows that were accompanied by currents (Seilacher and Hemleben 1966; Brett and Seilacher 1991). These microturbidites or distal gradient currents (Brett and Seilacher 1991) sometimes transported organisms for some distance (displaced obrution, *sensu* Brett and Seilacher 1991). The consensus is, however, that this was a within-habitat displacement (Bartels and Brassel 1990; Bartels 1995). The taxonomic composition of the fossil assemblages indicates no fundamental differences between event beds and the underlying and overlying beds (Bartels and Brassel 1990; Bartels 1995). Once embedded, the skeletons were not disturbed by burrowing animals (Bartels and Brassel 1990). With the exception of *Chondrites,* there are no traces in the Hunsrück Slate produced by deep burrowing animals (Seilacher and Hemleben 1966; Sutcliffe, Briggs, and Bartels 1999).

At present, it is unclear what caused the death of the fauna in the event beds. According to Brett and Seilacher (1991), the gradient currents eroded sulfide-rich sediment (Koenigswald 1930a), and this poisonous mudflow killed not only all the benthic organisms but also nektobenthic fish and arthropods. According to other authors, it was not poisoning of the organisms by sulfide that killed them, but the dense covering by fine-grained sediment that clogged the gills of the arthropods and especially the ambulacral system of the echinoderms (Bartels and Brassel 1990; Bartels 1995). This would lead to the observed predominance of echinoderms and explain why they could not escape even a thin sediment cover (Bartels and Brassel 1990).

Pyritization of the fossils started very early, almost immediately after they were buried (Kott and Wuttke 1987; Bartels and Wuttke 1994a; Briggs et al. 1996). The processes involved in early diagenetic pyritization are relatively well known. Sulfate, which is abundant in seawater, diffuses into the pore water of the sediment and is used as an energy source by anaerobic sulfate-reducing bacteria (Kott and Wuttke 1987; Bartels and Wuttke 1994a). The degree of pyritization is in part limited by the amount of available iron ions in the pore water (Bartels and Wuttke 1994a; Briggs et al. 1996). In the Hunsrück Slate, a principal source of mobile iron ions was layers of volcanic tuff (Bartels and Wuttke 1994a). As they are variably developed both regionally and stratigraphically (Wollanke and Zimmerle 1990), the degree of pyritization is also variable (Bartels and Brassel 1990; Bartels and Wuttke 1994a; Bartels, Briggs, and Brassel 1998).

The growth of sulfate reducers is limited by the available organic material in the sediment. If the surrounding sediment is rich in organic carbon, only disseminated pyrite will form. However, if the sediment has a low content of organic carbon, pyritization is confined largely to enclosed carcasses (Bartels and Wuttke 1994b; Briggs et al. 1996). The original content of organic carbon in the Hunsrück Slate (before metamorphism) was an estimated 1 to 1.5 percent (Briggs et al. 1996) and therefore quite low (Bartels and Wuttke 1994b).

In the first step of pyrite formation, sulfate reducers produce hydrogen sulfide, which is then released through the cell walls (Kott and Wuttke 1987; Bartels and Wuttke 1994a). These microbes, whose crucial role in pyritization was already suspected by earlier authors (Koenigswald 1930a; Kutscher 1931), thrive in the anoxic part of the sediment profile or in anoxic microenvironments near the oxygenated–anoxic interface (Bartels and Wuttke 1994a). The sulfide then reacts on the cell walls of the bacteria with Fe^{2+}, Fe^{3+} in the pore water to form iron monosulfide (Bartels and Wuttke 1994a).

In the second phase of pyrite formation, aerobic or facultatively anaerobic bacteria oxidize sulfide to elementary sulfur, which then reacts with iron monosulfide to form framboidal pyrite (Kott and Wuttke 1987; Bartels and Wuttke 1994a). The latter step requires the presence of at least minimal amounts of free oxygen (Bartels and Wuttke 1994a; Briggs et al. 1996). Early diagenetic pyrite formation was thus largely restricted to the oxic–anoxic interface in the sediment. Because the Hunsrück Slate was originally deposited as a fine-grained sediment, these conditions were met only in the top few centimeters of the seafloor (Bartels and Brassel 1990; Bartels and Wuttke 1994a; Briggs et al. 1996). It is significant in this context that bacteria were indeed involved in the production of the pyrite sulfur in the Hunsrück Slate, as can be shown by geochemical investigations (Briggs et al. 1996). At least in the horizons which show nearly complete pyritization of fossils, the background sedimentation rate after the sudden burial event was probably low (Stürmer and Bergström 1973). Therefore, embedded carcasses spent a long time at the oxic–anoxic interface.

PALEOBIOLOGY AND PALEOECOLOGY

For more than a century, geologists and paleontologists have been working on fossils from the Hunsrück Slate. Over time, an impressive number of taxa have been described and, up to now, more than 250 species and subspecies of animal fossils are known from this formation (Bartels 1995). In contrast, very few plant remains occur in the Hunsrück Slate. Apart from more than 40 types of spores (Holtz 1969; Karathanasopoulos 1975, cited in Bartels and Brassel 1990), only six species of terrestrial plant, one species of red algae (*Prototaxites*), and one species of receptaculitid (*Receptaculites*) have been found (Stürmer and Schaarschmidt 1980; Bartels and Brassel 1990; Bartels 1995; Bartels, Briggs, and Brassel 1998). Only *Receptaculites* sp., which is believed to be a calcareous green alga (Rietschel 1969), seems to be autochthonous (Bartels and Brassel 1990; Bartels and Blind 1995). This species has been found at only one locality and in one horizon (Bartels and Brassel 1990).

A semiquantitative analysis of the Hunsrück Slate fauna clearly shows that there is a sheer predominance of benthic organisms (Bartels and Brassel 1990; Bartels 1994, 1995). The presence of all age classes, as in crinoids and ophiuroids, clearly points to an autochthonous origin of this benthic fauna (Bartels and Brassel 1990). Nektonic and planktonic organisms are rare and present with only a few species of each. Among the organisms with swimming abilities, cephalopods and tentaculites dominate; they may have been forms that were nektobenthic rather than truly pelagic (Bartels and Brassel 1990). The vertebrates were mostly nektobenthic, as is evident from the flattened body forms of *Drepanaspis* and *Gemuendina* (Bartels and Brassel 1990; Bartels, Briggs, and Brassel 1998).

Among the benthos, various lifestyles are represented. The majority were filter feeders and deposit feeders, but scavengers and predators were also present (Stürmer and Bergström 1973; Bartels and Brassel 1990; Bartels and Wuttke 1994a). Especially among the suspension feeders, various taxa show adaptations to moderate current intensities (Kott and Wuttke 1987; Bartels and Brassel 1990). The Hunsrück Slate fossils thus represent a normal benthic fauna that does not show any signs of environmental stress. Only the ichnofauna, which lacks (with the exception of *Chondrites*) deep burrowers, can be cited as indicating somewhat lowered oxygen levels in the bottom-water (Savrda and Bottjer 1991).

The fauna of the Hunsrück Slate shows some notable variations among different regions and different beds (Bartels and Brassel 1990). The following account can, therefore, give only a very general picture and is valid for only the central region around Bundenbach. Complete listings of the Hunsrück Slate fossils can be found in Mittmeyer (1980b) and Bartels (1994). The most diverse group, which also contains some of the most common species, is the echinoderms. Asteroids include 27 described species (Lehmann 1957; Bartels and Brassel 1990). Among them are some large and multiarmed forms, such as *Helianthaster rhenanus* and *Palaeosolaster gregoryi* (Lehmann 1957; Bartels and Brassel 1990; Bartels 1994). However, with the exception of the common *Urasterella asperula*, asteroids are rare in the Hunsrück Slate (Bartels and Brassel 1990; Bartels 1994).

Ophiuroids, though, include some of the most common species of the Hunsrück Slate (Figure 8.2). Especially abundant is *Furcaster palaeozoicus*, with other frequently found ophiuroids including *Euzonosoma tischbeiniana* and *Taeniaster beneckei* (Lehmann 1957; Kutscher 1976a; Bartels and Brassel 1990). A total of 23 ophiuroid species are known (Bartels 1994).

The most diverse of the echinoderms are the crinoids, which include 65 species and subspecies (Schmidt 1934; Kutscher 1976b; Bartels and Brassel 1990). Delicate branching forms such as *Hapalocrinus, Imitatocrinus, Taxocrinus,* and *Parisangulocrinus* are common (Figure 8.3), whereas robust species such as *Codiacrinus* are much rarer (Schmidt 1934; Bartels and Brassel 1990). Decidedly rare are members of the cystoids (one species), blastoids (two species), and homalozoans (two species) (Bartels and Brassel 1990; Bartels 1994). The remainder of the echinoderms are known primarily from single specimens (Bartels and Brassel 1990). They include one species of edrioasteroid (Dehm 1961a; Kutscher and Sieverts-Doreck 1968), one species of holothurian (Seilacher 1961; Kutscher and Sieverts-Doreck 1977), and two species of echinoids (Dehm 1961b; Bartels and Brassel 1990).

After the echinoderms, the arthropods are the second most important group of fossils in the Hunsrück Slate. Especially common are the trilobites, of which 10 genera are known (Kutscher 1978; Bartels 1994). A few species are very abundant, especially the up to 16 cm long *Chotecops* (formerly *Phacops*) *ferdinandi*, which is the most common macrofossil in the Hunsrück Slate (Struve 1985; Bartels and Brassel 1990). However, most of the remains are considered to be exuviae (Bartels and Brassel 1990). Other important genera include *Asteropyge, Odontochile, Parahomalonotus,* and *Rhenops* (Bartels and Brassel 1990). Detailed radiographic investigations reveal many details of the anatomy of *Chotecops* and *Asteropyge* (Stürmer and Bergström 1973) (Figure 8.4). These trilobites are now among the best known in the world (Stürmer and Bergström 1973).

Three distant relatives of the trilobites (trilobitomorphs) have been described from the Hunsrück Slate, including the rare *Cheloniellon* (Stürmer and Bergström 1978) and *Vachonisia*, with its unusual bivalved carapace (Stürmer and Bergström 1976; Bartels and Brassel 1990). Much more common is the small *Mimetaster hexagonalis*, which has a small carapace that shows six projecting spines (Stürmer and Bergström 1976). It is notable that the closest known relative of this species seems to be *Marrella splendens* from the Cambrian Burgess Shale (Stürmer and Bergström 1976). Other arthropods that are also reminiscent of Cambrian forms include the newly described *Cambronatus* and *Wingertshellicus* (Briggs and Bartels 2001).

Two genera of phyllocarid crustaceans have been described, but only *Nahecaris stuertzi* is relatively common (Bergström et al. 1987; Bartels and Brassel 1990). *Heroldina rhenana* reaches a length of up to 60 cm and is thus one of the largest invertebrates of the Hunsrück Slate (Bartels and Brassel 1990). In addition, five other phyllocarid crustaceans have been found (Bergström et al. 1989).

Other arthropods that occur in the Hunsrück Slate include the very rare xiphosuran (horseshoe crab) *Weinbergina* (Lehmann 1956b; Stürmer and Bergström 1981); the very rare, small eurypterid *Rhenopterus* (only disarticulated remains, which are now lost) (Lehmann 1956a, 1956b; Kutscher 1975); and three species of pycnogonids (Bergström, Stürmer, and Winter 1980). Of the last, *Palaeoisopus problematicus* is unusually large and attains a leg span of 40 cm (Bergström, Stürmer, and Winter 1980) (Figure 8.3). In addition to these arthropods, a single specimen of a scorpion was found (Kutscher 1971); it most probably was swept into the depositional environment from land (Bartels and Brassel 1990).

Molluscs are also an important part of the Hunsrück Slate fauna. Because of their poor preservation, however, they have not been collected as systematically as echinoderms and arthropods (Bartels and Brassel 1990). Among the bivalves, of which some 13 species are known (Kutscher 1966), only *Buchiola* is common in certain horizons (Bartels and Brassel 1990). Gastropods are more typical of the silty and sandy facies, but some species occur in shaley horizons (Bartels and Brassel 1990). Cephalopods are much more common than the other mollusc groups. There are several forms of orthoceratids, but they are usually not preserved well enough to allow any species designation (Bartels and Brassel 1990). The goniatites are noteworthy because they comprise some of the earliest ammonoids, especially *Anetoceras* and *Mimagoniatites* (Kutscher 1969b). In the course of x-radiographic examination of fossiliferous slabs, rare teuthoid cephalopods were also detected (Bandel, Reitner, and Stürmer 1983; Stürmer 1985).

Tentaculites (*Viriatellina fuchsi* and some undescribed taxa) are very common in certain horizons (Bartels and Brassel 1990). Most authors have regarded these fossils as pteropods or molluscs of uncertain affinity. In the course of a detailed investigation of their shell, Blind (1969) concluded that tentaculites were cephalopods, and radiographs of tentaculites with preserved soft parts were thought to support this view (Brassel, Kutscher, and Stürmer 1971; Blind and Stürmer 1977), but the evidence is equivocal (Yochelson 1989). In the Hunsrück Slate, a single specimen of hyolithid has been found so far (Houbrick, Stürmer, and Yochelson 1988). As with the tentaculites, the systematic position of hyolithids is unresolved. They may be molluscs, but most workers place them into a separate phylum of unknown affinities (Bartels, Briggs, and Brassel 1998).

Brachiopods, like the molluscs, are usually poorly preserved (Bartels and Brassel 1990). They are quite common in the Hunsrück Slate, although not as common as in the sandy strata of the surrounding region (Bartels and Brassel 1990; Bartels 1994). At least two species of bryozoans (probably more) do occur in the Hunsrück Slate, but members of this group are rare (Bartels and Brassel 1990).

Corals, again more typical of the shallow-water strata of the adjacent region, have been found in some numbers (Bartels and Brassel 1990; Bartels 1994), and include small solitary rugose corals (*"Zaphrentis"*) as well as small colonies of tabulate corals (*Pleurodictyum, Aulopora*) (Bartels and Brassel 1990). Conularians, considered by most paleontologists to belong to the scyphozoans (cnidarians), are rare in the Hunsrück Slate (Hergarten 1994). Surprising results concerning the soft-part anatomy of these conularians were presented by Steul (1984). In radiographs, Steul (1984) believed he saw evidence for the presence of a phosphatic axial structure, a segmented body, and lens eyes. A chordate affinity was therefore postulated for the conularians (Steul 1984). However, this seems to be a case of misinterpretation of radiographs (Otto 1994). Furthermore, some of the investigated specimens were not conularians, but cephalopods (Hergarten 1994). Sponges are very rare, and only four species have been named (Kott and Wuttke 1987; Bartels 1994).

Not surprisingly, the entirely soft-bodied velellid medusa *Plectodiscus discoideus* (Yochelson, Stürmer, and Stanley 1983) (Figure 8.5) and two species of ctenophores (Stanley and Stürmer 1983, 1987) are exceedingly rare or known from single specimens. The same is true for the few polychaete annelids that have been figured so far (Bartels and Brassel 1990; Bergström 1990; Bartels and Blind 1995). In addition, one specimen of a questionable flatworm has been reported (Fauchald, Stürmer, and Yochelson 1988). Conspicuous by their absence are the conodonts.

Some of the vertebrates that occur in the Hunsrück Slate are quite common. They include the jawless fish *Drepanaspis,* of which several hundred specimens have been recovered, and the rhenanid placoderm *Gemuendina* (Bartels and Brassel 1990). Other agnathans (*Pteraspis*) and placoderms (six genera of arthrodirans, four other species of rhenanids) (Kutscher 1973) are very rare and mostly known from fragmentary remains (Bartels and Brassel 1990). Isolated spines of acanthodians are common only in the region around Mayen (Bartels and Brassel 1990).

The trace fossils of the Hunsrück Slate are still not satisfactorily known. Most abundant are shallow burrows of *Planolites,* whose pyritized fillings are often visible on radiographs (Bartels and Brassel 1990; Sutcliffe 1995). *Chondrites,* which is also quite common, seems to indicate somewhat lowered oxygen levels during deposition (Sutcliffe 1995). These traces, together with the various arthropod tracks (Seilacher and Hemleben 1966) and ophiuroid traces (Sutcliffe 1997), are best interpreted as post-depositional flysch-type traces (Sutcliffe 1995). However, elaborate mining traces or pre-depositional graphoglyptid burrows are absent, with the exception of *Heliochone* (Seilacher and Hemleben 1966).

CONCLUSIONS

In a global context, the Hunsrück Slate offers by far the best glimpse of Early Devonian marine life. Not only is a wide range of taxa documented, but, as a result of early diagenetic pyritization, many fossils also show, in an exceptional way, details of their soft-part anatomy. This circumstance has attracted many specialists from around the world, and the fauna of the Hunsrück Slate is now well known. Furthermore, the processes involved in early diagenetic pyritization and the role of bacteria in this process have been clarified through study of this fauna.

There are still many open questions concerning the depositional environment, however, and a comprehensive paleoecological analysis of the whole fauna has never been done. The main problem in performing such studies is the lack of a detailed stratigraphic framework that would allow a subdivision of the Hunsrück Slate. Without such a subdivision, it is not possible to compare the fauna and the facies of different localities beyond the anecdotal. But given the poor exposures, the tectonic deformation, and the rarity of fossils of stratigraphic value, it is doubtful if a finer stratigraphic subdivision is practically feasible in the next decades.

ACKNOWLEDGMENTS

The chapter was critiqued by J. Bergström and M. Wuttke, both of whom helped to clarify some important points. Additional references were provided by C. Bartels.

REFERENCES

Bandel, K., J. Reitner, and W. Stürmer. 1983. Coleoids from the Lower Devonian Black Slate ("Hunsrück-Schiefer") of the Hunsrück (West Germany). *Neues Jahrbuch für Geologie und Paläontologie, Abhandlungen* 165:397–417.

Bartels, C. 1994. Weltberühmt: Die "Bundenbacher Fossilien" des Hunsrückschiefers. *Schiefer-Fachverband in Deutschland e.V., Schriftenreihe* 3:11–85.

Bartels, C. 1995. Die unterdevonischen Dachschiefer von Bundenbach. In K. W. Weidert, ed., *Klassische Fundstellen der Paläontologie*, vol. 3, pp. 38–55. Korb: Goldschneck-Verlag.

Bartels, C., and W. Blind. 1995. Röntgenuntersuchung pyritisch vererzter Fossilien aus dem Hunsrückschiefer (Unter-Devon, Rheinisches Schiefergebirge). *Metalla* (Bochum) 2:79–100.

Bartels, C., and G. Brassel. 1990. *Fossilien im Hunsrückschiefer: Dokumente des Meereslebens im Devon.* Idar-Oberstein: Museum Idar-Oberstein.

Bartels, C., D. E. G. Briggs, and G. Brassel. 1998. *The Fossils of the Hunsrück Slate.* Cambridge: Cambridge University Press.

Bartels, C., and M. Wuttke. 1994a. Fossile Überlieferung von Weichkörperstrukturen und ihre Genese im Hunsrückschiefer (Unter-Ems, Rheinisches Schiefergebirge): Ein Forschungsbericht. *Giessener geologische Schriften* 51:25–61.

Bartels, C., and M. Wuttke. 1994b. Nachtrag zu: Fossile Überlieferung von Weichkörperstrukturen und ihre Genese im Hunsrückschiefer (Unter-Ems, Rheinisches Schiefergebirge): Ein Forschungsbericht. *Giessener geologische Schriften* 51:329–333.

Bergström, J. 1990. Hunsrück Slate. In D. E. G. Briggs and P. R. Crowther, eds., *Palaeobiology: A Synthesis*, pp. 277–279. Oxford: Blackwell Science.

Bergström, J., and G. Brassel. 1984. Legs in the trilobite *Rhenops* from the Lower Devonian Hunsrück Slate. *Lethaia* 17:67–72.

Bergström, J., D. E .G. Briggs, E. Dahl, W. D. I. Rolfe, and W. Stürmer. 1987. *Nahecaris stuertzi*, a phyllocarid crustacean from the Lower Devonian Hunsrück Slate. *Paläontologische Zeitschrift* 61:273–298.

Bergström, J., D. E. G. Briggs, E. Dahl, W. D. I. Rolfe, and W. Stürmer. 1989. Rare phyllocarid crustaceans from the Devonian Hunsrück Slate. *Paläontologische Zeitschrift* 63:319–333.

Bergström, J., W. Stürmer, and G. Winter. 1980. *Palaeoisopus, Palaeopantopus* and *Palaeothea*, pycnogonid arthropods from the Lower Devonian Hunsrück Slate, West Germany. *Paläontologische Zeitschrift* 54:7–54.

Blind, W. 1969. Die systematische Stellung der Tentakuliten. *Palaeontographica,* A 133:101–145.

Blind, W., and W. Stürmer. 1977. *Viriatellina fuchsi* Kutscher (Tentaculoidea) mit Sipho und Fangarmen. *Neues Jahrbuch für Geologie und Paläontologie, Monatshefte* 1977:513–522.

Brassel, G., F. Kutscher, and W. Stürmer. 1971. Erste Funde von Weichteilen und Fangarmen bei Tentaculiten. *Abhandlungen des hessischen Landesamtes für Bodenforschung* 60:44–50.

Brett, C. E., and A. Seilacher. 1991. Fossil Lagerstätten: A taphonomic consequence of event sedimentation. In G. Einsele, W. Ricken, and A. Seilacher, eds., *Cycles and Events in Stratigraphy*, pp. 283–297. Berlin: Springer-Verlag.

Briggs, D. E. G., and C. Bartels. 2001. New arthropods from the Lower Devonian Hunsrück Slate (Lower Emsian, Rhenish Massif, western Germany). *Palaeontology* 44:275–303.

Briggs, D. E. G., D. H. Erwin, and F. J. Collier. 1994. *The Fossils of the Burgess Shale.* Washington, D.C.: Smithsonian Institution Press.

Briggs, D. E. G., R. Raiswell, S. H. Bottrell, D. Hatfield, and C. Bartels. 1996. Controls on the pyritization of exceptionally preserved fossils: An analysis of the Lower Devonian Hunsrück Slate of Germany. *American Journal of Science* 296:633–663.

Chlupáč, I. 1976. The oldest goniatite faunas and their stratigraphical significance. *Lethaia* 9:303–315.

Conway Morris, S. C., and D. H. Collins. 1996. Middle Cambrian ctenophores from the Stephen Formation, British Columbia, Canada. *Philosophical Transactions of the Royal Society of London, B* 351:279–308.

Dehm, R. 1961a. Über *Pyrgocystis* (*Rhenopyrgus* nov. subgen.) *coronaeformis* Rievers aus dem rheinischen Unter-Devon. *Mitteilungen der bayrischen Staatssammlung für Paläontologie und historische Geologie* 1:13–17.

Dehm, R. 1961b. Ein zweiter Seeigel, *Porechinus porosus* nov. gen. nov. spec. aus dem rheinischen Unter-Devon. *Mitteilungen der bayrischen Staatssammlung für Paläontologie und historische Geologie* 1:1–8.

Fauchald, K., W. Stürmer, and E. L. Yochelson. 1988. Two worm-like organisms from the Hunsrück Slate (Lower Devonian), West Germany. *Paläontologische Zeitschrift* 62:205–215.

Fauchald, K., and E. L. Yochelson. 1990. Correction: A major error in Fauchald, K., Stürmer, W. & Yochelson, E. L., 1988. *Paläontologische Zeitschrift* 64:381.

Frech, F. 1889. Über das rheinische Unterdevon und die Stellung des "Hercyn." *Zeitschrift der deutschen geologischen Gesellschaft* 41:1–175.

Gürich, G. 1931. *Mimetaster hexagonalis*, ein neuer Kruster aus dem unterdevonischen Bundenbacher Dachschiefer. *Paläontologische Zeitschrift* 13:204–238.

Hergarten, B. 1994. Conularien des Hunsrückschiefers (Unter-Devon). *Senckenbergiana lethaea* 74:273–290.

Herrgesell, G. 1978. Geologische Untersuchungen im Raume Gemünden/Hunsrück (Rheinisches Schiefergebirge). Diploma thesis, University of Freiburg im Breisgau.

Holtz, S. 1969. Sporen im Hunsrückschiefer des Wispertales (Rheingaukreis, Hessen). *Notizblatt des hessischen Landesamtes für Bodenforschung* 97:389–390.

Houbrick, R. S., W. Stürmer, and E. L. Yochelson. 1988. Rare Mollusca from the Lower Devonian Hunsrück Slate of southern Germany. *Lethaia* 21:395–402.

Jaekel, O. 1895. Beiträge zur Kenntnis der paläozoischen Crinoiden Deutschlands. *Paläontologische Abhandlungen,* n.s., 3:1–176.

Karathanasopoulos, S. 1975. Die Sporenvergesellschaftungen in den Dachschiefern des Hunsrücks (Rheinisches Schiefergebirge, Deutschland) und ihre Aussage zur Stratigraphie. Ph.D. diss., University of Mainz.

Koenigswald, R. von. 1930a. Die Fauna des Bundenbacher Schiefers in ihren Beziehungen zum Sediment. *Zentralblatt für Mineralogie, Geologie und Paläontologie, B* 1930:241–247.

Koenigswald, R. von. 1930b. Die Arten der Einregelung im Sediment bei den Seesternen und Seelilien des unterdevonischen Bundenbacher Schiefers. *Senckenbergiana* 12:338–360.

Kott, R., and M. Wuttke. 1987. Untersuchungen zur Morphologie, Paläökologie und Taphonomie von *Retifungus rudens* Rietschel 1870 aus dem Hunsrückschiefer (Bundesrepublik Deutschland). *Geologisches Jahrbuch Hessen* 115:11–27.

Kutscher, F. 1931. Zur Entstehung des Hunsrückschiefers am Mittelrhein und auf dem Hunsrück. *Jahrbuch des nassauischen Vereins für Naturkunde* 81:177–232.

Kutscher, F. 1933. Fossilien aus dem Hunsrückschiefer I. *Jahrbuch der preussischen geologischen Landesanstalt* 54:628–641.

Kutscher, F. 1966. Lamellibranchiaten des Hunsrückschiefers. *Notizblatt des hessischen Landesamtes für Bodenforschung* 94:27–39.

Kutscher, F. 1969a. Aus der Frühgeschichte der Untersuchung von Hunsrückschiefer-Fossilien. *Decheniana* (Bonn) 122:15–20.

Kutscher, F. 1969b. Die Ammonoideen-Entwicklung im Hunsrückschiefer. *Notizblatt des hessischen Landesamtes für Bodenforschung* 97:46–64.

Kutscher, F. 1971. *Palaeoscorpius devonicus*, ein devonischer Skorpion. *Jahrbuch des nassauischen Vereins für Naturkunde* 99:82–88.

Kutscher, F. 1973. Zusammenstellung der Agnathen und Fische des Hunsrückschiefer-Meeres. *Notizblatt des hessischen Landesamtes für Bodenforschung* 101:46–79.

Kutscher, F. 1975. *Rhenopterus diensti*, ein Eurypteride im Hunsrückschiefer. *Notizblatt des hessischen Landesamtes für Bodenforschung* 103:37–42.

Kutscher, F. 1976a. Die Asterozoen des Hunsrückschiefers. *Geologisches Jahrbuch Hessen* 104:25–37.

Kutscher, F. 1976b. Die Crinoideen-Arten des Hunsrückschiefers. *Geologisches Jahrbuch Hessen* 104:9–24.

Kutscher, F. 1978. Über Trilobiten des Hunsrückschiefers. *Geologisches Jahrbuch Hessen* 106:23–52.

Kutscher, F., and H. Sieverts-Doreck. 1968. *Pyrgocystis*-Arten im Hunsrückschiefer und mittelrheinischen Unterdevon. *Notizblatt des hessischen Landesamtes für Bodenforschung* 96:7–17.

Kutscher, F., and H. Sieverts-Doreck. 1977. Über Holothurien im Hunsrückschiefer. *Geologisches Jahrbuch Hessen* 105:47–55.

Lehmann, W. M. 1934. Röntgenuntersuchungen von *Asteropyge* sp. Broili aus dem rheinischen Unterdevon. *Neues Jahrbuch für Mineralogie, Geologie und Paläontologie* 72:1–14.

Lehmann, W. M. 1938. Die Anwendung der Röntgenstrahlen in der Paläontologie. *Jahresbericht und Mitteilungen des oberrheinischen geologischen Vereins*, n.s., 27:16–24.

Lehmann, W. M. 1956a. Kleine Kostbarkeiten in Dachschiefern. 3. Sonderheft: Vom Hunsrück zum Westrich. Zur Geologie des oberen Nahegebietes um Idar-Oberstein. *Aufschluss* 1956:63–74.

Lehmann, W. M. 1956b. Beobachtungen an *Weinbergina opitzi* (R. und E. Richter) (Merostoma, Devon). *Senckenbergiana lethaea* 37:67–77.

Lehmann, W. M. 1957. Die Asterozoen in den Dachschiefern des rheinischen Unterdevons. *Abhandlungen des hessischen Landesamtes für Bodenforschung* 21:1–160.

Lehmann, W. M. 1958. Eine Holothurie zusammen mit *Palaeonectria devonica* und einem Brachiopoden in den devonischen Dachschiefern des Hunsrücks durch Röntgenstrahlen entdeckt. *Notizblatt des hessischen Landesamtes für Bodenforschung* 86:81–86.

Mittmeyer, H. G. 1980a. Zur Geologie des Hunsrückschiefers. In W. Stürmer, F. Schaarschmidt, and H. G. Mittmeyer, *Versteinertes Leben im Röntgenlicht*, pp. 26–33. Kleine Senckenberg-Reihe 11. Frankfurt: Kramer.

Mittmeyer, H. G. 1980b. Vorläufige Gesamtliste der Hunsrückschiefer-Fossilien. In W. Stürmer, F. Schaarschmidt, and H. G. Mittmeyer, *Versteinertes Leben im Röntgenlicht*, pp. 34–39. Kleine Senckenberg-Reihe 11. Frankfurt: Kramer.

Mosebach, R. 1952. Zur Petrographie der Dachschiefer des Hunsrückschiefers. *Zeitschrift der deutschen geologischen Gesellschaft* (Hannover) 103:368–376.

Opitz, R. 1932. *Bilder aus der Erdgeschichte des Nahe-Hunsrück-Landes Birkenfeld*. Birkenfeld: Selbstverlag.

Otto, M. 1994. Zur Frage der "Weichteilerhaltung" im Hunsrückschiefer. *Geologica et Palaeontologica* 28:45–63.

Raiswell, R., C. Bartels, and D. E. G. Briggs. 2001. Hunsrück Slate. In D. E. G. Briggs and P. R. Crowther, eds., *Palaeobiology II,* pp. 346–348. Oxford: Blackwell Science.

Rauff, H. 1939. *Palaeonectris discoidea* Rauff, eine siphonophore Meduse aus dem rheinischen Unterdevon nebst Bemerkungen zur umstrittenen *Brooksella rhenana* Kinkelin. *Paläontologische Zeitschrift* 21:194–213.

Reineck, H. E., and I. B. Singh. 1980. *Depositional Sedimentary Environments*. 2nd ed. Berlin: Springer-Verlag.

Richter, R. 1931. Tierwelt und Umwelt im Hunsrückschiefer: Zur Entstehung eines schwarzen Schlammsteins. *Senckenbergiana* 13:299–342.

Richter, R. 1935. Marken und Spuren im Hunsrückschiefer 1. Gefliessmarken. *Senckenbergiana* 17:244–263.

Richter, R. 1936. Marken und Spuren im Hunsrückschiefer 2. Schichtung und Grundleben. *Senckenbergiana* 18:215–244.

Richter, R. 1941. Marken und Spuren im Hunsrückschiefer 3. Fährten als Zeugnisse des Lebens auf dem Meeresgrunde. *Senckenbergiana* 23:218–260.

Richter, R. 1954. Marken und Spuren im Hunsrückschiefer 4. Marken von Schaumblasen als Kennmale des Auftauch-Bereichs im Hunsrückschiefer-Meer. *Senckenbergiana lethaea* 35:101–106.

Rietschel, S. 1969. Die Receptaculiten: Eine Studie zur Morphologie, Organisation, Ökologie und Überlieferung einer problematischen Fossil-Gruppe und die Deutung ihrer Stellung im System. *Senckenbergiana lethaea* 50:465–517.

Rievers, J. 1961. Zur Entstehung des Bundenbacher Dachschiefers und seiner Versteinerungen. *Mitteilungen der bayrischen Staatssammlung, Paläontologie und historische Geologie* 1:19–23.

Roemer, C. F. 1862–1864. Neue Asteriden und Crinoiden aus devonischen Dachschiefer von Bundenbach bei Birkenfeld. *Palaeontographica* 9:143–152.

Savrda, C. E., and D. J. Bottjer. 1991. Oxygen-related biofacies in marine strata: An overview and update. In R. V. Tyson and T. H. Pearson, eds., *Modern and Ancient Continental Shelf Anoxia*, pp. 201–219. Special Publication, no. 58. London: Geological Society.

Schäfer, W. 1972. *Ecology and Palaeoecology of Marine Environments*. Chicago: University of Chicago Press.

Schmidt, W. E. 1934. Die Crinoideen des rheinischen Devons. I. Teil: Die Crinoideen des Hunsrückschiefers. *Abhandlungen der preussischen geologischen Landesanstalt*, n.s., 163:1–149.

Seilacher, A. 1960. Strömungsanzeichen im Hunsrückschiefer. *Notizblatt des hessischen Landesamtes für Bodenforschung* 88:88–106.

Seilacher, A. 1961. Holothurien im Hunsrückschiefer (Unterdevon). *Notizblatt des hessischen Landesamtes für Bodenforschung* 89:66–72.

Seilacher, A., and C. Hemleben. 1966. Spurenfauna und Bildungstiefe der Hunsrückschiefer (Unterdevon). *Notizblatt des hessischen Landesamtes für Bodenforschung* 94:40–53.

Seilacher, A., W. E. Reif, and F. Westphal. 1985. Sedimentological, ecological and temporal patterns of fossil Lagerstätten. *Philosophical Transactions of the Royal Society of London, B* 311:5–23.

Solle, G. 1950. Obere Siegener Schichten, Hunsrückschiefer, tiefes Unterkoblenz und ihre Einstufung ins rheinische Unterdevon. *Geologisches Jahrbuch* 65:299–380.

Solle, G. 1970. Die Hunsrück-Insel im oberen Unterdevon. *Notizblatt des hessischen Landesamtes für Bodenforschung* 98:50–80.

Stanley, G., Jr., and W. Stürmer. 1983. The first fossil ctenophore from the Lower Devonian of West Germany. *Nature* 303:518–520.

Stanley, G., Jr., and W. Stürmer. 1987. A new fossil ctenophore discovered by x-rays. *Nature* 327:61–63.

Stemvers-van Bemmel, J. 1989. Fossiliseren en pyritiseren in de Hunsrück. *Gea* 22:18–20.

Steul, H. 1984. Die systematische Stellung der Conularien. *Giessener geologische Schriften* 37:1–117.

Struve, W. 1985. Phacopinae aus den Hunsrück-Schiefern (Unterdevon des rheinischen Gebirges). *Senckenbergiana lethaea* 66:393–432.

Stürmer, W. 1969. Pyrit-Erhaltung von Weichteilen bei devonischen Cephalopoden. *Paläontologische Zeitschrift* 43:10–12.

Stürmer, W. 1970. Soft parts of cephalopods and trilobites: Some surprising results of x-ray examination. *Science* 170:1300–1302.

Stürmer, W. 1980. Röntgenstrahlen erforschen die Urzeit. In W. Stürmer, F. Schaarschmidt, and H. G. Mittmeyer, *Versteinertes Leben im Röntgenlicht*, pp. 3–18. Kleine Senckenberg-Reihe 11. Frankfurt: Kramer.

Stürmer, W. 1985. A small coleoid cephalopod with soft parts from the Lower Devonian discovered by using radiography. *Nature* 318:53–55.

Stürmer, W., and J. Bergström. 1973. New discoveries on trilobites by x-rays. *Paläontologische Zeitschrift* 47:104–141.

Stürmer, W., and J. Bergström. 1976. The arthropods *Mimetaster* and *Vachonisia* from the Devonian Hunsrück Shale. *Paläontologische Zeitschrift* 50:78–111.

Stürmer, W., and J. Bergström. 1978. The arthropod *Cheloniellon* from the Devonian Hunsrück Shale. *Paläontologische Zeitschrift* 52:57–81.

Stürmer, W., and J. Bergström. 1981. *Weinbergina*, a xiphosuran arthropod from the Devonian Hunsrück Slate. *Paläontologische Zeitschrift* 55:237–255.

Stürmer, W., and F. Schaarschmidt. 1980. Pflanzen im Hunsrückschiefer. In W. Stürmer, F. Schaarschmidt, and H. G. Mittmeyer, *Versteinertes Leben im Röntgenlicht*, pp. 19–25. Kleine Senckenberg-Reihe 11. Frankfurt: Kramer.

Stürmer, W., F. Schaarschmidt, and H. G. Mittmeyer. 1980. *Versteinertes Leben im Röntgenlicht*. Kleine Senckenberg-Reihe 11. Frankfurt: Kramer.

Südkamp, W. H. 1997. Discovery of soft parts of a fossil brachiopod in the "Hunsrückschiefer" (Lower Devonian, Germany). *Paläontologische Zeitschrift* 71:91–95.

Sutcliffe, O. E. 1995. The Hunsrück Slate Fossil-Lagerstätte: Depositional setting and taphonomy. *Palaeontology Newsletter* 28:22–23.

Sutcliffe, O. E. 1997. An ophiuroid trackway from the Lower Devonian Hunsrück Slate, Germany. *Lethaia* 30:33–39.

Sutcliffe, O. E., D. E. G. Briggs, and C. Bartels. 1999. Ichnological evidence for the environmental setting of the Fossil-Lagerstätten in the Devonian Hunsrück Slate, Germany. *Geology* 27:275–278.

Wollanke, G., and W. Zimmerle. 1990. Petrographic and geochemical aspects of fossil embedding in exceptionally well preserved fossil deposits. *Mitteilungen aus dem Geologisch-Paläontologischen Institut der Universität Hamburg* 69:77–97.

Yochelson, E. L. 1989. Reconsideration of possible soft parts in dacryoconarids (incertae sedis) from the Hunsrück-Schiefer in Western Germany. *Senckenbergiana lethaea* 69:381–390.

Yochelson, E. L., W. Stürmer, and G. Stanley Jr. 1983. *Plectodiscus discoideus* Rauff: A redescription of a Chondrophorine from the Early Devonian Hunsrück Slate, West Germany. *Paläontologische Zeitschrift* 57:39–68.

Zeiss, A. 1969. Weichteile ectocochleater paläozoischer Cephalopoden in Röntgenaufnahmen und ihre paläontologische Bedeutung. *Paläontologische Zeitschrift* 43:13–27.

Ziegler, P. A. 1990. *Geological Atlas of Western and Central Europe*. 2nd ed. The Hague: Shell International Petroleum Maatschappij.

Zimmerle, W. 1992. Hunsrück-Slate. In W. Zimmerle and B. Stribrny, eds., *Organic Carbon-Rich Pelitic Sediments in the Federal Republic of Germany*, p. 43. Courier Forschungsinstitut Senckenberg, no. 152. Frankfurt: Senckenberg Museum.

9
Bear Gulch: An Exceptional Upper Carboniferous Plattenkalk

James W. Hagadorn

THE UPPER MISSISSIPPIAN BEAR GULCH BEDS OF CENTRAL Montana constitute one of the lesser-known conservation Lagerstätten, but are of importance because they contain one of the most diverse fossil fish assemblages in the world, as well as a range of exquisitely preserved soft-bodied and skeletonized nektonic and benthic organisms. Soft tissues, phosphatic fossils, cartilaginous fossils, and molds of carbonate skeletal elements are preserved in this deposit. The Bear Gulch Beds are a lensoidal unit of fine-grained limestone, or Plattenkalk, very similar in nature to the famous Solnhofen Limestone of the Bavarian Jurassic (Chapter 18), but differing somewhat in depositional style. Sediments in this unit grade laterally and downward into black shales and were likely deposited in a quiescent, oxygenated shallow basin or estuary with poorly oxygenated fine-grained sediments (Williams 1983; Factor and Feldmann 1985). The Bear Gulch basin was periodically subjected to microturbidite sedimentation and, perhaps, rapid shifts in salinity—factors that may have killed resident and vagrant Bear Gulch faunas. For example, the best-preserved vertebrate fossils occur in the inferred depocenter of the basin, where rapid burial by fine-grained carbonate, restriction of bioturbation, and minimal transport are thought to have mitigated sealing of soft tissues in alternating thin and thick flat tabular limestones. The preservation of veins, skin pigments, organ pigments, gut contents, sexual organs, and other soft tissues has allowed reconstruction of the life habits, feeding strategies, sexual dimorphism, trophic structure, and evolutionary history of numerous taxa from this deposit.

Given the exceptional soft-tissue preservation and diversity of faunas preserved in this deposit, it is surprising that it has not received wider public attention. Bear Gulch's lack of notoriety may stem from its relatively short history of study—the first fossil fish were not discovered until 1967 (Melton 1969b)—coupled with limited outcrop accessibility on private land. Nevertheless, the abundance and diversity of well-preserved fossils in this classic obrution deposit make Bear Gulch a spectacular conservation Lagerstätte.

GEOLOGICAL CONTEXT

The Bear Gulch Beds are part of a late Mississippian transgressive sequence exposed on Potter Creek Dome, about 30 km northeast of the Big Snowy Mountains in Fergus County, Montana (Melton 1969b; Smith and Gilmour 1979). This sequence was deposited about 12° north of the equator in a series of open-ended en-echelon basins within the Big Snowy Trough, a narrow east–west-trending embayment that connected the Big Snowy Basin to the east with the Cordilleran Miogeosyncline to the west (Mallory 1972; Williams 1981; Horner 1985; Witzke 1990) (Figure 9.1). At the base of this sequence are terrestrial red beds of the Kibbey Formation overlain by supratidal and shallow subtidal green shales, thin gypsum deposits, and limestones of the Otter Formation (Harris 1972; summaries in Smith and Gilmour 1979; Feldman et al. 1994). The Heath Formation overlies the green shales of the Otter Formation and consists of black shale, sandstone, and three lensoidal limestone beds—the Becket, Bear Gulch, and Surenough Beds (Figure 9.2)

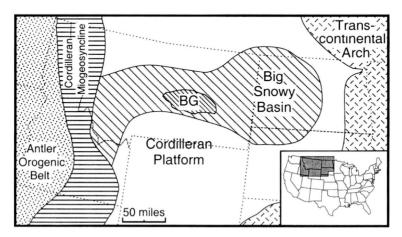

FIGURE 9.1 Generalized paleogeographic and locality map for the Bear Gulch region during the Carboniferous. Stippled area labeled BG indicates surface and subsurface extent of the Bear Gulch Limestone within the Big Snowy Trough. (Modified from Williams 1983 and Sando, Gordon, and Dutro 1975)

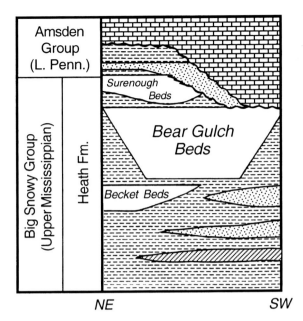

FIGURE 9.2 Generalized stratigraphic section and lateral facies relationships of Middle Carboniferous strata of central Montana. (Modified from Williams 1983)

—the last of which is nearly 115 km² in lateral extent and is separated from the overlying Pennsylvanian Tyler Formation by an erosional unconformity (Williams 1981; Horner 1985). Although well-preserved fossils occur in the Heath Formation (Schram and Schram 1979; Zidek 1980) and the lithologically similar Becket and Surenough Beds, only the Bear Gulch Beds have been systematically quarried for fossils (Melton 1985), largely because many of the fossils are articulated and exhibit soft-tissue preservation. Biostratigraphic analyses of ammonoids, conodonts, palynomorphs, bryozoans, and fish from the Bear Gulch Beds and surrounding Heath Formation shales suggest a late Chesterian (latest Mississippian) age for the unit (Scott 1973; Cox 1986; Landman and Davis 1988; Lund et al. 1993).

The Bear Gulch Beds consist of alternating massive argillaceous silty dolomitic micritic limestone beds (up to 25 cm thick) and platy clayey dolomitic micritic beds (up to 30 cm thick) (Williams 1981, 1983). Such rhythmically alternating thinner, argillaceous limestone beds and thick, nonfissile limestone units are typical of the Flinz and Fäule style of bedding made famous in other fine-grained lithographic limestones, such as Solnhofen (Chapter 18; Hemleben and Swinburne 1991). Bear Gulch Flinz (massive) beds are characterized by normally graded laminations with sharp bottoms, and individual beds are laterally continuous for distances that may exceed 1 km (Feldman et al. 1994). Graded laminae are composed of skeletal fragments, microspar, and quartz silt grading upward into micrite, with occasional argillaceous caps (Williams 1983; Feldman et al. 1994). Bear Gulch Fäule (platy) beds have similar, but less well graded laminations, as well as a higher clay content (Feldman et al. 1994).

PALEOENVIRONMENTAL SETTING

Minimal lithologic variation and limited outcrop exposure (except within deeply incised canyons) has hampered attempts to identify small-scale facies variations within the Bear Gulch Beds. However, by analyzing available sedimentary structures, faunal distributions, thickness patterns, and sedimentologic relationships, Williams (1983) recognized marginal, slope, and basin center facies within the Bear Gulch Beds, suggesting that deposition occurred within a shallow basin with gently sloping margins.

Lack of evidence for syndepositional subsidence suggests that the present-day section thickness approximates the original basin depth. Assuming that the thickest part of the Bear Gulch limestone lens represents the basin depocenter, and after consideration of compaction effects, the basin center was likely about 40 m deep (Williams 1983). Beds are thickest (~24 m thick) in the center of the inferred 15 km long depositional basin, and they thin laterally into platy black phosphatic shales along the margins (Lund et al. 1993). Absence of erosional features, tool marks, channels, or current-oriented skeletal debris in the basin facies suggests that it was likely characterized by generally quiescent conditions, periodically interrupted by waning sediment pulses. The graded Flinz beds, for example, are a typical sedimentologic by-product of waning bottom currents. Taphonomic evidence also supports this paleoenvironmental interpretation. For example, fish carcasses in this region are not significantly disarticulated, except perhaps by scavenging, and scale-filled coprolites are often preserved with an intact halo of skeletal debris (Feldman et al. 1994).

Along the inferred basin slopes, beds contain small normal faults and flame structures, indicative of rapid sedimentation followed by early dewatering, as well as tightly folded but unfractured large (3 to 4 m) slump features (Williams 1983; Feldman et al. 1994). Axes of larger slumps are oriented toward the basin depocenter (Williams 1983) and, together with thinning of overlying beds in response to slump topography (Feldman et al. 1994), suggest deposition on an inclined but unstable slope. In some places, beds with repetitive normal faulting are overlain and underlain by continuously laminated beds, suggesting that bed movements occurred shortly after deposition (Feldman et al., 1994). Surface lineations on these beds also suggest down-dip movement of sediment toward the basin center. Together with observed gradual facies transitions,

Williams (1983) used this evidence to hypothesize slopes of approximately 1° between the basin margins and the basin center.

Toward the basin margins, beds exhibit vague to poor rhythmic bedding and are characterized by peloidal laminations thought to represent rip-up clasts of organically bound mud. Marginal sediments also contain rare ripples, common gastropods, and benthic algae and are typically bound by networks of cyanobacterial fibers (Williams 1983). The large cyanobacterial population in these facies led Williams (1983) to propose that it was important in producing oxygen and stimulating benthic activity along basin margins. In particular, these latter two factors may have attracted mobile vertebrate opportunists, predators, grazers, and scavengers to the basin.

The basin margins were likely characterized by brackish surface waters and terrestrial sediment input. Marginal strata are closely associated with iron-stained sandstones bearing land-plant fragments and black shales bearing coals, brackish water ostracods, and acanthodians. Acanthodian fish, which exist in brackish water environments, are common in adjacent black shales of the Heath Formation, decrease in abundance from marginal to slope facies, and are almost absent in Bear Gulch basin facies (Colbert 1980; Zidek 1980; Lund and Poplin 1999). Near the top of the shallowing-upward Bear Gulch Beds, terrestrial palynomorphs and other plant foliage have been documented and may reflect windblown debris (Cox 1986; Lund et al. 1993). Proximity to nonmarine and brackish water deposits, association with a freshwater to brackish water fauna, and presence of several rafted *Lepidodendron* logs are evidence for a strong nonmarine influence on this basin.

All three facies (basin, slope, and margin) are finely laminated by either peloids or micrite alternating with coarser-grained laminae of sparry calcite, skeletal fragments, and quartz silt (Williams 1983). Microscopic algal filaments and alternating Flinz and Fäule bedding are also common to all three Bear Gulch facies. Deposition of the Flinz beds is thought to reflect rapid microturbidite deposition and periods of increased turbidity, which inhibited algal growth and benthic colonization. Deposition of the Fäulen in the slope and basin plain facies is thought to reflect periods of background sedimentation and decreased turbidity, conditions that may have been more amenable to benthic colonization (Williams 1983).

Lateral lithologic variations among the three Bear Gulch facies are subtle, but all shoal upward into shallow-water micrites of the marly facies, suggesting that after establishment, the basin was filled in continuously with sediment. For example, in the basin depocenter, the Bear Gulch Beds grade downward into phosphatic black platy shales, and grade upward into shallower marly facies characterized by stromatolites, intraformational conglomerates, burrows, ripple marks, and rare mud cracks (Williams 1983) (Figure 9.2). Rhythmic bedding and alternating

Flinze and Fäule bedding occur less frequently in the upper portions of the unit, supporting the idea that water depth decreased toward the top of the Bear Gulch Beds (Williams 1983; Feldman et al. 1994).

TAPHONOMY

Although a systematic taphonomic analysis of the deposit has not yet been published (but see Feldman et al. 1992), fossils are generally well preserved throughout the three Bear Gulch facies, and provide a rich source of information about the communities that inhabited, visited, and died in the Bear Gulch basin. The deposit includes a variety of vertebrate, invertebrate, plant, and bacterial fossils, including benthic, pelagic, mobile, and sessile forms. Fossils are typically flattened and preserved at lamination interfaces, except for large fish and large cephalopods (10 to 40 cm in diameter) that project through laminations, and the branching sponge *Arborispongia*, which is typically flattened on several layers. Even small coiled cephalopods are typically preserved as flattened impressions (Feldman et al. 1994).

Among skeletonized invertebrates, skeletal hashes are relatively rare, occurring sparsely in only a few beds. In dense invertebrate assemblages, complete sponges tens of centimeters in diameter, large 10 to 20 cm long conulariids, and spiny productid brachiopods are preserved fully articulated, with all ornamentation and features intact (Lutz-Garihan 1985; Rigby 1985; Babcock and Feldman 1986; Feldman et al. 1994). Kelp-like algal fronds also occur and are sometimes covered by dense associations of epibiont bivalves and brachiopods (McRoberts and Stanley 1989). Delicate arborescent sponges are sometimes preserved with attached bivalves and associated shrimp and worms intact (Rigby 1985). Ammonoids are also found covered with epibionts such as *Sphenothallus* or encrusting bryozoa (Lund 1990). Fully articulated asteroids with as many as 35 arms and visible madreporites are preserved as compressed casts and molds. All asteroid arms are outstretched and attached (i.e., not curled up, twisted, or inverted), suggesting burial and smothering of the fauna while in life position (Welch 1984). Although soft-tissue preservation is common among such mobile fauna, it has not been studied in the cephalopod specimens (Feldman et al. 1994). However, carbon impressions of cephalopod mandibles are commonly preserved *in situ* with ammonoids and less commonly with orthocone nautiloids (Mapes 1987; Cox, personal communication, in Mapes 1987). This type of preservation is extremely rare in shallow oxygenated marine deposits, yet is common in the Bear Gulch basin (which is still relatively shallow at 40 m depth) (Mapes 1987).

Unskeletonized invertebrates are also preserved, including a diverse worm fauna dominated by nemertines, nematodes, and polychaetes

(Schram 1979b). Worms are preserved as external body molds, casts, actual organic remains, and "color differences" in the rock. Several of the polychaetes are preserved with their jaw apparatuses, denticles, parapodia, acicula, and gut contents intact.

The widest range of taphonomic information available for this deposit comes from analysis of the fish, because workers from the Carnegie Museum have systematically collected specimens along with stratigraphic data since 1968 (Melton 1985). Although much of this work remains unpublished, Feldman et al. (1994) hint at the extraordinary taphonomic information available from the fish taxa when they note that approximately 70 percent of specimens in the museum collection are complete or nearly complete. Furthermore, many of the fish specimens exhibit only minor disarticulation resulting from gut rupture or minor scavenging. Dark skin color patterns, intact venous systems, as well as liver, spleen, and eye pigments also occur in many specimens (Grogan and Lund 1997). More rarely, muscles are phosphatized (Lund et al. 1993).

Taphonomic features of the fish fauna allow further insight into the mode of death and burial of the Bear Gulch fauna. For example, distinctive postmortem features, such as preferential preservation of superficial vascular structures and distended gill covers, suggest a physiological response to asphyxia-induced stresses. Spectacular preservation of abdominal vessel pathways allowed Grogan and Lund (1997) to determine that postmortem diffusion of blood pigments was greatly limited in the Bear Gulch fish, a condition that would be expected only if minimal time existed between death and burial. Such physiological response could occur if fish were asphyxiated and buried by the detached turbidity currents thought to frequently blanket the basin center (Grogan and Lund 1997). Similarly, the variety of fish tissues preserved in the Bear Gulch basin is unusual when considered in light of the tropical setting of the basin during the Carboniferous—a setting in which basin waters were almost certainly quite warm. Without rapid burial, such warm waters would have led to rapid carcass decay and disarticulation (Lund et al. 1993), again suggesting that articulated fish preservation (as opposed to scavenged carcasses) was mediated by rapid burial in the Bear Gulch basin.

For many of the articulated faunas, soft-bodied forms, and mixed faunal–floral communities, rapid burial is almost a taphonomic prerequisite for preservation. However, time from death to burial likely varied during Bear Gulch deposition, ranging from short intervals represented by the articulated fish and buried arborescent sponge communities, to longer periods of seafloor exposure reflected by scavenged fish carcasses. Although it appears that many faunas were rapidly buried, general lack of alignment of articulated skeletons on bedding planes, coprolite halos, and other taphonomic features suggest minimal current transport or erosive scouring occurred.

Styles and patterns of fossil preservation also provide insights into the paleoenvironmental conditions near the sediment–water interface in the Bear Gulch basin. For example, preservation of chitinous and phosphatic fossils increases from the black shales toward the basin facies, and calcareous invertebrates occur mostly in marginal facies, whereas in the basin facies the latter are typically preserved as casts and molds. Fish abundance and diversity also increase from shallow to deeper portions of the basin, and increases toward the eastern mouth (Lund and Poplin 1999); in the adjacent black shales, only patchy invertebrate distributions and dwarf assemblages occur (Sando, Gordon, and Dutro 1975). Because carbonate content also increase toward the basin center, Williams (1983) hypothesized that buffering of pore-water pH mediated these taphonomic patterns. In particular, buffering may have caused increased preservation potential of phosphatic materials and thus increased dissolution of calcareous shells (relative to micrite) toward the center of the basin.

Williams (1983) also attempted to gain insights into oxygen levels in the basin, suggesting that oxygen in the water column may have been consumed by respiration in surface waters and by oxidation of large amounts of organic matter and detritus produced along basin margins, thus decreasing oxygen levels in sediment pore waters. Reduced pore-water oxygen levels would account for the lack of bioturbation and sessile benthic organisms in basin facies and would have allowed preservation of the observed laminated sediments, organic carbon, and phosphatic and chitinous fossils (Williams 1983). Although the sediments may have been anaerobic or dysaerobic, the presence of mobile benthic vertebrates (including forms specialized for bottom feeding as well as shallow-sediment burrowing) and rare benthic invertebrates suggests that bottom-waters were oxygenated (Factor and Feldmann 1985).

In summary, these taphonomic and paleoecologic features collectively suggest that faunas in the Bear Gulch basin were rapidly buried and sealed by waning fine-grained turbid flows. Although many of the articulated and soft-bodied forms likely died shortly before or concomitant with burial, rafted and scavenged carcasses were also incorporated into the deposit and reflect a longer duration of preburial taphonomic exposure. Both mobile and sessile forms, as well as vagrant, rafted, and resident faunas, were incorporated into the deposit. Although the role of pore-water chemistry and sediment oxygenation has not been fully explored, it appears that variations in sediment alkalinity affected the preservation of calcareous skeletal and soft-tissue elements. Similarly, restriction of pore-water oxygen levels inhibited bioturbation in the basin center and slopes, ensuring the sealing of carcasses within the sediment.

PALEOBIOLOGY AND PALEOECOLOGY

Although best known for the spectacular preservation of fish (Figures 9.3 and 9.4), the Bear Gulch Beds contain a variety of invertebrates, including limulines, shrimp, gastropods, asteroids, straight nautiloid cephalopods, a trilobite pygidium, crustaceans, bryozoans, conulariids, conodonts, phosphatic and calcareous brachiopods, and polychaete worms (Schram and Horner 1978; Schram 1979a, 1979b; Welch 1984; Factor and Feldmann 1985; Lutz-Garihan 1985; Rigby 1985; Mapes 1987; Landman and Davis 1988) (Figure 9.5). In addition to metazoans, algae, worm borings, and flora consisting of plant fragments, wood fragments, acritarchs, and spores occur and are typically preserved as carbon residues (Melton 1969a, 1969b; Scott 1973; Hill 1978; Cox 1986). Although discrete trace fossils have not been noted, bioturbated layers do occur in the uppermost layers and eastern fringes of the Bear Gulch lens, and are associated with bryozoans and productid brachiopods (Lund et al. 1993). The most abundant fossils in the Bear Gulch are small cephalopods (including straight-shelled and coiled nautiloids and ammonoids), followed by shrimp, fish, worms, and other taxa.

Within each of these groups, a variety of feeding types are present. Among the malocostracan arthropods, for example, all four feeding types known from the Paleozoic occur: carnivores, scavengers, filter feeders, and algal-detrital feeders (Schram 1981; Factor and Feldmann 1985). Plankton were also present in the basin, as documented by the filter-feeding shrimp *Crangopsis eskdalensis*, which is noteworthy because it has

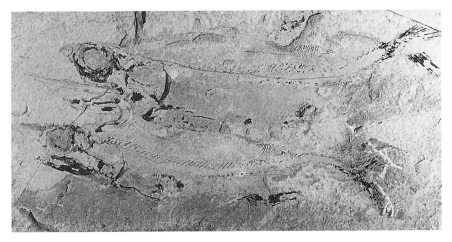

FIGURE 9.3 Female (*upper*) and male (*lower*) specimens of the stethacanthid shark *Falcatus falcatus*, likely preserved in a pre- or post-copulatory position. Note the absence of arched vertebral curling (indicating minimal preburial rigor mortis). Specimens are approximately 14.5 cm long. (Photo courtesy of H. Feldman and C. Maples, Indiana University, Bloomington)

FIGURE 9.4 The unusual coelacanth *Allenypterus montanus*. Specimen is approximately 15 cm long. (Photo courtesy of R. Lund, Adelphi University)

been found intact in the intestines of sharks and in fish coprolites (Factor and Feldmann 1985).

Many of the Bear Gulch invertebrates also provide insights into regional biogeography and paleoenvironments. In particular, many of the worms, crustaceans, and chelicerates found in the Bear Gulch Beds are similar to those found in concretions in the Pennsylvanian Mazon Creek fauna (Chapter 10), and are biostratigraphically intermediate between forms found in the Mazon Creek and in the lower Carboniferous Glencartholm deposits of Scotland (Schram and Horner 1978; Schram 1979a, 1979b). For example, limuline chelicerates such as *Paleolimulus* (Figure 9.6) and *Euproops* occur in both Bear Gulch and Mazon Creek. The latter form is important because it is typically found only in brackish water deposits of Mazon Creek, and rarely in its nearshore marine deposits, supporting suggestions that brackish water conditions existed in parts of the Bear Gulch basin.

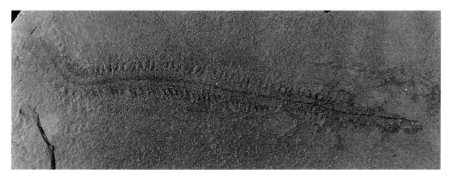

FIGURE 9.5 The fleshy polychaete worm cf. *Carbosesostris megaliphagon*, characterized by abundant well-preserved parapodia. Specimen is approximately 13 cm long. (Photo courtesy of H. Feldman and C. Maples, Indiana University, Bloomington)

FIGURE 9.6 The malacostracan crustacean *Paleolimulus longispinus.* Specimen is approximately 5.1 cm wide. (Photo courtesy of H. Feldman and C. Maples, Indiana University, Bloomington)

Among the many soft-bodied organisms found in this deposit, one in particular stimulated early interest in Bear Gulch. Shortly after the discovery of fish fossils in the Bear Gulch Beds, Richardson (1969) reported what were thought to be fossils of the conodont animal (Melton 1972; Melton and Scott 1973). Although this discovery generated widespread interest among the paleontological community, Rhodes (1973), Lindström (1974), and Conway Morris (1976) subsequently debunked these claims, instead suggesting that "complete" conodont assemblages were actually conodont clusters within the gut of a lancelet-like organism. Conway Morris (1990) reexamined the original specimens collected by Melton (1972), and although he did not pinpoint the phylogenetic affinity of this keeled wormlike organism, he renamed it *Typhloesus,* suggesting that it was an actively swimming nonchordate predator or scavenger containing conodont apparatuses, fish, and worm teeth in its gut. Ironically, Briggs, Clarkson, and Aldridge (1983) later discovered the real conodont animal from another poorly known lower Carboniferous Lagerstätte—the Granton Shrimp Beds.

Over the past 30 years, 108 fish species have been documented from the Bear Gulch Beds, including numerous sharks (Figure 9.3), skates, platysonids, paleoniscids, coelacanths, a dorypterid, and a tarrassiid (Melton 1969a, 1969b; Lund and Zangerl 1974; Lund 1977; Di Canzio 1978; Zidek 1980) (Figure 9.4). Among the more than 4,800 specimens collected from Bear Gulch, the most abundant fish taxon is the coelacanth *Caridosuctor populosum;* yet the most diverse group, comprising about 60 percent of the taxa, are the Chondrichthyes (Lund and Poplin 1999). Over half of the fauna are known from only a few specimens. For example, most of the chondrichthyans are rare or uncommon in Bear Gulch; only 12 of the 55 species are known from 10 or more specimens (Lund 1990). Among the rare taxa, many are thought to be opportunists or vagrants, and include large predators and fish with shell-crushing dentition (Lund 1990). Among the common taxa, larval forms, juveniles, fetuses, and mature adults have been documented, allowing for extraction of information about the life histories of faunal elements that inhabited or visited this basin. In addition to life history information, preservation of sexual organs and evidence of internal fertilization allowed Lund (1990, and references therein) to assemble information about sexual dimorphism for the more common fish taxa. For example, male:female ratios for taxa within the basin range from 11:1 for *Damocles serratus* to 1:10 for *Harpagofututor volsellorhinus* (Lund 1990).

Based on the paleoenvironmental distribution of fossils, four general paleocommunities have been documented within the Bear Gulch basin: nektonic, benthic, algal, and terrestrial-dominated assemblages (Di Canzio 1978; Lund et al. 1993). The first community, a nektonic assemblage, is thought to include small cephalopods, shrimp, fish, and soft-bodied invertebrates. The assemblage is sparsely distributed and typically occurs on bed surfaces lacking other fossil debris. Although uncommon, fish with fusiform bodies (such as the ray-finned Osteichthyes) are found in this assemblage, and are thought to reflect forms that lived in the middle to top of the water column (Lund et al. 1993). Even rarer are fish with specialized shell-crushing dentition, such as the holocephalan Chondrichthyes. The rarity of these forms is thought to stem from the paucity of shelled invertebrates (food sources) available to nektonic faunas in the basin (Lund et al. 1993). Flattened bottom-lurking benthic fish also occur throughout the basin, but are most abundant in the basin center. Isolated teeth, large and bone-filled coprolites, beheaded (but otherwise perfectly preserved) fish carcasses, and rare partial carcasses also point to the presence of large predatory sharks in this part of the Bear Gulch ecosystem (Melton 1969b; Zidek 1980; Lund 1990).

The benthic assemblage includes branching sponges, brachiopods, bivalves, algae, and conulariids. Fossils in this community typically occur as dense assemblages on bedding surfaces, and more rarely as isolated colonies. For example, the Bear Gulch bay mouth contains extensive bioherms of arborescent sponge colonies (Rigby 1985), which are commonly found preserved with several types of bivalves and conulariids in life position, as well as associated algae, worms, shrimp, and fish. Shelly fossils constitute another important component of this benthic assemblage, but likely reflect transported forms, either from floating nektonic communities or from basin margins (Lund et al. 1993). For example, *Sargassum*-like algal mats, which hosted diverse invertebrate communities, may have been floating in the basin and have periodically fallen to the basin floor. Evidence for such assemblages includes bivalves found attached to an algal thallus, as well as accumulations of spiny productid brachiopods that not only are unabraded, but lack a preferred orientation (i.e., pedicle valve up or down) relative to bedding planes, suggesting that they were not transported downslope but were dropped into the basin center (McRoberts and Stanley 1989; Lund et al. 1993). Such occurrences are more common near the top of the Bear Gulch limestone lens and along the eastern and northern edges of the limestone, in areas where shallowing is inferred (Lund et al. 1993). Larval fish concentrated at the margins of the Bear Gulch basin may form another component of the benthic community. For example, a number of Chondrichthyes and Osteichthyes in the Bear Gulch Beds have morphologies adapted to cryptic, defensive, or evasive swimming behavior—morphologies indicative of a shelter-dwelling mode of life. Typically, these forms are also found in close association with the arborescent sponge– and filamentous algae–rich horizons, suggesting a link to higher productivity and shelter associated with sponge-algal communities characteristic of marginal facies (Lund et al. 1993).

The third assemblage is dominated by dense concentrations of fine filamentous algae, primarily along the southern margin of the inferred basin. This assemblage also includes small productid brachiopods found within the mats and may reflect deposition during a low-energy regime (Lund et al. 1993).

Finally, Lund et al. (1993) identified a fourth assemblage dominated by terrestrial plants, lycopsid logs, leaves, and other plant material. With the exception of the rafted logs, which occur in the center of the basin, most floral elements occur in the upper layers of the Bear Gulch lens (Lund et al. 1993). Such a terrestrial floral component would be expected to occur in the upper portions of this deposit, as they represent shallowing of the Bear Gulch basin.

In summary, species richness of the fish, together with the other invertebrate and floral evidence, suggests that the Bear Gulch basin contained a complex and high-diversity ecosystem similar to that of a modern estuary or bay (Lund et al. 1993). Fish diversity, together with occurrence of arborescent sponge communities, was generally concentrated along the eastern margin of the Bear Gulch basin, where it likely connected to the Williston Basin in the Carboniferous. Access to this larger marine environment probably allowed development of a relatively complex food web in the Bear Gulch basin that attracted migratory and larger predatory faunas. At the top of the Bear Gulch food chain were large fish and vagrant predatory sharks that probably fed on fish, larger invertebrates, and mobile shelled invertebrates (Lund 1990). The last community likely grazed on the sponge and algal communities, worms, algal and bacterial mats, and plankton found in the basin (Di Canzio 1978; Zidek 1980). Despite the diversity of resident and vagrant faunas within the basin, the Bear Gulch assemblage is still quite unique in the Carboniferous because it generally lacks foraminifera, ostracods, trilobites, and stenohaline forms, such as stalked echinoderms, corals, and byrozoans (Scott 1973; Factor and Feldmann 1985).

CONCLUSIONS

The Bear Gulch Beds were likely deposited in a shallow nearshore marine basin or estuary that was characterized by periodically brackish surface waters and oxygenated bottom-waters. Although the role of salinity variations has not been fully explored, it appears that many of the articulated faunas were asphyxiated and rapidly buried along with scavenged carcasses by fine-grained waning turbiditic flows. Both mobile and sessile faunas, floral elements, and vagrant and resident communities were incorporated into alternating thickly and thinly bedded pure micritic Plattenkalk-style limestones. Decreased pore-water oxygen concentrations probably restricted bioturbation in these beds, thus mediating sealing of carcasses in the sediment. Spectacular preservation of these faunas within this small Lagerstätte has thus far afforded us rare insights into ancient fish physiology, life histories, sexual dimorphism, and taphonomy, as well as insights into the paleocommunity structure of a highly diverse Carboniferous bay ecosystem. With further work, there remains much more to learn about the geochemical, paleoceanographic, taphonomic, and paleoecologic conditions that favored preservation in this exceptional deposit.

ACKNOWLEDGMENTS

I wish to thank H. Feldman, A. Fischer, E. Grogan, and C. Maples for their constructive reviews. I also thank C. Maples, H. Feldman, R. Lund, and F. Schram for providing unpublished data and observations that helped improve this contribution.

REFERENCES

Babcock, L. E., and R. M. Feldman. 1986. Devonian and Mississippian conulariids of North America. Part B. *Paraconularia, Reticulaconularia,* new genus, and organisms rejected from Conulariida. *Annals of the Carnegie Museum* 55:411–479.

Briggs, D. E. G., E. N. K. Clarkson, and R. J. Aldridge. 1983. The conodont animal. *Lethaia* 16:1–14.

Colbert, E. H. 1980. *Evolution of the Vertebrates.* New York: Wiley.

Conway Morris, S. 1976. A new Cambrian lophophorate from the Burgess Shale of British Columbia. *Palaeontology* 19:199–222.

Conway Morris, S. 1990. *Typhloesus wellsi* (Melton and Scott, 1973), a bizarre metazoan from the Carboniferous of Montana, U.S.A. *Philosophical Transactions of the Royal Society of London, B* 327:595–624.

Cox, R. S. 1986. Preliminary report on the age and palynology of the Bear Gulch Limestone (Mississippian, Montana). *Journal of Paleontology* 60:952–956.

Di Canzio, J. 1978. Ecomorphology of the Osteichthyes from the Bear Gulch Limestone. Master's thesis, Adelphi University.

Factor, D., and R. Feldmann. 1985. Systematics and paleoecology of malacostracan arthropods in the Bear Gulch Limestone (Namurian) of central Montana. *Annals of the Carnegie Museum* 54:319–356.

Feldman, H. R., R. Lund, C. G. Maples, and A. W. Archer. 1992. Taphonomy of the Bear Gulch Limestone, Mississippian, Montana. *Geological Society of America Abstracts with Programs* 24:345.

Feldman, H. R., R. Lund, C. G. Maples, and A. W. Archer. 1994. Origin of the Bear Gulch Beds (Namurian, Montana, USA). *Geobios* 27:283–291.

Grogan, E. D., and R. Lund. 1997. Soft tissue pigments of the upper Mississippian chondrenchelyid *Harpagofututor volsellorhinus* (Chondrichthyes, Holocephalie) from the Bear Gulch Limestone, Montana, USA. *Journal of Paleontology* 71:337–342.

Harris, W. L. 1972. Upper Mississippian–Pennsylvanian stratigraphy of central Montana. Ph.D. diss., University of Montana.

Hemleben, C., and N. M. H. Swinburne. 1991. Cyclical deposition of the Plattenkalk facies. In G. Einsele, W. Ricken, and A. Seilacher, eds., *Cycles and Events in Stratigraphy*, pp. 572–591. Berlin: Springer-Verlag.

Hill, V. 1978. *Spenothallus* cf. *S. angustifolius* from the Mississippian Bear Gulch Limestone of central Montana. B.A. thesis, Princeton University.

Horner, J. R. 1985. The stratigraphic position of the Bear Gulch Limestone (lower Carboniferous) of central Montana. *Compte Rendu, Neuvième Congrès International de Stratigraphie et de Géologie du Carbonifère* 5:427–436.

Landman, N. H., and R. A. Davis. 1988. Jaw and crop preserved in an orthoconic nautiloid cephalopod from the Bear Gulch Limestone (Mississippian, Montana). *New Mexico Bureau of Mines and Mineral Resources, Memoir* 44:103–107.

Lindström, M. 1974. The conodont apparatus as a food-gathering mechanism. *Palaeontology* 17:729–744.

Lund, R. 1977. *Echinochimaera meltoni*, new genus and species (Chimaeriformes), from the Mississippian of Montana. *Annals of the Carnegie Museum* 46:195–221.

Lund, R. 1990. Chondrichthyan life history styles as revealed by the 320 million years old Mississippian of Montana. *Environmental Biology of Fishes* 27:1–19.

Lund, R., H. Feldman, W. L. Lund, and C. G. Maples. 1993. The depositional environment of the Bear Gulch Limestone, Fergus County, Montana. In L. D. V. Hunter, ed., *Energy and Mineral Resources of Central Montana: 1993 Field Conference Guidebook*, pp. 87–96. Billings: Montana Geological Society.

Lund, R., and C. Poplin. 1999. Fish diversity of the Bear Gulch Limestone, Namurian, lower Carboniferous of Montana, USA. *Geobios* 32:285–295.

Lund, R., and R. Zangerl. 1974. *Squatinactis caudispinatus*, a new elasmobranch from the upper Mississippian of Montana. *Annals of the Carnegie Museum* 45:43–55.

Lutz-Garihan, A. B. 1985. Brachiopods from the upper Mississippian Bear Gulch Limestone of Montana. *Compte Rendu, Neuvième Congrès International de Stratigraphie et de Géologie du Carbonifère* 5:457–467.

Mallory, W. W. 1972. Regional synthesis of the Pennsylvanian system. In W. W. Mallory, ed., *Geologic Atlas of the Rocky Mountain Region*, pp. 111–127. Denver: Rocky Mountain Association of Geologists.

Mapes, R. H. 1987. Upper Paleozoic cephalopod mandibles: Frequency of occurrence, modes of preservation, and paleoecological implications. *Journal of Paleontology* 61:521–538.

McRoberts, C. A., and G. D. Stanley. 1989. A unique bivalve–algae life assemblage from the Bear Gulch Limestone (upper Mississippian) of central Montana. *Journal of Paleontology* 63:578–581.

Melton, W. G. 1969a. The Bear Gulch fauna from central Montana. In *Proceedings of the North American Paleontological Convention*, pp. 1202–1207. Lawrence, Kans.: Allen Press.

Melton, W. G. 1969b. A new Dorypterid fish from central Montana. *Northwest Science* 43:196–206.

Melton, W. G. 1972. The Bear Gulch Limestone and the first conodont-bearing animals. In *Montana Geological Society, 21st Annual Field Conference*, pp. 65–68. Billings: Montana Geological Society.

Melton, W. G. 1985. A brief history of exploration for the Bear Gulch fauna of central Montana. *Compte Rendu, Neuvième Congrès International de Stratigraphie et de Géologie du Carbonifère* 5:425–427.

Melton, W. G., and H. W. Scott. 1973. Conodont-bearing animals from the Bear Gulch Limestone, Montana. *Geological Society of America Special Paper* 141:31–65.

Rhodes, F. H. T. 1973. Conodont research: Programs, progress and priorities. *Geological Society of America Special Paper* 141:277–286.

Richardson, E. S. 1969. The conodont animal. *Earth Science* 6:256–257.

Rigby, J. K. 1985. The sponge fauna of the Mississippian Heath Formation of central Montana. *Compte Rendu, Neuvième Congrès International de Stratigraphie et de Géologie du Carbonifère* 5:443–456.

Sando, W. J., M. Gordon, and T. J. Dutro. 1975. *Stratigraphy and Geologic History of the Amsden Formation (Mississippian and Pennsylvanian) of Wyoming*. United States Geological Survey Professional Paper, no. 848A. Washington, D.C.: Department of the Interior, Geological Survey.

Schram, F. R. 1979a. Limulines of the Mississippian Bear Gulch Limestone of central Montana, USA. *Transactions of the San Diego Society of Natural History* 19:67–74.

Schram, F. R. 1979b. Worms of the Mississippian Bear Gulch Limestone of central Montana, USA. *Transactions of the San Diego Society of Natural History* 19:107–120.

Schram, F. R. 1981. Late Paleozoic crustacean communities. *Journal of Paleontology* 55:126–137.

Schram, F. R., and J. Horner. 1978. Crustacea of the Mississippian Bear Gulch Limestone of central Montana. *Journal of Paleontology* 52:394–406.

Schram, J. M., and F. R. Schram. 1979. *Joanellia lundi* sp. nov. (Crustacea: Malacostraca) from the Mississippian Heath Shale of central Montana. *Transactions of the San Diego Society of Natural History* 19:53–56.

Scott, H. W. 1973. New Conodontochordata from the Bear Gulch Limestone (Namurian, Montana). *Publications of the Michigan State Museum of Paleontology Series* 1:85–99.

Smith, D. L., and E. H. Gilmour. 1979. *The Mississippian and Pennsylvanian (Carboniferous) Systems in the United States—Montana*. United States Geological Survey Professional Paper, no. 1110–X. Washington, D.C.: Department of the Interior, Geological Survey.

Welch, J. R. 1984. The asteroid *Lepidasterella montanensis* n. sp., from the upper Mississippian Bear Gulch Limestone of Montana. *Journal of Paleontology* 58:843–851.

Williams, L. A. 1981. The sedimentational history of the Bear Gulch Limestone (Carboniferous, central Montana)—An explanation of "How Them Fish Swam Between Them Rocks." Ph.D. diss., Princeton University.

Williams, L. A. 1983. Deposition of the Bear Gulch Limestone: A Carboniferous Plattenkalk from central Montana. *Sedimentology* 30:843–860.

Witzke, B. J. 1990. Paleoclimatic constraints for Paleozoic paleolatitudes of Laurentia and Euramerica. In W. S. McKerrow and C. R. Scotese, eds., *Paleozoic Paleogeography and Biogeography*, pp. 57–73. Geological Society Memoir, no. 12. London: Geological Society.

Zidek, J. 1980. *Acanthodes lundi*, new species (Acanthodii), and associated coprolites from uppermost Mississippian Heath Formation of central Montana. *Annals of the Carnegie Museum* 49:69–78.

10
Mazon Creek: Preservation in Late Paleozoic Deltaic and Marginal Marine Environments

Stephen A. Schellenberg

IN THE MAZON CREEK AREA OF NORTHEASTERN ILLINOIS (Figure 10.1), fossiliferous concretions of the Francis Creek Shale provide an unparalleled window into late Pennsylvanian life in deltaic and coastal environments. The exquisitely preserved, commonly soft-bodied fauna spans a wide range of life modes (flying to infaunal) and environments (terrestrial to freshwater to full marine) (Richardson and Johnson 1971), while the flora is among the most diverse known from North America (Darrah 1970; Horowitz 1979). Some Mazon Creek specimens represent the earliest or only known fossil occurrence of their higher taxa; a few remain systematic mysteries altogether (Nitecki 1979a). Many Mazon Creek organisms passed relatively unscathed through the fossilization barrier, starting with rapid burial within low-oxygen sediments, followed by early diagenetic encapsulation within iron-rich carbonate (siderite) concretions (Baird et al. 1986). Rapid fossilization is apparent by the preservation of ink sacs in cephalopods, yolk sacs in hatchling fish, and articulated setal hairs in polychaete worms. In some cases, preservation is three-dimensional and color patterns are preserved. Thus, Mazon Creek concretions are considered to be an extreme form of conservation Lagerstätte (Seilacher, Reif, and Westphal 1985).

The first reported Mazon Creek fossils were collected by the local populace during the mid-nineteenth century from natural outcrops of the Francis Creek Shale. Subsequent development of regional strip and shaft mining of the underlying Colchester (No. 2) Coal (Ledvina 1997)

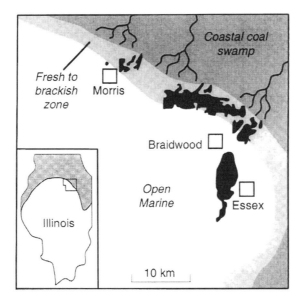

FIGURE 10.1 Major strip mines (black areas) in the Mazon Creek area overlain by interpreted late Pennsylvanian paleoenvironments. Coastal coal swamp and deltaic paleoenvironments predominated in the northeast, with an increasingly open-marine paleoenvironment toward the southwest. Pit Eleven is directly west of the town of Essex. Inset map of Illinois shows inferred maximum incursion of the late Pennsylvanian epicontinental seaway. (Modified from Baird et al. 1986)

produced immense spoil piles of the Francis Creek Shale rich in concretions. Avid collectors passed specimens on to the Illinois Geological Survey, and formal scientific study of this Lagerstätte commenced in 1864 with James Dana's description of two extinct insect orders. Dana's publication, along with extensive paleobotanical work by Charles Lesquereux, heralded the beginning of a productive, if uneven, study of this Lagerstätte (Nitecki 1979a). An important event in the history of Mazon Creek study was the opening in 1945 of an extensive strip mine, including the now famous Pit Eleven, by the Peabody Coal Company near Essex, Illinois (Figure 10.1). In contrast to those from previous collection sites, concretions from Pit Eleven revealed a highly diverse and abundant fauna of rare soft-bodied estuarine invertebrates. The discovery of this fauna, coupled with an improved understanding of the regional distribution of different taxa, led to the recognition of two distinct assemblages: the Braidwood biota, composed mostly of plants with terrestrial and freshwater invertebrates and vertebrates, and the Essex biota, formed predominately of soft-bodied estuarine invertebrates most common around Pit Eleven (Richardson and Johnson 1971).

Through many years of collecting by amateur and professional paleontologists, the Field Museum of Natural History in Chicago has become the major repository for Mazon Creek fossils, housing well over 50,000 cataloged specimens. Description and interpretation of the biota, paleo-environment, and preservational processes have resulted in more than 100 published papers, numerous symposia at professional meetings, and two edited books (Nitecki 1979b; Shabica and Hay 1997). In recent years, Gordon Baird and colleagues have played a major role in deciphering the Pennsylvanian coastal paleocommunity preserved in the fossils and sediments of the Mazon Creek area, but many systematic, paleoenviron-mental, and taphonomic questions remain.

GEOLOGICAL CONTEXT

Throughout the Pennsylvanian Period, terrigenous and marine sedi-ments were deposited on the subsiding structural margin of the north-eastern Illinois Basin, the general location of the Mazon Creek Lager-stätte. During the middle Pennsylvanian, this region was within 10° of the paleoequator (Zeigler et al. 1979; Rowley et al. 1985) and had a warm, wet climate (Schopf 1979; Phillips and Dimichele 1981; Phillips, Peppers, and Dimichele 1985). To the east, active Alleghenian tecton-ism produced abundant clastic sediments that were transported basin-ward by major fluvial systems (Potter and Pryor 1961). Deposition of these terrigenous and associated marine sediments along the basin's margin was strongly controlled by sea-level fluctuations of the epiconti-nental seaway to the southwest (Figure 10.1). These sea-level fluctua-tions, in turn, reflected the net effect of local delta progradation, re-gional plate flexing from the Laurasia–Gondwana collision, and eustatic (global) sea-level oscillations driven by southern Gondwana's glaciation (Heckel 1986). The combination of these local- to global-scale processes produced cyclothems, repeated cycles of marine, deltaic, and terrestrial sedimentation totaling over 1 km in thickness (Wanless and Weller 1932; Shabica 1979; Klein and Willard 1989; Klein and Kupperman 1992). The Mazon Creek Lagerstätte is preserved within the Francis Creek Shale Member of the Carbondale Formation, which forms the lower portion of the Liverpool Cyclothem. The formation is middle Pennsylvanian in age (~296 Ma) and is assigned to the Desmoinesian Series of North America and Westphalian D Stage of Europe (Wanless 1975; Baird 1979; Pfefferkorn 1979). This chapter concentrates on the three lower mem-bers of the Carbondale Formation: the Colchester (No. 2) Coal, Francis Creek Shale, and Mecca Quarry Shale (Figure 10.2).

The basal Colchester (No. 2) Coal is a roughly 1 m thick remnant of a considerably thicker (10 to 20 m), lycopod-dominated peat formed in a coastal swamp paleoenvironment (Pfefferkorn 1979; Phillips, Peppers, and Dimichele 1985; Winston 1986) (Figure 10.3). This economically important coal seam is one of the most extensive known from the geo-logic record, extending from western Indiana to Oklahoma (Wright

1979). Within the Mazon Creek area, the Colchester (No. 2) Coal is over-lain by the Francis Creek Shale, a lobate unit dominated by gray mud-stones and siltstones deposited along a river-influenced coastline (Fig-ure 10.3). Overlying the Francis Creek Shale is the marine Mecca Quarry Shale, a fissile to sheety black shale containing abundant brachiopods, fish remains, and phosphatic nodules (Zangerl and Richardson 1963; Zangerl 1980). Toward the southwest, the Francis Creek Shale thins irregularly and pinches out, while the marine Mecca Quarry Shale

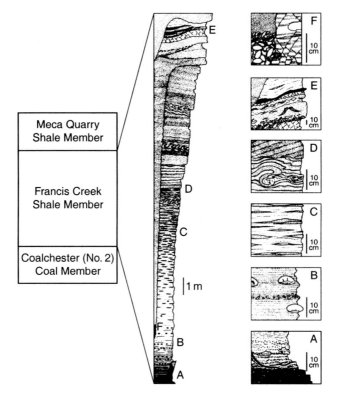

FIGURE 10.2 Stratigraphy of the lower Carbondale Formation and composite mi-crofacies sequence of the Francis Creek Shale Member in northeastern Illinois. The Colchester (No. 2) Coal Member is overlain, and occasionally cut, by the Francis Creek Shale Member, an upward-coarsening sequence with local sandstone chan-nels. Thinner outcrops of the Francis Creek Shale Member are commonly overlain by the marine Mecca Quarry Shale Member. The composite microfacies sequence of the Francis Creek Shale Member is (A) basal mudstone with abundant plant debris, concretions, small burrows, and rare *in situ* tree trunks; (B) finely laminated silty mudstone reflecting rapid, regular ebb-and-flood tidal events and containing abun-dant siderite concretions; (C) foresets of silty mudstone/muddy siltstone with uni-directional (current-produced) and, occasionally, bidirectional (tide-produced) cur-rent ripples; (D) thinly bedded siltstone with climbing ripples and synsedimentary deformation, commonly associated with sandstone channels of various sizes; (E) complex uppermost interval containing thin coals and small channels with peat, shale, and underclay debris and fragments (suggesting levee facies); and (F) a chan-nel sandstone with basal lag accumulations of siderite concretions, underclay and peat fragments, and exotic clasts. (Modified from Baird and Sroka, 1990)

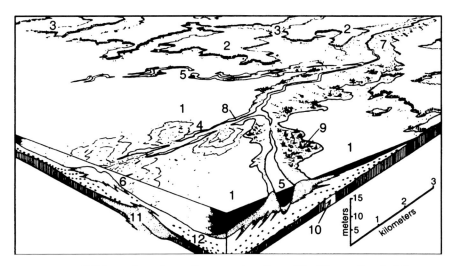

FIGURE 10.3 Oblique-view schematic reconstruction of the Mazon Creek delta complex. Features include (1) interdistributary bays, (2) coastal swamp, (3) waterways within swamp, (4) active distributary bar finger during flood, (5) abandoned distributary channel with erosional bank, (6) outflow of muddy freshwater, (7) overbank flood sedimentation, (8) subaerial levee, (9) subaerial levee with established flora, (10) underlying peat deposit (proto-Colchester Coal), (11) sand-filled fluvial channel cutting through peat deposit, and (12) laminar, silty Francis Creek mud deposits containing concretions (prominently inclined proximal to distributaries). (Modified from Baird, Shabica, et al. 1985)

thickens and eventually directly overlies the Colchester (No. 2) Coal (Baird 1979).

Although the Francis Creek Shale contains significant variation in thickness and internal facies, a general stratigraphic sequence exists (Figure 10.2). The base of the shale is mudstone with 0.1 to 10 mm thick alternating laminations of silt and clay. These laminated couplets grade upward into silty mudstones with ripple cross-stratification, which, in turn, coarsen into silty sandstones with climbing ripples. The member also contains sandstone distributary channels with trough cross-stratification, plane-bedding, and bed-loads of concretions and peat clumps; adjacent to these channels are off-lapping foreset beds of mudstone inclined over 10° (Shabica 1979). The shale is thickest (~26 m) and most variable in sediment texture and structure in the northeast and thins unevenly to the southwest, with increased mudstone dominance (Baird 1979). The vast majority of the fossiliferous siderite concretions are found parallel to the fine-grained bedding within the lower 3 to 5 m of the Francis Creek Shale (Baird and Sroka 1990). The thinner Francis Creek Shale sections to the south and west typically contain fewer fossiliferous concretions and a more poorly preserved, normal marine fauna (e.g., brachiopods, corals, bryozoans, trilobites) in highly bioturbated sediments (Smith et al. 1970).

PALEOENVIRONMENTAL SETTING

Building on the foundation of earlier studies (Richardson 1956; Richardson and Johnson 1971; Shabica 1979), Baird, Shabica, et al. (1985) interpret the lower Carbondale Formation as a coastal estuarine environment recording ongoing transgression with subsequent deltaic progradation (Figure 10.3). The coastal water table supported a spatially complex and abundant assemblage of ferns, sphenopsids, lycopods, and other flora typical of Carboniferous swamps (Peppers and Pfefferkorn 1970; Taylor 1981). As the transgression continued, the progressive northeastward rise of the coastal water table and deposition of plant material produced a laterally extensive, southwestward-sloping peat unit of irregular paleorelief, the parent material of the Colchester (No. 2) Coal Member.

The Francis Creek Shale Member represents the progradation of a delta complex on this transgressive peat base in an estuarine environment. The delta complex consisted of distributary channels, interdistributary bays, and a proximal prodelta (Shabica 1979). During progradation of the delta complex, laminated prodelta muds were deposited distally in the depressions and swales of the peat unit, producing the topographically irregular contact between the Colchester (No. 2) Coal and Francis Creek Shale Members (Baird and Shabica 1980). The general coarsening-upward sequence within the Francis Creek Shale reflects the progressive filling of the estuary by sediments carried seaward by distributary channels. Additional support for this paleoenvironmental reconstruction includes the member's varying thickness and geometry and the northeastward and up-section transition from predominately marine to estuarine to terrestrial facies and assemblages (Richardson and Johnson 1971; Broadhurst 1985). Concurrent with this local delta progradation, continued transgression led to regional deposition of the open-marine Mecca Quarry Shale on the flooded Colchester peat and thinner, topographically lower portions of the Francis Creek Shale.

TAPHONOMY

The Mazon Creek fossils are preserved within siderite concretions that generally occur parallel to bedding, either in mudstones of the episodically, torrentially deposited distributary foresets or in the rapid, regularly deposited prodelta. These concretions are generally smooth and flat, discoidal to elongate in shape, and less than 20 cm in maximum diameter (Nitecki 1979a). Most carbonate and cuticle skeletal remains (e.g., mollusc shells, polycheate setae, horseshoe crab exoskeletons) are preserved as three-dimensional molds, some with degraded cuticle or partial to complete casts of sphalerite, kaolinite, calcite, and, rarely, quartz or galena (Schopf 1979; Baird and Sroka 1990). In contrast, most soft tissues are preserved as two-dimensional or highly compressed, light-on-dark impressions that detail both body sides and, occasionally, internal features; tissues are rarely replaced. The taphonomic pathway that produced these fossiliferous concretions consisted of two distinct processes: the physical events burying the biota and the geochemical reactions entombing them within siderite concretions.

Limited postmortem transport and rapid burial of the Mazon Creek biota are indicated by the fine preservation of delicate invertebrates and minimal deformation of soft materials, such as coprolites. Additional evidence includes the pronounced biofacies boundary between the Braidwood and the Essex biota as well as the unsuccessful upward escape burrows that accompany many bivalve and worm fossils (Figure 10.2 [microfacies D]). Preservational quality generally decreases southwestward within the Mazon Creek area, reflecting an increased taphonomic overprint attributed to slower sedimentation rates, increased bioturbation, and greater aerobic decay within an increasingly distal, open-marine environment (Baird, Shabica, et al. 1985).

Baird et al. (1986) focused on two sedimentary microfacies within the basal portion of the Francis Creek Shale where high sedimentation rates led to rapid burial and concretions are most common. The first microfacies is composed of large-scale mudstone foresets adjacent to distributary channels. These foresets are approximately 10 to 12 m in vertical thickness, are inclined up to 20°, and contain climbing ripples, dewatering features, and slump structures—all features supporting episodic, torrential sedimentation. Further evidence for rapid foreset sedimentation is the occasional preservation of complete lycopod trunks in life position and individual leaves that crosscut inclined bedding planes (Baird, Shabica, et al. 1985). These episodic pulses of sedimentation, likely related to river flooding or distributary switching, rapidly buried transported remains and smothered benthic organisms *in situ.*

The second microfacies is the laminated distal mudstone of the prodelta (Figure 10.2 [microfacies B]). The laminations consist of alternating thin (0.1 to 2 mm) and thick (0.5 to 10 mm) layers of siltstone, each separated by a very thin clay parting. Broadhurst (1985) interpreted these alternating laminae to reflect flood and ebb tides, respectively, and the intercalated clays to represent finer-grained sedimentation during intervening slack-water intervals. Assuming that the original layers consisted of greater than 70 percent water, as in modern deltaic environments, pre-dewatering accumulation rates in these subtidal depths would have been 3 to 4 mm per day (Baird et al. 1986). This substantial, but highly regular, prodelta sedimentation provided for rapid burial of transported and native remains.

Following their burial in these environments, the remains of Mazon Creek organisms were relatively rapidly entombed by siderite precipitation. Pore waters surrounding the newly deposited, organic-rich sediments were relatively fresh due to seaward and upward groundwater flow from the adjacent coastal aquifer (Woodland and Stenstrom 1979). This freshwater flow, together with high sedimentation rates and minimal bioturbation, limited infiltration of seawater rich in oxygen and sulfate, thereby minimizing aerobic oxidation as well as sulfate reduction and subsequent pyrite production. Instead, organic detritus was broken down largely by bacterial methanogenesis, which increased pore-water concentrations of ammonia (increasing pH) and bicarbonate (increasing ΣCO_2) to levels favoring interstitial precipitation of siderite around buried organisms (Berner 1971; Woodland and Stenstrom 1979). Pyrite halos are occasionally present and likely precipitated very early as sulfide was released by initial decay of the entombed organism. Overall rapid formation of the concretions is supported by preservation of organisms that quickly decay or deform, such as jellyfish, and by precipitated cements that form up to 80 percent of concretion volume, indicating formation when sediments were still extremely water-rich. This general scenario is consistent with the predominance of siderite concretions in modern freshwater to brackish water deposits, where high iron:calcium ratios further favor precipitation of siderite over calcium carbonate (Curtis and Spears 1968; Pye et al. 1990).

Although Mazon Creek preservation is exceptional, it is not a complete record of local biotic diversity; taphonomic biases do exist. The dearth of larger invertebrates, fishes, and amphibians may relate to their greater ability to escape episodic deluges of sediment. The low abundance of these larger organisms, together with relatively small organisms such as ostracodes, may also reflect maximum and minimum limits on the volume around which siderite concretions could form. The scarcity of small organisms is also an artifact of sampling; many nodules fail to break along the surface containing minute fossils (Baird et al. 1986).

PALEOECOLOGY AND PALEOBIOLOGY

Many paleoecological insights into the Mazon Creek Lagerstätte are based on an extensive fossil census study carried out by Baird, Shabica, et al. (1985). The spatial distributions of 150 fossil taxa and trace fossils were obtained from analyzing over 285,000 concretions from more than 270 sites. Analysis of this census data quantitatively reinforced and refined the observations of Richardson and Johnson (1971), who divided the biota into almost mutually exclusive assemblages: the Braidwood (plants and terrestrial to freshwater animals) and the Essex (predominately soft-bodied marine invertebrates). Mixing between the assemblages appears unidirectional: Braidwood plants, coprolites, and animal body fossils are occasionally found within the Essex assemblage. However, the abundance and preservation of Braidwood-type fossils decreases seaward with increasing transport from their native habitat through riverine flow.

The Braidwood biota includes over 200 animal taxa, but is volumetrically dominated (>90 percent of specimens) by more than 250 plant taxa, including pteridosperm tree ferns, arborescent lycopods, and horsetails (Darrah 1970; Peppers and Pfefferkorn 1970). The terrestrial fauna consists largely of insects, arachnids, millipedes, and myriapods (Figures 10.4 and 10.5), while the freshwater fauna includes bivalves, syncarid shrimps, lungfish scales, shark teeth, insect larvae, coprolites, paleoniscoid fish, and amphibians. One of the rarer, but more intriguing, members of the Braidwood fauna is *Euproops danae,* a xiphosuran (horseshoe crab) (Figures 10.6 and 10.7). Xiphosurans are generally considered exclusively aquatic animals; however, Fisher (1979) provides numerous independent lines of evidence favoring extensive subaerial activity in *Euproops,* probably within moist leaf litter or floating mats of vegetation. Evidence for this life mode in *Euproops* includes (1) its nearly exclusive restriction to the

FIGURE 10.4 Ventral view of *Bicarinitarbus* sp., a phalangiotarbid arachnid. Specimen is approximately 18 mm in length. (Courtesy of National Museum of Natural History, Smithsonian Institution [NMNH 3841150])

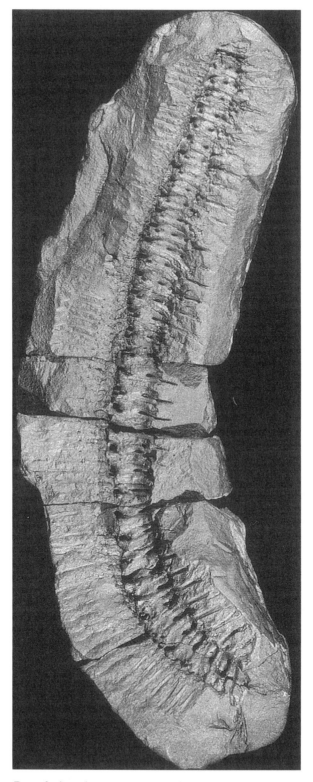

FIGURE 10.5 Dorsal view (*anterior to left*) of *Myriacantherpestes hystricosus*, a euphoberiid millipede, with left-lateral spines prominently preserved along upper portion of specimen and left series of legs folded under body. The two series of dark pits parallel along the body are molds of dorsal paramedian spines. Specimen is approximately 200 mm in length. (Courtesy of National Museum of Natural History, Smithsonian Institution [Holotype, USNM 38038])

FIGURE 10.6 Dorsal view of *Euproops danae,* a xiphosuran crustacean. Specimen is approximately 14 mm in length. (Photo courtesy of D. Fisher, University of Michigan)

Braidwood assemblage, as Essex assemblage specimens are poorly preserved and consist largely of disarticulated tagmata, reflecting postmortem transport; (2) its preservation within coprolitic material containing exclusively terrestrial organisms (e.g., scorpions, millipedes), indicating a land-based predator; (3) its appendage-functional morphology, consistent with movement along plant stems and branches; and (4) the striking visual similarity between *Euproops* spines and lycopod foliage, interpreted as a camouflage adaptation (Figure 10.7).

Compared with the Braidwood biota, the Essex biota is dominated by marginal marine organisms that generally lacked heavily calcified skeletons (Johnson and Richardson 1969; Baird et al. 1986). The biota includes hydrozoans, larval fish, sea anemones, cephalopods, xiphosurids, echiurans, chaetognaths, priapulids, nemerteans, polychaetes (Figure 10.8), and polyplacophorans (Figures 10.9 and 10.10) (for a broad multiauthor review of taxa, see Nitecki 1979b; Shabica and Hay 1997). Significantly, many taxa considered typical of late Paleozoic open-marine environments (e.g., crinoids, sponges, foraminifera, brachiopods, cephalopods) are absent to extremely rare in Mazon Creek

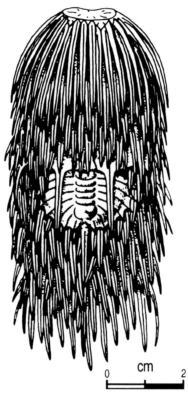

cm

0   2

FIGURE 10.7 Interpretive drawing of the xiphosuran *Euproops danae* hiding or foraging in lycopod litter. (Modified from Fisher 1979)

FIGURE 10.8 Dorsal view of *Palaeocampa anthrax*, a polycheate annelid of uncertain taxonomic affinity. Ten body segments are outlined through radiating articulated setae. Specimen is approximately 40 mm in length. (Courtesy of National Museum of Natural History, Smithsonian Institution [Holotype, USNM 38032])

FIGURE 10.9 Dorsal view of *Glaphurochiton concinnus*, a polyplacophoran mollusc with all eight valves articulated. Specimen is approximately 28 mm in length. (Photo by E. Richardson, Field Museum of Natural History [FMNH PE31878])

FIGURE 10.10 Tail plate impression of *Glaphurochiton concinnus* showing spiculate girdle margin; apparent variation in spicule length is a result of individual spicules' orientation to the plane of breakage. Photo field of view is 2 mm. (Photo by E. Richardson, Field Museum of Natural History [FMNH PE31946])

concretions, a pattern consistent with a deltaic paleoenvironment subject to large fluctuations in, or generally low, salinity.

Perhaps the most interesting and frustrating to systematists, however, are those organisms not easily placed within the established phylogeny of life. An excellent example is *Tullimonstrum gregarium*, first described by Richardson (1966) from a specimen collected in Pit Eleven by Francis Tully (*T. gregarium* translates as "abundant monster of Tully") (Figure 10.11). Ranging from 3 to 35 cm in length, the creature had a proboscis with a terminal jaw or claw, an anterior rod with terminal visual apparatuses, a dorsoventrally flattened trunk with transverse segments, and a tail with flexible lateral fins. Based on its stalked eyes, distinct proboscis, transverse muscle bands, and within-tail intestine, Foster (1979) compared *T. gregarium* with living bathypelagic nemerteans, pelagic polychaetes, and heteropod gastropods. In contrast to these groups, however, *T. gregarium* lacks segmentation, paired appendages, swimming fins, obvious statocysts, and a distinct visceral mass. Thus, Foster (1979) tentatively interpreted *T. gregarium* as an early prosobranch gastropod that actively fed on plankton (perhaps exclusively jellyfish), a niche later exploited by heteropods. More recently, Beall (1991) reinterpreted the posterior fins to be dorsoventral with a slight asymmetry and expanded the possible functions of the transverse segments to include copulatory assistance, highly modified sensory setae, or substrate support structures. Beall (1991) also explored the phylogenetic position of *T. gregarium* using cladistic analyses, which supported both a close affinity with the Conodonta and Foster's (1979) phylogenetic interpretations. Thus, the position of *Tullimonstrum*'s phylogenetic twig on the "tree of life" remains enigmatic. It may well be a life-form quite disparate from anything known.

Other soft-bodied organisms whose natures remain enigmatic include *Etacystis communis* (Figure 10.12a) and *Escumasia roryi* (Figure 10.12b). *Etacystis communis* averages 9.5 cm in diameter and is shaped like an uppercase "H" with a sac extending from the crossbar (Nitecki

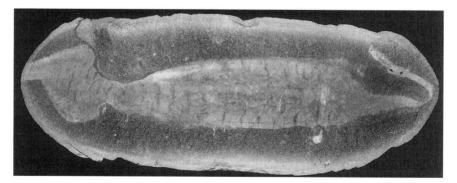

FIGURE 10.11 *Tullimonstrum gregarium* in concretion. Specimen is 6 cm in length. (Photo courtesy of A. Hay, Northeastern Illinois University [NEIU MCP 418])

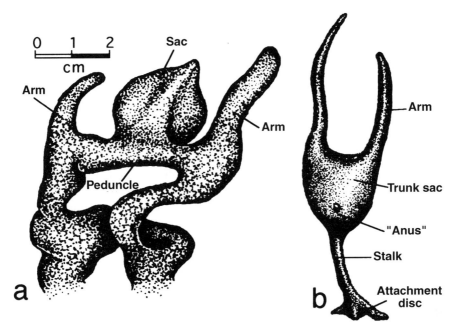

FIGURE 10.12 (a) Composite reconstruction of *Etacystis communis*, a problematic fossil form of possible hemichordate affinity, approximately 35 mm in height (Modified from Nitecki and Schram 1976 and Foster 1979, after Sroka 1997); (b) composite reconstruction of *Escumasia roryi*, approximately 85 mm in height. (Modified from Nitecki and Solem 1973, after Sroka and Richardson 1997).

and Schram 1976). On many specimens, one bar of the H exhibits a regularly patterned papillose texture, interpreted as suckers or clumps of nematocysts. The other bar is interpreted as the portion of a stolon-like organ connecting numerous individuals, suggesting a colonial life with clonal or asexual reproduction. A small opening at the stalk–sac junction is interpreted as a mouth and/or an anus by Nitecki and Schram (1976), who classified *Etacystis* as a passive filter-feeding echiuroid or perhaps an extinct hemichordate. *Escumasia roryi* ranges from 7.5 to 20.5 cm in length and is bilaterally symmetrical, with a smooth body surface (Figure 10.12b). At the base of its long, slender arms is a slit interpreted as a mouth, while a large pore on one side of the main trunk sac is interpreted as an anal opening (Nitecki and Solem 1973). The trunk sac is attached to a circular basal disk by a slender featureless stalk. Nitecki and Solem (1973) propose *Escumasia* as a dead end in cnidarian evolution, with large prey taken in by the mouth and solid waste excreted from the anal opening.

CONCLUSIONS

The Mazon Creek Lagerstätte is a truly exceptional window into late Paleozoic coastal paleoenvironments, paleoecology, and paleobiology. Similar, although less studied, locales with Braidwood- and Essex-type fossils are known from putative stratigraphic equivalents of the Illinois Basin in Missouri, Oklahoma, and other regions of Illinois (Baird, Sroka, et al. 1985). In examining these locales and the Mississippian Glencartholm and Bear Gulch (Chapter 9) Lagerstätten, Schram (1979) concluded that their similar faunal composition and paleoenvironment reflect a relatively stable ecological association throughout the Carboniferous Period. Furthermore, the presence of Essex-type fauna in the medial Triassic Grès à Voltzia Lagerstätte (Chapter 11) implies that this ecological assemblage existed from the late Paleozoic into the early Mesozoic (Baird, Sroka, et al. 1985).

Much is known about Mazon Creek's ancient environment and the life, death, and preservation of its myriad organisms, but many questions remain. These range from the relative importance of sea-level fluctuations versus delta switching in the formation of the Mazon Creek deposits, to the exact geochemical processes responsible for concretion formation and exquisite soft-bodied preservation. The latter can be addressed by an improved understanding of where and how siderite concretions form in modern coastal, deltaic, and estuarine environments. Appreciating such biases in preservation, both environmental and biological, will help elucidate how completely late Paleozoic coastal communities are preserved through this exceptional window into past life.

ACKNOWLEDGMENTS

I am grateful to M. Nitecki for reviewing an earlier version of this chapter, and thank E. Yochelson for discussion of certain aspects of the fauna.

REFERENCES

Baird, G. C. 1979. Lithology and fossil distributions, Francis Creek Shale in northeastern Illinois. In M. H. Nitecki, ed., *Mazon Creek Fossils*, pp. 41–67. New York: Academic Press.

Baird, G. C., and C. W. Shabica. 1980. The Mazon Creek depositional event: Examination of Francis Creek and analogous facies in the midcontinent region. In R. L. Langeheim and C. J. Mann, eds., *Middle and Late Pennsylvanian Strata on Margin of Illinois Basin* [guidebook], pp. 79–92. Danville, Ill.: Society of Economic Paleontologists and Mineralogists Great Lakes Section.

Baird, G. C., C. W. Shabica, J. L. Anderson, and E. S. Richardson, Jr. 1985. Biota of a Pennsylvanian muddy coast: Habitat within the Mazonian Delta Complex, northeastern Illinois. *Journal of Paleontology* 59:253–281.

Baird, G. C., and S. D. Sroka. 1990. Geology and geohistory of Mazon Creek area fossil localities, Illinois. In W. Hammer and D. F. Hess, eds., *Geology Field Guidebook: Current Perspectives on Illinois Basin and Mississippi Arch Geology*, pp. C1–C70. Macomb: Western Illinois University.

Baird, G. C., S. D. Sroka, C. W. Shabica, and T. L. Beard. 1985. Mazon Creek-type fossil assemblages in the U.S. midcontinent Pennsylvanian: Their recurrent character and palaeo-environmental significance. In H. B. Whittington and S. Conway Morris, eds., *Extraordinary Fossil Biotas: Their Ecological and Evolutionary Significance,* pp. 87–98. London: Royal Society of London.

Baird, G. C., S. D. Sroka, C. W. Shabica, and G. J. Kuecher. 1986. Taphonomy of middle Pennsylvanian Mazon Creek area fossil localities, northeast Illinois: Significance of exceptional fossil preservation in syngenetic concretions. *Palaios* 1:271–285.

Beall, B. S. 1991. The Tully Monster and a new approach to analyzing problematica. In A. M. Simonetta and S. Conway Morris, eds., *The Early Evolution of Metazoa and the Significance of Problematic Taxa,* pp. 271–285. New York: Cambridge University Press.

Berner, R. A. 1971. *Principles of Chemical Sedimentology.* New York: McGraw-Hill.

Broadhurst, F. M. 1985. Discussion addendum to Baird et al. In H. B. Whittington and S. Conway Morris, eds., *Extraordinary Fossil Biotas: Their Ecological and Evolutionary Significance,* pp. 98–99. London: Royal Society of London.

Curtis, C. D., and D. A. Spears. 1968. The formation of sedimentary iron minerals. *Economic Geology* 63:257–270.

Darrah, W. C. 1970. *A Critical Review of the Upper Pennsylvanian Floras of the Eastern United States with Notes on the Mazon Creek Flora of Illinois.* Gettysburg, Pa.: Gettysburg Press.

Fisher, D. C. 1979. Evidence for subaerial activity of *Euproops danae* (Merostomata, Xiphosurida). In M. H. Nitecki, ed., *Mazon Creek Fossils,* pp. 379–447. New York: Academic Press.

Foster, M. W. 1979. A reappraisal of *Tullimonstrum.* In M. H. Nitecki, ed., *Mazon Creek Fossils,* pp. 269–302. New York: Academic Press.

Heckel, P. H. 1986. Sea-level curve for Pennsylvanian eustatic marine transgressive–regressive depositional cycles along mid-continent outcrop belt, North America. *Geology* 14:330–334.

Horowitz, A. S. 1979. The Mazon Creek flora: Review of research and bibliography. In M. H. Nitecki, ed., *Mazon Creek Fossils,* pp. 143–158. New York: Academic Press.

Johnson, R. G., and E. S. Richardson. 1969. Pennsylvanian invertebrates of the Mazon Creek area, Illinois: The morphology and affinities of *Tullimonstrum.* *Fieldiana: Geology* 12:119–149.

Klein, G., and J. B. Kupperman. 1992. Pennsylvanian cyclothems: Methods of distinguishing tectonically induced changes in sea level from climatically induced changes. *Geological Society of America Bulletin* 104:166–175.

Klein, G., and D. A. Willard. 1989. Origin of the Pennsylvanian coal-bearing cyclothems of North America. *Geology* 17:152–155.

Ledvina, C. T. 1997. History of coal mining in northeastern Illinois. In C. W. Shabica and A. A. Hay, eds., *Richardson's Guide to the Fossil Fauna of Mazon Creek,* pp. 3–16. Chicago: Northeastern Illinois Press.

Nitecki, M. H. 1979a. Mazon Creek fauna and flora: A hundred years of investigation. In M. H. Nitecki, ed., *Mazon Creek Fossils,* p. 1–11. New York: Academic Press.

Nitecki, M. H., ed. 1979b. *Mazon Creek Fossils.* New York: Academic Press.

Nitecki, M. H., and F. R. Schram. 1976. *Etacystis communis,* a fossil of uncertain affinities from the Mazon Creek fauna (Pennsylvanian of Illinois). *Journal of Paleontology* 50:1157–1161.

Nitecki, M. H., and A. Solem. 1973. A problematic organism from the Mazon Creek (Pennsylvanian) of Illinois. *Journal of Paleontology* 47:903–907.

Peppers, R. A., and H. W. Pfefferkorn. 1970. A comparison of the floras of the Colchester (No. 2) Coal and Francis Creek Shale. In W. H. Smith, N. B. Nance, M. E. Hopkins, R. G. Johnson, and C. W. Shabica, eds., *Depositional Environments in Parts of the Carbondale Formation—Western and Northern Illinois,* pp. 61–72. Guidebook Series / Geological Survey of America, no. 8. Urbana: Illinois State Geological Survey.

Pfefferkorn, H. W. 1979. High diversity and stratigraphic age of the Mazon Creek flora. In M. H. Nitecki, ed., *Mazon Creek Fossils,* pp. 129–142. New York: Academic Press.

Phillips, T. L., and W. A. Dimichele. 1981. Paleoecology of middle Pennsylvanian age coal swamps in southern Illinois/Herrin Coal Member at Sahara Mine No. 6. In K. J. Niklas, ed., *Paleobotany, Paleoecology, and Evolution,* pp. 231–284. New York: Praeger.

Phillips, T. L., R. A. Peppers, and W. A. Dimichele. 1985. Stratigraphic and interregional changes in Pennsylvanian coal-swamp vegetation: Environmental inferences. *International Journal of Coal Geology* 5:43–109.

Potter, P. E., and W. A. Pryor. 1961. Dispersal centers of Paleozoic and later clastics of the Upper Mississippi Valley and adjacent areas. *Geological Society of America Bulletin* 71:1195–1250.

Pye, K., J. A. D. Dickson, N. Schiavon, M. L. Coleman, and M. Cox. 1990. Formation of siderite-Mg-calcite-iron sulphide concretions in intertidal marsh and sandflat sediments, North Norfolk, England. *Sedimentology* 37:325–343.

Richardson, E. S., Jr. 1956. Pennsylvanian invertebrates of the Mazon Creek area, Illinois. *Fieldiana: Geology* 12:3–76.

Richardson, E. S., Jr. 1966. Wormlike creature from the Pennsylvanian of Illinois. *Science* 151:75–76.

Richardson, E. S., Jr., and R. G. Johnson. 1971. The Mazon Creek faunas. In E. Yochelson, ed., *Proceedings of the North American Paleontological Convention,* vol. 2, pp. 1222–1235. Lawrence, Kans.: Allen Press.

Rowley, D. M., A. Raymond, J. T. Parrish, A. L. Lottes, C. R. Scotese, and A. M. Ziegler. 1985. Carboniferous paleogeography, phytogeography, and paleoclimate reconstructions. *International Journal of Coal Geology* 5:17–42.

Schopf, J. M. 1979. Pennsylvanian paleoclimate in the Illinois Basin. In J. E. Palmer and R. R. Dutcher, eds., *Deposition and Structural History of the Pennsylvanian System of the Illinois Basin.* Part 2, *Invited Papers,* pp. 1–13. Guidebook Series / Geological Survey of America, no. 15. Urbana: Illinois State Geological Survey.

Schram, F. R. 1970. Isopod from the Pennsylvanian of Illinois. *Science* 169:854–855.

Schram, F. R. 1973. Pseudosoelomates and a nemertine from the Illinois Pennsylvanian. *Journal of Paleontology* 47:985–989.

Schram, F. R. 1979. The Mazon Creek biotas in the context of a Carboniferous faunal continuum. In M. H. Nitecki, ed., *Mazon Creek Fossils,* pp. 159–190. New York: Academic Press.

Seilacher, A., W. E. Reif, and F. Westphal. 1985. Sedimentological, ecological, and temporal patterns of fossil Lagerstätten. *Philosophical Transactions of the Royal Society of London, B* 311:5–23.

Shabica, C. W. 1979. Pennsylvanian sedimentation in northern Illinois: Examination of delta models. In M. H. Nitecki, ed., *Mazon Creek Fossils*, pp. 13–40. New York: Academic Press.

Shabica, C. W., and A. A. Hay, eds. 1997. *Richardson's Guide to the Fossil Fauna of Mazon Creek*. Chicago: Northeastern Illinois Press.

Smith, W. H., R. B. Nance, M. E. Hopkins, R. G. Johnson, and C. W. Shabica. 1970. Depositional environments in parts of the Carbondale Formation—Western and northern Illinois [field trip road log]. In W. H. Smith, N. B. Nance, M. E. Hopkins, R. G. Johnson, and C. W. Shabica, eds., *Depositional Environments in Parts of the Carbondale Formation—Western and Northern Illinois*, pp. 1–300. Guidebook Series / Geological Survey of America, no. 8. Urbana: Illinois State Geological Survey.

Taylor, T. N. 1981. *Paleobotany: An Introduction to Fossil Plant Biology*. New York: McGraw-Hill.

Wanless, H. R. 1975. Distribution of Pennsylvanian coal in the United States. *United States Geological Survey Professional Paper* 853:33–47.

Wanless, H. R., and J. M. Weller. 1932. Correlation and extent of Pennsylvanian cyclothems. *Geological Society of America Bulletin* 43:1003–1016.

Willman, H. B., and J. N. Payne. 1942. Geology and mineralogy of the Marseilles, Ottawa and Streater Quadrangles. *Illinois Geological Survey Bulletin* 66:1–388.

Winston, R. B. 1986. Characteristic features and compaction of plant tissues traced from permineralized peat to coal in Pennsylvanian coals (Desmoinesian) from the Illinois Basin. *International Journal of Coal Geology* 6:21–41.

Woodland, B. G., and R. C. Stenstrom. 1979. The occurrence and origin of siderite concretions in the Francis Creek Shale (Pennsylvanian) of northeastern Illinois. In M. H. Nitecki, ed., *Mazon Creek Fossils*, pp. 69–104. New York: Academic Press.

Wright, C. R. 1979. Depositional history of the Pennsylvanian System in the Illinois Basin—A summary of work by Dr. Harold R. Wanless and associates. In J. E. Palmer and R. R. Dutcher, eds., *Deposition and Structural History of the Pennsylvanian System of the Illinois Basin*. Part 2, *Invited Papers*, pp. 21–27. Guidebook Series / Geological Survey of America, no. 15. Urbana: Illinois State Geological Survey.

Zangerl, R. 1980. The Pennsylvanian black, carbonaceous, sheety shales of the Illinois and Forest City Basins. In R. L. Langeheim and C. J. Mann, eds., *Middle and Late Pennsylvanian Strata on Margin of Illinois Basin* [guidebook], pp. 225–238. Danville, Ill.: Society of Economic Paleontologists and Mineralogists Great Lakes Section.

Zangerl, R., and E. S. Richardson. 1963. *The Paleoecological History of Two Pennsylvanian Black Shales*. Fieldiana Geology Memoir, vol. 4. Chicago: Chicago Natural History Museum.

Zeigler, A. M., C. R. Scotese, W. S. McKerrow, M. E. Johnson, and R. K. Bambach. 1979. Paleozoic paleogeography. *Annual Review of Earth and Planetary Sciences* 7:473–502.

11
Grès à Voltzia: Preservation in Early Mesozoic Deltaic and Marginal Marine Environments

Walter Etter

THE GRÈS À VOLTZIA (VOLTZIA SANDSTONE) OF NORTH-eastern France, named after the abundant remains of the conifer *Voltzia heterophylla*, has become famous as a Lagerstätte in which a large number of invertebrates–including jellyfish, annelid worms, spiders, insects, and crustaceans–exhibit a high degree of soft-bodied preservation. The Grès à Voltzia is the uppermost formation of the Buntsandstein (Triassic) of continental Europe, and its depositional environment records the gradual change of a deltaic setting to a marginal marine environment (Gall 1971, 1985; Gall, Grauvogel-Stamm, and Papier 1995). The lower part of the Grès à Voltzia, the Grès à meules, represents mainly deltaic environments, whereas the upper Grès à Voltzia, the Grès argileux, marks the beginning of the marine Muschelkalk transgression (Gall 1985; Gall, Grauvogel-Stamm, and Papier 1995). Soft-part preservation of fossils is restricted to certain layers within the Grès à meules, and this chapter, therefore, concentrates on the lower Grès à Voltzia.

The complex depositional environment of the Grès à Voltzia is expressed in several distinct sedimentary facies, each with its characteristic floral and faunal associations (Gall and Grauvogel-Stamm 1984; Gall 1985) These include (1) large sandstone lenses containing poorly preserved plant remains and bones of stegocephalid amphibians and representing rapid accumulation of deposits in fluvial channels; (2) thin intercalated lenses of green silty claystone deposited in stagnant waters of abandoned river channels, temporary ponds, and pools; and (3) carbonate sediments, now represented mainly by brecciated dolomitic intercalations, containing a restricted marine fauna and originating in the coastal littoral mudflats bordering the delta.

It is only in the silty-clay layers of the Grès à meules that the fossils show extraordinary preservation (Gall 1971, 1985; Grauvogel-Stamm 1978; Gall and Grauvogel-Stamm 1984; Gall, Grauvogel-Stamm, and Papier 1995). The aquatic fauna was probably killed by depletion of oxygen in stagnant pools, associated with a decrease in the size of these pools through evaporation (Briggs and Gall 1990). The assemblages are preserved *in situ* at the base of thin, graded laminae, and this Lagerstätte therefore represents a mixture of stagnation and obrution deposits. Furthermore, sealing by microbial mats may have been an important factor in fossil preservation (Gall 1990). Hence, it is not possible to clearly assign the Grès à Voltzia to one of the major conservation regimes of Seilacher, Reif, and Westphal (1985).

The Grès à Voltzia has a long history of intensive research. Systematic collection of fossils has been done for more than 50 years by Louis Grauvogel (1947a, 1947b, 1947c) and for 30 years by Jean-Claude Gall from the Louis Pasteur University of Strasbourg, France. Moreover, since 1965, Léa Grauvogel-Stamm, Louis Grauvogel's daughter, has participated in their study. Therefore, a large amount of data has been accumulated over the years, and the Grès à Voltzia is a well-investigated Lagerstätte, in terms of paleoecology, sedimentology, and taphonomy. Because of the remarkable state of preservation, this deposit probably allows the most detailed insight into a marginal continental ecosystem of Triassic times.

GEOLOGICAL CONTEXT

The Buntsandstein, the lower part of the tripartite Trias of continental Europe, consists mainly of red, gray, and green sandstones and siltstones deposited under more or less semiarid climates in vast alluvial plains. In eastern France and southwestern Germany, the Grès à Voltzia is the upper formation of the Buntsandstein, and its age has been determined as Anisian (earliest Middle Triassic) (Briggs and Gall 1990). Unlike the lower Buntsandstein, the Grès à Voltzia was deposited in a deltaic setting bordering the sea, and it actually records the transition from the fluvial sedimentary environment of the Buntsandstein to the marine environment of the Middle Triassic Muschelkalk. Although this formation is quite widespread in eastern France and southwestern Germany, it is only in the northern Vosges region of eastern France that the Grès à Voltzia has yielded spectacular fossils. In this region, the sandstones of the Grès à Voltzia are still quarried and used for construction (formerly for mill-

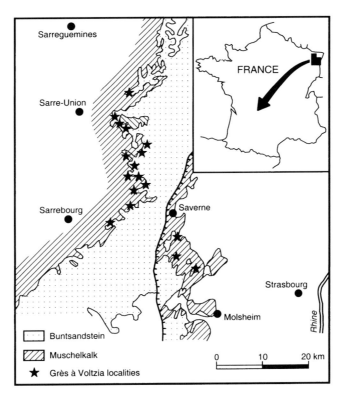

FIGURE 11.1 Northern Vosges and geographic location of the most important out-crops in the Grès à Voltzia. (Modified from Gall 1985)

stones, French *meules*). The principal outcrops are located 30 to 40 km northwest of the city of Strasbourg, just west of the Rhine Graben (Gall 1985) (Figure 11.1).

In the northern Vosges, the Grès à Voltzia is approximately 20 m thick (Gall 1971, 1985). Its lower part, the Grès à meules, which has yielded the exceptionally preserved fossils, is 5 to 13 m thick (average 12 m). The Grès à meules consists of three distinct facies: the sandstone facies, the silt-clay facies, and the carbonate intercalation facies (Figure 11.2).

Sandstone Facies

Two types of sandstones have been distinguished in the Grès à meules (Gall 1971, 1985). Pure, fine-grained gray, red, or variously colored sandstones, forming lenticular bodies with a diameter of several tens of meters, are built up of subhorizontal laminae a few millimeters thick. This type of sandstone is the most common lithology of the Grès à meules. The bases of the large lenses show various flute and groove casts, whereas the upper surfaces are covered by current ripple marks. These fine, pure sandstones do not contain any fossils except for some carbonate inter-

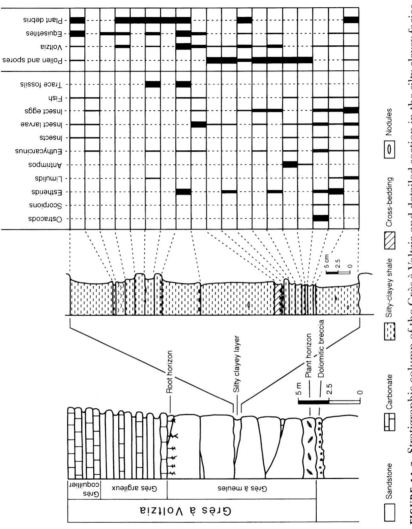

FIGURE 11.2 Stratigraphic column of the Grès à Voltzia and detailed section in the silty-clayey facies with indication of the fossil content in different layers. Note the general impoverishment of the fauna toward the top of the section. (Modified from Gall 1971)

calations where marine faunal elements (foraminifers, molluscs) may be preserved (Gall 1985).

Coarser gray sandstones are present as lenses of a few meters in diameter. They show an erosive base covered with flat pebbles reworked from the underlying substratum. These smaller sandstone lenses are rich in floral remains, which are preserved mainly as wood fragments but occasionally as intact remains (Gall and Grauvogel-Stamm 1984). The coarser plant-bearing sandstones contain disarticulated and fragmented bones of stegocephalid amphibians (Gall and Grauvogel-Stamm 1984), although only one complete skull has been found (Kamphausen and Morales 1981).

Silt-Clay Facies

Green and, rarely, red silty-clayey lenses up to several decimeters thick are intercalated between the sandstones (Figure 11.2). They are either homogeneous in appearance and devoid of fossils or finely laminated, consisting of graded laminae a few millimeters thick. It is in these laminated fine-grained deposits that the remarkably well preserved fossils occur (Gall 1971, 1985; Gall and Grauvogel-Stamm 1984; Gall, Grauvogel-Stamm, and Papier 1995). The fossils include terrestrial animals (scorpions, spiders, myriapods, insects), continental plants (horsetails, ferns, conifers), and aquatic animals (jellyfish, lingulid brachiopods, bivalves, annelids, limulids, crustaceans, fishes). In general, the silty-clayey lenses show a progressive impoverishment in species numbers from the bottom toward the upper layers. The tops of the silty-clayey lenses commonly show desiccation cracks and plant-root traces (Gall 1971, 1985).

Carbonate Intercalation Facies

Carbonate intercalations are present most frequently as intraformational breccias and, rarely, as autochthonous lenticular banks of dolomitic sandstones, calcareous sandstones, or sandy dolomites. These carbonate deposits, which are on average only a few centimeters thick (Figure 11.2), contain a restricted to fully marine fauna (Gall and Grauvogel-Stamm 1984; Gall 1985). Occasionally, carbonate clasts up to 60 cm in diameter have been observed (Gall and Grauvogel-Stamm 1984; Gall 1985).

Termination of the Deltaic Facies

The uppermost Grès à meules contains many plant roots *in situ* and is overlain by the Grès argileux. In the Grès argileux, the individual sandstones and silty-clayey beds are thinner than in the Grès à meules and have a larger horizontal extension. Toward the top of the Grès à meules, calcareous sandstones become increasingly frequent, and the fauna becomes fully marine, with an increasing proportion of stenohaline taxa (e.g., echinoderms, cephalopods) (Gall 1985).

Paleoenvironmental Setting

The different facies occurring in the lower Grès à Voltzia and their close interfingering clearly indicate deposition in a complex deltaic setting bordering the sea. Furthermore, the evolution of the facies testifies to the gradual development of an alluvial plain into a deltaic environment, which finally was drowned by the transgressing Muschelkalk Sea (Gall 1985; Gall, Grauvogel-Stamm, and Papier 1995) (Figure 11.3).

The sandstones, predominant in the lower Grès à Voltzia, represent the channel facies of the delta (Figure 11.3). The finer, well-sorted pure sandstones formed as channel bars during normal fluvial sedimentation (Gall 1971, 1985; Gall and Grauvogel-Stamm 1984) or, as is indicated by the carbonate intercalations containing a marine fauna, as stream mouth bars in the vicinity of the sea (Gall 1985). The coarser plant-bearing sandstones, though, are the result of sudden ruptures of riverbanks during high flood stages (Gall and Grauvogel-Stamm 1984; Gall 1985).

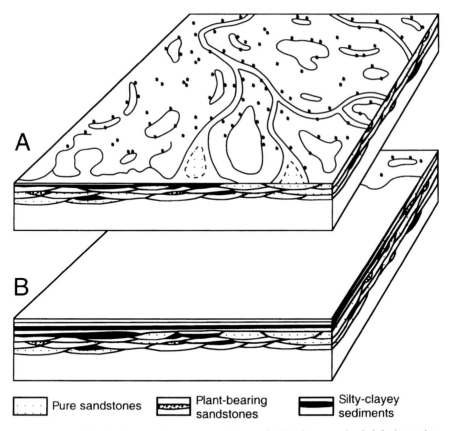

| | Pure sandstones | | Plant-bearing sandstones | | Silty-clayey sediments |

FIGURE 11.3 Block diagram showing evolution of (A) the marginal deltaic setting in the Grès à meules to (B) the shallow sea in the Grès argileux. (Modified from Gall 1985)

Plants growing on the banks bordering the river channels were buried under a thick layer of sand and may be preserved as large intact remains. However, if the plants were transported farther downstream, the sandstones would contain only fragmented plant remains (Gall and Grauvogel-Stamm 1984).

Fine-grained sediments were deposited between the river channels in temporary ponds and pools in the overbank plain (interdistributary plain) and subordinately in abandoned channels (Gall 1971, 1985) (Figure 11.3). Homogeneous lenses formed by slow and continuous settling of the suspension-load in standing waters (Gall 1985). In the finely laminated lenses where the fossils occur, the individual laminae are usually several millimeters thick and grain size is graded (Gall 1990). The sediment was introduced by overspilling of floodwater from the river channels or as a result of very strong tides. The presence of a brackish water fauna indicates that salinity in these temporary ponds and pools was elevated. Salinity could have even reached levels of supersaturation, as is indicated by the occurrence of salt casts and the high boron content of the clay minerals (Gall 1971, 1985). The individual ponds and pools probably existed for only a short time, estimated as a few weeks to several seasons (Gall 1985). The tops of the fine-grained lenses usually show signs of emergence, such as mud cracks and plant-root traces (Gall 1971, 1985).

The carbonate intercalations in the Grès à meules are the result of marine incursions (Gall 1971, 1985; Gall and Grauvogel-Stamm 1984). They are present either as intraformational breccias or, rarely, as autochthonous lenticular banks of dolomitic or calcareous sandstones. The carbonate beds originated in the distal domain of the delta plain near the shoreline along the sea, either on littoral mudflats or in brackish to euhaline lagoonal ponds (Gall 1985). During exceptional storms, carbonate clasts were eroded from the mudflats and transported far into the delta proper, where they accumulated as breccias (Gall and Grauvogel-Stamm 1984; Gall 1985).

Taphonomy

Because the fossil preservation in the sandstones and the carbonate intercalations does not require any special taphonomic circumstances, only the taphonomy of the fossils in the laminated silty-clayey lenses is discussed here. In these deposits, various groups of organisms occur (Gall 1971, 1985; Gall and Grauvogel-Stamm 1984). The plant remains (pteridophytes: horsetails and ferns, gymnosperms) are very well preserved, sometimes with anatomical structures preserved under a crust of iron oxides (Grauvogel-Stamm 1978; Gall and Grauvogel-Stamm 1984). The quality and abundance of certain plants (*Aethophyllum, Voltzia*) have even permitted the reconstruction of their entire life cycle, from seed to adults with preserved reproductive organs (Grauvogel-Stamm 1978; Gall and Grauvogel-Stamm 1984).

The aquatic fauna is represented by medusae, lingulid brachiopods, bivalves, annelids, limulids, crustaceans, and fish. The terrestrial fauna includes scorpions, spiders, myriapods, and insects. Some of these fossils show a remarkable state of preservation, with soft parts of animals preserved (Gall 1971, 1990; Gall and Grauvogel-Stamm 1984).

The fully articulated animals show minimal signs of decay, and their soft parts are preserved as organic material or as a thin crust of iron oxides (Gall 1990). Occasionally, what appear to be the initial stages of sideritic concretion formation can be observed around fossils (Gall and Grauvogel-Stamm 1984). However, all the calcareous shell material is dissolved, and only the thin periostracum is still present (Gall 1971, 1983). Most spectacular is the preservation of the medusae *Progonionemus,* whose gonads and stinging cells are visible (Grauvogel and Gall 1962; Gall 1990). Other examples of exceptional preservation include polychaetes with fossilized jaws and parapodia (Gall and Grauvogel 1967b) (Figure 11.4) and egg clutches of insects and fish (Grauvogel 1947d; Gall

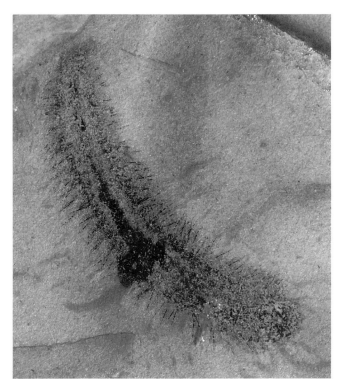

FIGURE 11.4 The polychaete *Homaphrodite speciosa,* showing soft-part preservation. Length of specimen is 7.5 mm. (Photo courtesy of J.-C. Gall, Université Louis Pasteur, Strasbourg)

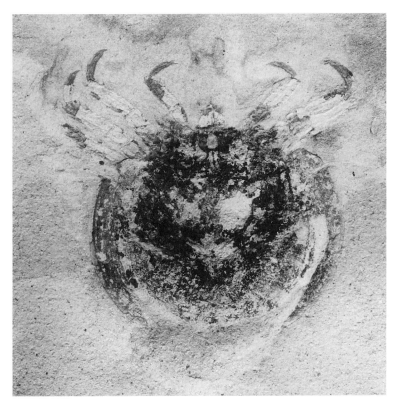

FIGURE 11.5 The small crustacean *Halycine ornata*. Width of shield is 15 mm. (Photo courtesy of J.-C. Gall, Université Louis Pasteur, Strasbourg)

and Grauvogel 1966; Gall 1985). The insect clutches show not only the individual eggs but their internal structure as well, and even the mucilagenous cover that originally bound them together is still visible (Gall and Grauvogel 1966). Crustaceans and terrestrial arthropods are preserved fully articulated and show remarkable details of their morphology, sometimes including the preserved digestive tract. For example, the splendid preservation of the small enigmatic crustacean *Halycine* has allowed, for the first time, a detailed reconstruction of this genus (Gall and Grauvogel 1967a) (Figure 11.5).

Aquatic organisms are concentrated near the tops of individual laminae, immediately beneath the base of the next lamina (Gall 1971, 1985, 1990; Gall and Grauvogel-Stamm 1984). Sometimes the organisms are locally enriched in places that seem to have been residual ponds in very shallow water before desiccation (Gall 1971, 1985). The fossiliferous laminae have an organic-rich top, with up to 37.9 mg extractable organic matter per 100 g of sediment (Briggs and Gall 1990; Gall 1990). Although detailed geochemical analyses of these organic-rich horizons have not been made, the investigation of their microfabric shows that they are actually fossilized microbial mats (Gall 1990).

There are clear indications that the aquatic fauna is autochthonous and that the terrestrial organisms are derived from the immediate vicinity of the stagnant pools. The evidence includes the preservation of certain organisms in life position (*Lingula*), the absence of any directional orientation, the presence of both larval and adult forms in the same assemblage, and the concentration of organisms in former residual ponds (Gall 1971, 1985; Gall and Grauvogel-Stamm 1984). The bed-by-bed analysis of the silty-clayey lenses shows that with every new influx of water a new assemblage was established (Figure 11.2). Some time thereafter, the entire fauna perished and was essentially preserved *in situ* (Gall 1971, 1985).

The following scenario has been proposed for the fossilization of the aquatic fauna of the ponds and pools (Gall 1990):

1. An overspill of water from river channels and/or brackish water during high tides produced a freshwater to brackish water environment in the stagnant ponds. The newly introduced aquatic fauna diversified.
2. With increasing evaporation, the restricted pool lost water. In the residual pond, the entire fauna was killed by severe oxygen depletion and increasing salinity of the water.
3. In this hostile environment, a microbial community proliferated, resulting in the rapid establishment of microbial mats covering and sealing the carcasses in a reducing environment and producing a mechanical as well as geochemical envelope.
4. A renewed overspill of floodwater buried the microbial mats and the enclosed carcasses under a sediment cover, and a new aquatic community became established.

Earlier explanations of the exceptional style of preservation in the Grès à Voltzia emphasized the role of desiccation of the carcasses before burial (Grauvogel and Gall 1962; Gall and Grauvogel 1966, 1967b; Gall 1971, 1985; Gall and Grauvogel-Stamm 1984). Mummification was thought to be a crucial process for the splendid preservation of soft-part tissues. More recent discussions of the taphonomy of the Grès à Voltzia have abandoned this earlier notion and suggest that all the taphonomic processes took place under water (Briggs and Gall 1990; Gall 1990). Instead of desiccation, dehydration and partial conservation in a brine, sealing by a microbial mat, and blanketing by a sediment cover are now thought to account for the exceptional preservation. Indeed, remains of these microbial mats were preserved themselves (Briggs and Gall 1990; Gall 1990).

PALEOBIOLOGY AND PALEOECOLOGY

Analysis of plant-bearing sandstones, carbonate intercalations, and the fossils of the fine-grained lenses of the Grès à Voltzia allows a very detailed reconstruction of the life of the Grès à Voltzia delta. According to these three facies, three different faunal and floral assemblages can be recognized.

The Taphocoenoses of the Sandstone Facies

The fossils found in the plant-bearing sandstone lenses allow a partial reconstruction of the flora and fauna of the proximal domain of the delta (Gall 1985). In this part, the delta platform was intersected by river channels lined by vegetation consisting of horsetails (*Equisetites, Schizoneura*), ferns (*Anomopteris, Neuropteridium*), and gymnosperms (*Aethophyllum, Voltzia, Yuccites, Albertia*). Many of the plants appear to have been highly opportunistic, fast-growing colonizers (Rothwell, Grauvogel-Stamm, and Mapes 2000). The levees were inhabited by stegocephalid amphibians (*Eocyclotosaurus lehmani*), which needed freshwater for their reproduction and consequently are not encountered in the brackish water environment of the distal part of the delta.

The Biocoenoses of the Silt-Clay Facies

Although in the fine-grained laminated sediments, fossil content can sometimes differ markedly from one lamina to the next, detailed investigation of these deposits allows the reconstruction of a generalized picture of the aquatic fauna of the brackish water ponds and the terrestrial fauna and flora living in the immediate neighborhood. The aquatic fauna is dominated by benthic arthropods (Gall 1971, 1985). The horseshoe crab *Limulitella bronni* is very common, sometimes with juvenile and adult specimens as well as exuviae and locomotion traces preserved in the same layer. The most common members of the fauna are, however, crustaceans (Bill 1914), including branchiopods (*Triops,* estherids), ostracodes, *Euthycarcinus,* syncarides, mysidaceans, isopods, decapods (*Antrimpos, Clytiopsis*), and the enigmatic *Halycine* (Figure 11.5). The estherids (Figure 11.6), which can occur in huge numbers, are represented by six species, and numerous individuals contain eggs. The abundance of estherids is characteristic of temporary standing waters (Gall 1985). These crustaceans have a very short biological cycle and can therefore develop in environments that dry out rapidly. Their very resistant eggs are subsequently distributed by the wind.

At the base of the temporary water bodies lived annelids (*Eunicites triasicus, Homaphrodite speciosa*) as well as larvae of insects (ephemeropterids, odonatopterids, dipterids) (Gall 1971, 1985). Burrowing endobenthos, represented by only *Lingula tenuissima* and some bivalves (e.g., *Homomya impressa*), is rare (Gall 1971, 1985).

FIGURE 11.6 The estherid *Palaeolimnadia dictyonata.* Length of carapace is 6.25 mm. (Photo courtesy of J.-C. Gall, Université Louis Pasteur, Strasbourg)

Nektonic animals are represented mainly by juvenile specimens of fish (*Dipteronotus aculeatus, Dorsolepis virgatus, Pericentrophorus minimus,* coelacanths) (Gall 1985). Eggs of fish are commonly concentrated in residual ponds. Finally, a medusa (*Progonionemus vogesiacus*) has been preserved (Grauvogel and Gall 1962; Gall 1990).

The terrestrial flora and fauna preserved in the silty-clayey lenses belong to the same ecosystem. The excellent preservation of the plants and the abundant plant roots in life position indicates that these organisms lived immediately in the neighborhood of standing-water bodies (Gall 1985). The horsetails (*Equisetites, Schizoneura*) grew at the margins of ponds and pools. The gymnosperms, like *Voltzia, Aetophyllum,* and *Yuccites,* formed wooded areas consisting of bushes up to several meters high (Grauvogel-Stamm 1978). The terrestrial surface was inhabited by abundant insects (ephemeropterids, odonatopterids, blattopterids, coleopterids, dipterids, paratrichopterids, hemipteroids), as well as spiders, scorpions (Figure 11.7), and myriapods (Gall 1971, 1985; Papier and Grauvogel-Stamm 1995; Papier et al. 1997). The presence of reptiles is documented by locomotion tracks attributed to the ichnogenus *Chirotherium* (Gall 1985).

The composition of this biocoenosis resembles those of modern brackish ponds and lagoons, but also those of some fossil deposits, most notably the Carboniferous Mazon Creek fauna (Chapter 10) and some other Carboniferous deposits (Briggs and Gall 1990). In all cases, the biocoenoses are poor in number of species, but rich in individuals of several species adapted for living in environments with fluctuating conditions (salinity, oxygen levels, temperature). The composition of these restricted communities has remained very stable during much of the Phanerozoic (Briggs and Gall 1990).

The Assemblages of the Carbonate Intercalation Facies

Two associations can be recognized in the distal domain of the delta: the *Lingula*–pelecypod assemblage, and the littoral fauna (Gall 1985). In the first assemblage, which occurs predominantly in silty-clayey lenses of the upper part of the Grès à meules, the inarticulate brachiopod *Lingula tenuissima* and the bivalves *Homomya impressa* and *Myophoria vulgaris* dominate. They are accompanied by crustaceans (estherids, *Clytiopsis argentoratensis, Halycine ornata*) and burrows of the ichnogenus *Rhizocorallium*. Burrowing endobenthos is much more frequent than in the assemblages of the brackish ponds (Gall 1985). Plant remains are rare. This association reflects a brackish water to restricted marine environment.

The littoral fauna, which is found in the carbonaceous intercalations in the Grès à meules, reflects a marginal but fully marine environment. Here the fauna includes foraminifers (*Glomospirella, Glomospira, Agathammina, lagenids*), ostracodes, bivalves (*Gervilleia, Myophoria, Homomya,*

FIGURE 11.7 Scorpion. Length of body is 60 mm. (Photo courtesy of J.-C. Gall, Université Louis Pasteur, Strasbourg)

Pleuromya), and gastropods (*Naticopsis gaillardoti, Loxonema absoletum*). The composition of this association is already very similar to those of the assemblages found in the lower Muschelkalk and illustrates the first incursions of the Muschelkalk Sea (Gall 1985). Stenohaline taxa, however, make their first appearance only in the overlying Grès argileux (Gall 1985).

CONCLUSIONS

The Middle Triassic Grès à Voltzia consists of several distinct sedimentary facies, reflecting a very complex depositional environment in a deltaic to marginal marine setting. The Grès à Voltzia biota certainly allows the best insight into marginal terrestrial life of Triassic times. The taxonomic composition and the species abundance pattern of the stagnant pond assemblage preserved in fine-grained lenses of the lower Grès à Voltzia are very similar to those of communities of modern brackish ponds and lagoons. Furthermore, close parallels exist with several Carboniferous Lagerstätten that have formed in similar settings, highlighting the remarkable stability of these restricted communities (Briggs and Gall 1990).

The exquisitely preserved fossils found in the silty-clayey layers represent mass killings due to oxygen depletion and possibly hypersalinity in small stagnant ponds and pools (Briggs and Gall 1990; Gall 1990; Gall, Grauvogel-Stamm, and Papier 1995). The plants and animals, which now constitute the taphocoenosis, are a mixture of aquatic species embedded *in situ* and washed-in terrestrial life-forms (Gall 1985). Detailed bed-by-bed analyses of the silty-clayey lenses have revealed considerable differences even between individual laminae (Gall 1971, 1985). The sedimentologic and taphonomic processes that have been involved in the formation of these fossiliferous lenses include dehydration of the organisms in a brine, sealing of the carcasses by proliferating microbial mats, and subsequent blanketing by sediments (Gall 1990). Although the sedimentology, taphonomy, and paleoecology of the Grès à Voltzia are in general very well documented, a modern geochemical investigation could probably improve our understanding of the formation of this Lagerstätte.

ACKNOWLEDGMENTS

J.-C. Gall carefully reviewed the chapter and offered important suggestions.

REFERENCES

Bill, Ph. C. 1914. Über Crustaceen aus dem Voltziensandstein des Elsass. *Mitteilungen der geologischen Landesanstalt Elsass-Lothringen* 8:289–338.

Briggs, D. E. G., and J.-C. Gall. 1990. The continuum in soft-bodied biotas from transitional environments: A quantitative comparison of Triassic and Carboniferous Konservat-Lagerstätten. *Paleobiology* 16:204–218.

Gall, J.-C. 1971. *Faunes et paysages du Grès à Voltzia du nord des Vosges: Essai paléoécologique sur le Buntsandstein supérieur.* Mémoires du Service de la Carte Géologique d'Alsace et de Lorraine, no. 34. Strasbourg: Service de la Carte Géologique d'Alsace et de Lorraine.

Gall, J.-C. 1983. The Grès à Voltzia delta. In J.-C. Gall, *Ancient Sedimentary Environments and the Habitats of Living Organisms: Introduction to Palaeoecology,* pp. 134–148. Berlin: Springer-Verlag.

Gall, J.-C. 1985. Fluvial depositional environment evolving into deltaic setting with marine influences in the Buntsandstein of northern Vosges. In D. Mader, ed., *Aspects of Fluvial Sedimentation in the Lower Triassic Buntsandstein of Europe,* pp. 449–477. Lecture Notes in Earth Sciences, no. 4. Berlin: Springer Verlag.

Gall, J.-C. 1990. Les voiles microbiens: Leur contribution à la fossilisation des organismes au corps mou. *Lethaia* 23:21–28.

Gall, J.-C., and L. Grauvogel. 1966. Faune du Buntsandstein I: Pontes d'invertébrés du Buntsandstein supérieur. *Annales de Paléontologie (Invertébrés)* 52:155–161.

Gall, J.-C., and L. Grauvogel. 1967a. Faune du Buntsandstein II: Les Halycines. *Annales de Paléontologie (Invertébrés)* 53:1–14.

Gall, J.-C., and L. Grauvogel. 1967b. Faune du Buntsandstein III: Quelques annelides du Grès à Voltzia. *Annales de Paléontologie (Invertébrés)* 53:105–110.

Gall, J.-C., and L. Grauvogel-Stamm. 1984. Genèse des gisements fossilifères du Grès à Voltzia (Anisien) du nord des Vosges (France). *Géobios, Mémoire Special* 8:293–297.

Gall, J.-C., L. Grauvogel-Stamm, and F. Papier. 1995. Der Buntsandstein der Nordvogesen. *Jahresbericht und Mitteilungen des oberrheinischen geologischen Vereins* 77:155–165.

Grauvogel, L. 1947a. Contribution à l'étude du Grès à Voltzia. *Comptes rendus sommaires de la Societé Géologique de France* 1947:35–37.

Grauvogel, L. 1947b. Note préliminaire sur la flore du Grès à Voltzia. *Comptes rendus sommaires de la Societé Géologique de France* 1947:64–66.

Grauvogel, L. 1947c. Note préliminaire sur la faune du Grès à Voltzia. *Comptes rendus sommaires de la Societé Géologique de France* 1947:90–92.

Grauvogel, L. 1947d. Sur quelques types de pontes du Grès à Voltzia (Trias inférieur) des Vosges. *Comptes rendus de l'Academie de Science de France* 225:1165–1167.

Grauvogel, L., and J.-C. Gall. 1962. *Progonionemus vogesiacus* nov. gen. nov. sp., une méduse du Grès à Voltzia des Vosges septentrionales. *Bulletin du Service de la Carte Géologique d'Alsace et de Lorraine* 15:17–27.

Grauvogel-Stamm, L. 1978. *La Flore du Grès à Voltzia (Buntsandstein supérieur) des Vosges du Nord (France): Morphologie, anatomie, interprétations phylogénique et aléogéographique.* Mémoires des sciences géologiques, no. 50. Strasbourg: Université Louis Pasteur, Institut de Geologie.

Kamphausen, D., and M. Morales. 1981. *Eocyclotosaurus lehmani,* a new combination for *Stenotosaurus lehmani* Heyler, 1969 (Amphibia). *Neues Jahrbuch für Geologie und Paläontologie, Monatshefte* 1981:651–656.

Papier, F., and L. Grauvogel-Stamm. 1995. Les Blattodea du Trias: Le genre *Voltziablatta* n. gen. du Buntsandstein supérieur des Vosges (France). *Palaeontographica, A* 235:141–162.

Papier, F., A. Nel, L. Grauvogel-Stamm, and J.-C. Gall. 1997. La plus ancienne sauterelle Tettigoniidae, Orthoptera (Trias, NE France): Mimétisme ou exaptation? *Paläontologische Zeitschrift* 71:71–77.

Rothwell, G. W., L. Grauvogel-Stamm, and G. Mapes. 2000. An herbaceous fossil conifer: Gymnospermous ruderals in the evolution of Mesozoic vegetation. *Palaeogeography, Palaeoclimatology, Palaeoecology* 156:139–145.

Seilacher, A., W. E. Reif, and F. Westphal. 1985. Sedimentological, ecological and temporal patterns of fossil Lagerstätten. *Philosophical Transactions of the Royal Society of London, B* 311:5–23.

12
Monte San Giorgio: Remarkable Triassic Marine Vertebrates

Walter Etter

THE MIDDLE TRIASSIC LAGERSTÄTTE OF MONTE SAN Giorgio, located in the southernmost part of Switzerland, is well known among vertebrate paleontologists and is considered the most important locality for Triassic marine fish and reptiles in the world (Bürgin et al. 1989). Its organic-rich black bituminous shales and gray laminated dolomites have yielded thousands of fully articulated and exquisitely preserved vertebrate fossils. However, because invertebrates other than ammonites and thin-shelled bivalves are virtually absent, this Lagerstätte is practically unknown among invertebrate paleontologists and taphonomists outside central Europe. Monte San Giorgio has consequently not been mentioned in any of the recent publications summarizing the stratigraphic and environmental occurrence of conservation deposits (Allison and Briggs 1991, 1993), even though some fish and reptiles show soft-part preservation. This is due mainly to the absence of a modern study integrating sedimentologic, paleontologic, and taphonomic features.

The orthodox view about the depositional environment holds that the fossiliferous layers of Monte San Giorgio formed in a stagnant basin with permanent anoxic bottom-water conditions (Rieber and Sorbini 1983; Bernasconi 1991, 1994; Furrer 1995). The richly fossiliferous units are finely laminated throughout, trace fossils are entirely absent, and the invertebrate fauna consists mainly of taxa traditionally believed to represent pseudoplankton and nekton. The very rare undisputed benthic invertebrates were probably washed into the basin from a nearshore environment. The same is certainly true for the rare land plants and terrestrial vertebrates found in a few layers (Rieber 1973a). The Middle Triassic of Monte San Giorgio thus seems to be a much better model example for a stagnation deposit (Seilacher, Reif, and Westphal 1985) than the much-cited Posidonia Shale of southern Germany (Chapter 15).

The sedimentologic and paleontologic facts can also be interpreted in a different way, however. The pseudoplanktonic bivalves may have been epibenthic, and the lamination may, in part, be the product of microbial mats that grew on the seafloor with strongly dysoxic, but not anoxic, bottom-waters.

The first excavations in the richly fossiliferous Middle Triassic layers were done in the late nineteenth century by paleontologists from the Natural History Museum of Milan, Italy. They collected vertebrate fossils at a site near the Italian village of Besano at the Swiss border (Bassani 1886; Kuhn-Schnyder 1974). By that time, a mining industry had been established in Besano and in the Swiss village of Meride, where Middle Triassic bituminous shales were quarried and processed. Through heating of the organic-rich sediments, bitumen was distilled; it, in turn, was the basic ingredient for a pharmaceutical product called Saurol (Kuhn-Schnyder 1974).

These early sampling efforts yielded some 22 fish species and three reptile species (Bassani 1886; Kuhn-Schnyder 1974). However, it was not until the early twentieth century that the exceptional value of the Monte San Giorgio Lagerstätte was truly recognized. In 1924 the first of a series of major excavations was started by Bernhard Peyer (1944) from the Zoological Museum of Zurich, Switzerland. The success of this and the subsequent excavations was very much a consequence of a quarrying technique that exposed a very large surface area of the bedding planes (Kuhn-Schnyder 1974). Although the earlier excavations conducted by the Zurich Museum concentrated entirely on the exceptionally preserved vertebrate remains, a new major campaign, starting in 1950 and continuing until 1968, documented for the first time both the invertebrate fossils and the sedimentology of the deposit (Rieber 1973a, 1973b; Kuhn-Schnyder 1974). Since 1968, no more large-scale excavations have been undertaken, but preparation work on the many thousands of vertebrate specimens continues. The most beautiful and remarkable specimens are now on display in the Paleontological Institute and Museum of the University of Zurich, which, after renovation, reopened in 1991.

GEOLOGICAL CONTEXT

Monte San Giorgio (Mount Saint George), which rises 1,097 m above sea level, is located in the southernmost part of Switzerland, in the Italian-speaking canton of Ticino near the border with Italy (Figure 12.1). The mountain is composed of pre-Permian basement rocks and of

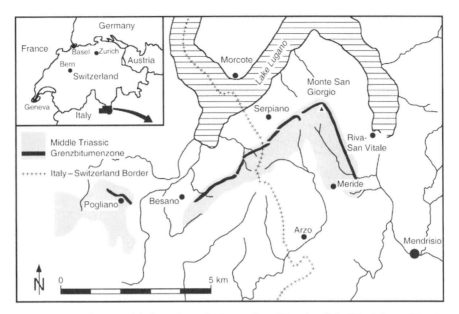

FIGURE 12.1 Geographic location of Monte San Giorgio. (Modified from Bürgin 1992)

Permian and Triassic volcanoclastics and sediments that dip toward the south-southwest. The northern slope of Monte San Giorgio borders on Lake Lugano. The richly fossiliferous Middle Triassic strata are best exposed on the southern and western slopes of the peak, where they extend into Italy (Kuhn-Schnyder 1974; Sander 1989; Bürgin 1992; Furrer 1995).

The Middle Triassic succession at Monte San Giorgio consists of several distinct units (Frauenfelder 1916; Kuhn-Schnyder 1974; Rieber and Sorbini 1983) (Figure 12.2). The lowermost formation is the lower Salvatore Dolomite, which is of Anisian age and consists of dolomitized microbial limestone. Overlying it is the so-called Grenzbitumenzone, a relatively thin alternation of bituminous shales and bituminous dolomites. The next lithological unit is the approximately 60 m thick San Giorgio Dolomite, which grades upward into the well-bedded, dark gray Meride Limestone, which reaches a total thickness of 600 m and is of Ladinian age (Rieber and Sorbini 1983; Furrer 1995).

The vast majority of the vertebrate fossils have been recovered from the Grenzbitumenzone, which at Monte San Giorgio is 5 to 16 m thick (Frauenfelder 1916). To a lesser extent, complete vertebrate skeletons have also been recovered from thin bituminous shaley layers of the lower 90 m of the Meride Limestone (Rieber and Sorbini 1983). In the Grenzbitumenzone, more or less bituminous dolomites alternate with thin bituminous black shales and intercalated volcanoclastic layers (Figure 12.2). The dolomites, which constitute 87 percent of the section, are

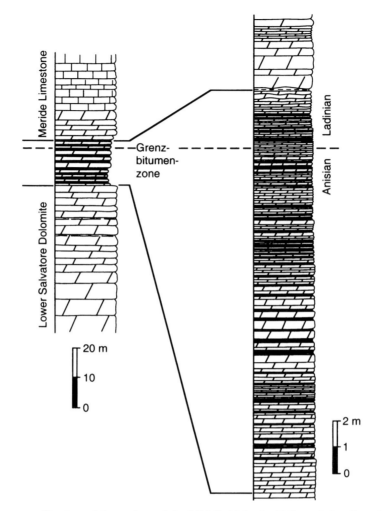

FIGURE 12.2 Stratigraphic section of the Middle Triassic (*left*) and the Grenzbitumenzone (*right*) at Monte San Giorgio. (Modified from Brack and Rieber 1993)

finely laminated and contain up to 20 percent organic matter (Rieber 1975). With increasing organic content, the lamination becomes much more visible. The thickness of the individual beds ranges from 3 to 30 cm, the average being 10 to 20 cm. A thinning upward of the beds is clearly visible. The intercalated bituminous shales, which account for 11 percent of the section, are 10 cm thick at most, and are usually much less. The organic matter content is unusually high in these layers and reaches values of up to 40 percent (Müller 1969; Bernasconi 1991; Bernasconi and Riva 1993). The strongly bituminous shales do not contain any carbonate (Rieber and Sorbini 1983). The thin volcanoclastic layers contribute 2 to 3 percent of the sections. Sedimentological features (Müller, Schmid, and Vost 1964), as well as their wide geographic distribution (Brack and Rieber 1993), indicate an aeolian transport of these volcanic tuffs (but see Zorn 1971, who argues for a subaqueous origin). The Grenzbitumenzone lies at the Anisian–Ladinian boundary, as has been determined by the succession of ammonoids (Rieber 1973b; Brack and Rieber 1993) and bivalves of the genus *Daonella* (Rieber 1968, 1969).

Monte San Giorgio is part of the austroalpine tectonic unit, which is the southernmost unit of the Alps. During the Middle Triassic, this unit was part of the western Tethys (Ziegler 1988), where both compressional and extensional tectonics were accompanied by intensive volcanism and led to a highly irregular seafloor topography, consisting of uplifted and subsiding blocks. The Grenzbitumenzone and the lower Meride Limestone are believed to have formed in an intraplatform depression surrounded by reefs and shallow-water platform carbonates (Brack and Rieber 1993; Furrer 1995; Mundil et al. 1996).

PALEOENVIRONMENTAL SETTING

The finely laminated sediments of the Grenzbitumenzone and the lower Meride Limestone have obviously been deposited below storm wave base (Bernasconi 1991) (Figure 12.3). There are absolutely no signs of bioturbation or physical reworking. Many laminae or individual beds show, however, characteristics of distal turbidites (Bernasconi 1991, 1994; Bernasconi and Riva 1993), and in the lower parts of the Grenzbitumenzone, slumps are quite frequent (Müller 1969) (Figure 12.3). The bituminous shales are so rich in organic matter that, once lighted, they will burn with a smoky flame (Rieber 1975). They also contain large concentrations of trace elements (Ti, V, Mn, Ni, Co, U) (Rieber 1975; Bernasconi 1991). Pyrite is present throughout the section, but only in small amounts, and shows a very negative sulfur isotope composition (Bernasconi 1991, 1994). This indicates that sulfide formation occurred within an open system in the presence of abundant sulfate, and that pyrite formation was limited by the amount of available reactive iron (Bernasconi 1994).

The accumulation and preservation of the organic matter and the timing of dolomitization of the sediments of the Grenzbitumenzone have only recently been thoroughly studied (Bernasconi 1991; Bernasconi and Riva 1993). A major proportion of the preserved organic material derives from bacterial biomass (Bernasconi and Riva 1993). The organic matter is characterized by a very high hopane concentration, up to 10 times the amount of steranes. Hopanoids are known from only prokaryotic organisms, in which they reinforce the cell membranes. Sterols, which degrade to steranes, have the same function in eukaryotic organisms (Bernasconi 1991). A predominant bacterial origin of the

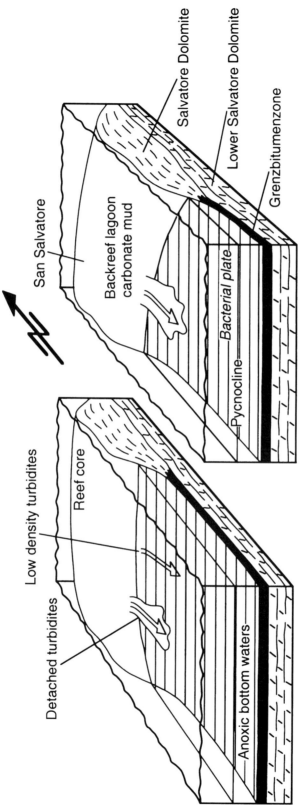

FIGURE 12.3 Reconstruction of the depositional environment of the Monte San Giorgio region at the time of Grenzbitumenzone deposition, according to the traditional stagnant basin model. In the new model proposed in this chapter, the dysoxic–anoxic interface is situated at the sediment surface where microbial mats were growing. (Modified from Bernasconi 1991)

biomass is supported by the fact that the $\delta^{13}C$ values are strongly negative (Bernasconi and Riva 1993). In similar modern environments, the bacterial biomass is depleted in ^{13}C by 10‰ to 20‰, compared with phytoplankton (Bernasconi 1991). Another remarkable feature of the preserved organic material in the Grenzbitumenzone is the very high concentration of porphyrins, which, as is supported by the correlation with other biomarkers, are derivatives of cyanobacterial chlorophyll (Bernasconi 1991, 1994).

Dolomitization of the carbonate-rich layers of the Grenzbitumenzone occurred during early diagenesis in the sulfate reduction zone and was facilitated by increasing alkalinity produced by the degradation of organic matter. Furthermore, a diagenetically open system and hence a low sedimentation rate is implied, allowing seawater magnesium to diffuse into the sediment for a prolonged time (Bernasconi 1991). Low sedimentation rates have long been hypothesized for the Grenzbitumenzone, and are based mainly on faunal evidence. Within only a few meters of the stratigraphic column, many different species of ammonoids (Rieber 1973b) and of the bivalve genus *Daonella* (Rieber 1968, 1969) succeed one another. Additional evidence for low sedimentation rates includes calculations of trace metal accumulation rates and of primary productivity. At present, the time span for deposition of the Grenzbitumenzone (~15 m) is estimated at 2 to 3 Ma (Bernasconi 1991). However, the bituminous shales probably represent times of much slower accumulation than the deposition of the dolomitic layers (Rieber 1968; Bernasconi 1991, 1994).

The preservation of laminations, the absence of trace fossils and undisputably autochthonous benthic invertebrates, the high content in organic matter, and the preservation of very unstable biomarkers such as porphyrines were traditionally thought to indicate deposition of the Grenzbitumenzone under completely anoxic conditions. The stagnant basin model for the Monte San Giorgio deposits dates back to Peyer (1944) and has especially been advanced by Rieber (1973a, 1975, 1982; Rieber and Sorbini 1983). Detailed reconstruction of the depositional environment is somewhat hampered by the fact that the lateral dimensions of the Grenzbitumenzone are not known. The maximum east–west extension is 20 km, but the north–south extension cannot be constrained because of the thick Cenozoic cover in the south (Bernasconi 1991). Most workers believe, however, that the richly fossiliferous Monte San Giorgio deposits formed in a rather small intraplatform basin. Just a few kilometers to the north of Monte San Giorgio, the time-equivalent sediments of the Grenzbitumenzone developed as thick dolomitized microbial reef carbonates, the Salvatore Dolomite (Figure 12.3); and at least 250 m of Salvatore Dolomite correspond to about 15 m of Grenzbitumenzone (Zorn 1971). Similar reef complexes are found farther to the west (Rieber and Sorbini 1983); and about 20 km to the east, contemporaneous sediments developed as shallow-water platform carbonates (Esino platform) (Brack and Rieber 1993). It seems, therefore, that the Middle Triassic Monte San Giorgio basin was surrounded by reef complexes and/or carbonate platforms. The presence of an abundant open-water fauna indicates, however, that the surface waters were normal marine and were connected to the Tethys, allowing an exchange of nektonic and planktonic animals.

According to the traditional model, a permanent water stratification developed in the silled Monte San Giorgio basin (Figure 12.3), although the underlying mechanism is unclear. Geochemical evidence indicates normal saline bottom-waters (biomarker distributions in modern hypersaline environments are markedly different from those of the Grenzbitumenzone) (Bernasconi 1991), and the rich pelagic life documents normal salinity of the surface waters as well. A sharp and stable depth-related temperature gradient (as opposed to a salinity stratification) seems unlikely, because during Middle Triassic times Monte San Giorgio was located in the tropical belt near the equator (Bernasconi 1991). It has been proposed that aerobic photosynthetic (cyanobacteria) and anaerobic chemoautotrophic bacteria formed a bacterial plate at the oxic–anoxic interface of the stratified water column (Bernasconi 1991, 1994; Bernasconi and Riva 1993) (Figure 12.3). Because the underlying lower Salvatore Dolomite as well as the overlying upper Meride Limestone represent shallow-marine environments, and there was no steep slope from the Salvatore reef complex to the Monte San Giorgio basin (no reef debris in the basin), the water depth during deposition of the Grenzbitumenzone has been estimated to be only between 30 and 100 m (Zorn 1971; Rieber and Sorbini 1983; Bernasconi 1991, 1994).

The alternation of black bituminous shales with dolomites, although not very regularly developed (Figure 12.2), may be the result of cyclic sea-level fluctuations (Bernasconi 1991, 1994). During sea-level low stands, carbonate production on the surrounding platforms was inhibited or minimal, and only clay minerals together with organic matter accumulated in the black shales. Sea-level high stands, on the contrary, promoted carbonate production in the shallow-water environments. It is assumed that the carbonate was produced on the Salvatore–Esino platform and was transported into the basin by normal and/or detached turbidity currents, which passed along the pycnocline at the oxic–anoxic boundary (Bernasconi 1991; Bernasconi and Riva 1993). The black shale–dolomite alternations of the Grenzbitumenzone, therefore, most likely represent variation in carbonate supply to the basin over a more or less constant background sedimentation of clay minerals.

For a reconstruction of the paleoenvironmental setting, the mode of life of the bivalve *Daonella* is crucial. If the bivalve was pseudoplanktonic, then there was no life at the seafloor and the traditional depositional model is correct; if *Daonella* was benthic, then a new depositional mode is required. Such an alternative model is favored here. The bottom-waters of the Monte San Giorgio basin were severely oxygen depleted in general, but not anoxic during deposition of the Grenzbitumenzone. The low oxygen content allowed the colonization of the seafloor by extremely tolerant benthic bivalves of the genus *Daonella*, but oxygen values were too low for the establishment of a more diverse and burrowing benthic fauna. Cyanobacteria and chemoautotrophic bacteria formed microbial mats on the sediment surface and were not floating at an oxic–anoxic pycnocline in the water column. Such a model, according to which the oxic–anoxic interface is situated at the sediment surface, circumvents the problem of maintaining stable water stratification during several million years, and would also better explain some of the geochemical data. In a diagram where the total sulfur content is plotted against the total organic carbon, the values from the Grenzbitumenzone of Monte San Giorgio are scattered rather well around a regression line intersecting the axes at the zero point (Bernasconi 1991). In Holocene sediments, such a relationship between total sulfur and total organic carbon is characteristic of sediments deposited under oxic or dysoxic conditions (Bernasconi 1991). During times of slower basin water renewal, the bottom-water may have become anoxic for a short time, but turbidites and improved water circulation during times of sea-level high stands would again lead to dysoxic conditions and renewed colonization of the seafloor by benthic bivalves.

TAPHONOMY

The Middle Triassic of Monte San Giorgio is most remarkable for the rich occurrences of completely articulated vertebrate skeletons. Isolated bones are much less common than articulated skeletons (Rieber and Sorbini 1983). The fish and reptile remains occurring in the bituminous shales, as well as in the dolomitic layers, are splendidly preserved and exhibit very fine details of bone, scale, and fin anatomy (Figures 12.4, 12.5, 12.7, 12.8). It has been discovered that rarely calcified cartilage of shark skeletons was preserved as well (Bürgin et al. 1989). Other examples of exceptionally well preserved vertebrates include embryos of the fish genus *Saurichthys* (Rieppel 1985) and one embryo of a nothosaur (Sander 1988). Soft-part preservation is very rare and is confined to a few occurrences of phosphatized digestive tracts of fishes (Rieppel 1985) and to preserved nonmineralized scales of nothosaurs (Sander 1989) and *Macrocnemus* (Premru 1991) (Figures 12.4 and 12.5). Although the bedding planes around vertebrate skeletons are usually enriched in organic matter (Rieber 1975), the skin outline is never preserved (compare with ichthyosaurs of the Posidonia Shale [Chapter 15]).

FIGURE 12.4 Skin preservation in *Macrocnemus bassanii*, showing an overview of the entire skeleton. Body length of specimen is 25 cm. (Photo courtesy of H. Rieber, University of Zurich)

The invertebrate fossils of Monte San Giorgio are not as spectacular as the vertebrate fauna, with the exception of a coleoid cephalopod with preserved armhooks, jaws, and ink sac (Rieber 1970) and of preserved conodont clusters (Rieber 1980). Bivalves and ammonoids are found only in the dolomitic layers, and their shells are invariably dissolved (Rieber 1968, 1973b). In only slightly bituminous dolomites, the ammonoid phragmocones are preserved as voids internally lined with dolomite and/or quartz crystals, and the body chambers are preserved as steinkerns (Rieber 1973b). With increasing bitumen content, the dolomite layers underwent increasing compaction, and the enclosed ammonoids are strongly flattened or are preserved as impressions. The fact that some ammonoids are lying obliquely to the bedding planes suggests quite a high sedimentation rate for the dolomite layers. In the carbonate-free bituminous black shales, no shelly fossils are preserved (Rieber 1973b).

Bivalves, which belong mainly to the genus *Daonella,* are predominantly present as single valves. In most beds, articulated bivalved shells account for less than 1 percent of all the specimens (Rieber 1968). The bivalves are never sorted by size, but in some layers the majority of the valves are oriented with the convex side up (Rieber 1968). These patterns indicate that the shells were reoriented and perhaps transported by gentle currents (Rieber 1968).

Based on the preservation style of the invertebrates and compaction features of slump folds, it can be concluded that compaction in the pure dolomites was only minor, whereas a compaction rate of 1:10 is estimated for the black bituminous shales (Bernasconi 1991). In the latter, vertebrate skeletons are usually strongly flattened (Bürgin et al. 1989).

According to the traditional depositional model described in the previous section, there were no benthic animals in the Monte San Giorgio basin. The carcasses of nektonic animals and the shells of presumably

FIGURE 12.5 Skin preservation in *Macrocnemus bassanii,* showing detail from pelvic region with preserved scales. Individual scales are 0.3 mm long. (Photo courtesy of H. Rieber, University of Zurich)

pseudoplanktonic bivalves (*Daonella*) simply sank to the bottom into anoxic water, where they eventually were covered with mud (Kuhn-Schnyder 1974; Rieber 1982; Rieber and Sorbini 1983). However, anoxia is not a prerequisite for excellent preservation. Severely dysoxic bottom-water and sealing of the carcasses by microbial mats would also lead to the observed preservational patterns (Chapter 15). In the absence of scavenging organisms and under quiet conditions, fully articulated skeletons accumulated in large numbers. In several layers, vertebrate remains, as well as phosphatized coprolites, are very frequent. The fact that vertebrate fossils are more common in the bituminous shales than in the dolomite layers is probably a consequence of the lower sedimentation rate of the black shales (Rieber 1968).

Most of the fossils show no preferred orientation. There are, however, some indications of gentle currents along the bottom. Evidence includes layers where the majority of the bivalve shells are oriented with the convex side upward (Rieber 1968), as well as unimodally dispersed skeletal elements in some fish (Tintori 1992). Weak bottom currents may have been coupled with renewal of the deep water, but strong oxygenation events never did occur. Trace fossils are absent throughout the section, and the rare benthic animals other than bivalves were most probably washed into the basin by storm events (Etter 1994). The same is certainly true for the few land plants and terrestrial vertebrates. In addition, many of the fish and reptile species are typical of nearshore shallow water and show various adaptations to reef-like environments (Bürgin et al. 1989; Bürgin 1992, 1996). Obviously, these animals were carried into the central part of the basin as well.

PALEOBIOLOGY AND PALEOECOLOGY

It is mainly the vertebrate fauna of Monte San Giorgio that has attracted the attention of numerous workers. Indeed, the variety of fish and reptile species from this Lagerstätte is richer than that from any other Triassic marine locality. Five species of sharks are known, of which four are rather small and have shell-crushing dentition (Rieppel 1982; Bürgin et al. 1989). More than 30 species of actinopterygian (ray-finned) fish have been described. A few of these (e.g., *Birgeria, Colobodus*) are large, predatory, open-water forms (Schwarz 1970; Bürgin et al. 1989). The different species of *Saurichthys*, one of the most common fish genera at Monte San Giorgio, range in adult size from 30 to 85 cm and were certainly predatory. Their habitat was probably the reef-like environment of the basin borders (Rieppel 1985, 1992). However, most of the ray-finned fish species are medium to small (30 to less than 5 cm) and include both open-water forms and reef inhabitants (Bürgin et al. 1989; Bürgin 1992,

1996). Actinistian (lobe-finned) fishes are known from only about 20 specimens, which probably belong to three species (Bürgin et al. 1989).

Among the reptiles, ichthyosaurs are one of the best-represented groups (Kuhn-Schnyder 1974; Brinkmann 1998). The most common species (adult size, 1 to 1.2 m) belong to the genus *Mixosaurus,* of which several embryos were recently discovered (Brinkmann 1996). *Mixosaurus* specimens have been found in densities of up to four specimens per 10 m² in one layer (Rieber 1973a) (Figure 12.6). Some of the specimens contain many cephalopod hooks between the ribs (Rieber 1970), and it can be concluded that cephalopods were an important part of their diet (Kuhn-Schnyder 1974). At least three other genera of the fully aquatic, open-water ichthyosaurs have been found in Monte San Giorgio and are currently under investigation (Brinkmann 1998).

The most abundant reptiles at Monte San Giorgio are the amphibious nothosaurs, common Triassic members of the order Sauropterygia (Bürgin et al. 1989). These animals probably fed on fish, but may have laid their eggs on land. Several hundred specimens of the small *Neusticosaurus* (adult size, ~30 cm) have been collected, although the majority of them do not come from the Grenzbitumenzone but from bituminous layers in the lower Meride Limestone. From *Neusticosaurus,* a complete growth series from embryo to adult can be documented, and it is now known that these small amphibious lizards reached sexual maturity at three to four years and that they lived to a maximum of six years (Sander 1988, 1989). Other nothosaurs found at Monte San Giorgio include species of the genera *Ceresiosaurus, Paranothosaurus,* and *Lariosaurus,* some of which attained an adult length of more than 3 m (Kuhn-Schnyder 1974) (Figure 12.7).

Two genera of placodonts, an entirely Triassic reptile group, are known from Monte San Giorgio. *Cyamodus* had turtle-like armor covering its back, whereas *Paraplacodus* possessed broadened and flattened dorsal as well as heavy ventral ribs (Peyer 1931; Kuhn-Schnyder 1974). Like all the placodonts, these specimens show very enlarged blunt teeth, which were certainly adapted for crushing shells. Most probably these animals fed on molluscs from the reefs surrounding the Monte San Giorgio basin (Bürgin et al. 1989). The enigmatic *Helveticosaurus,* known from only the three specimens of Monte San Giorgio, may belong to the placodonts as well.

Archosauromorphs are present with three genera. Perhaps the strangest reptile of this deposit is *Tanystropheus* (Figure 12.8), the so-called giraffe-neck saurian, which was first reconstructed as a pterosaur (Nopcsa 1923). This up to 6 m long animal is most remarkable because of its extremely elongated neck, which was more than twice the length of its trunk (Wild 1974; Tschanz 1988). The grotesque long neck was not the result of a multiplication of vertebrae numbers, but the consequence

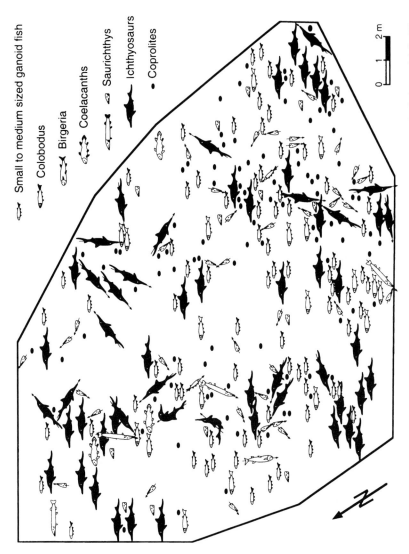

FIGURE 12.6 Fossil distribution of bed number 113, a black shale approximately 8.5 cm thick. The mapped grid is composed of 4 m² squares. Excavation is at Point 902 at Monte San Giorgio. Only the fossils with an arrow are shown in the correct orientation, the horizontally oriented fossils have not been measured. (Modified from Rieber and Sorbini 1983)

Small to medium sized ganoid fish

Colobodus

Birgeria

Coelacanths

Saurichthys

Ichthyosaurs

Coprolites

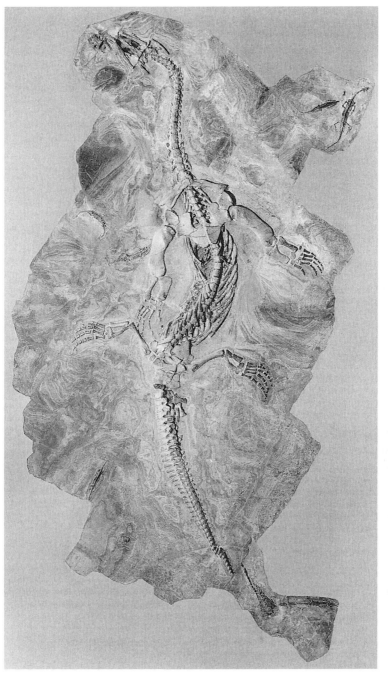

FIGURE 12.7 *Ceresiosaurus calcagnii.* Length of specimen is 2.3 m. (Photo courtesy of H. Rieber, University of Zurich)

FIGURE 12.8 Reconstruction of 5.5 m long adult *Tanystropheus longobardicus*. (Modified from Bürgin et al. 1989)

of strongly positive allometric growth of the individual vertebrae (Tschanz 1988). Closely related to *Tanystropheus*, but reaching only 80 cm in length, is *Macrocnemus*, which also possesses elongated neck vertebrae (Tschanz 1988; Premru 1991) (Figure 12.4). Whereas *Tanystropheus* probably lived most of the time in the sea and fed on fish, *Macrocnemus* is believed to have been mainly terrestrial. The latter is certainly the case for the only true archosaur of Monte San Giorgio: *Ticinosuchus* (Krebs 1965).

Three genera of the somewhat enigmatic thalattosaurs occur in the Middle Triassic of Monte San Giorgio (Rieppel 1987; Bürgin et al. 1989). *Askeptosaurus* was a fish-eating reptile about 2.5 m long, but the smaller *Clarazia* possessed shell-crushing teeth. The diet of *Hescheleria*, known from only an incomplete specimen, is not known with certainty, although this animal most probably also fed on molluscs.

The invertebrate fauna of Monte San Giorgio is restricted primarily to the dolomite layers of the Grenzbitumenzone. With the exception of only one echinoid specimen, no macroinvertebrates are known from the lower Meride Limestone (Kuhn-Schnyder 1974). In various layers, however, there is a quite diverse and probably autochthonous microfauna (foraminifers and ostracods, together with sponge spicules) (Wirz 1945), which suggests a somewhat different depositional history of the lower Meride Limestone with better-oxygenated bottom-water. The most common and conspicuous invertebrates of the Grenzbitumenzone are bivalves and cephalopods, but their abundances vary considerably from bed to bed and they are noticeably lacking in the bituminous black shales for diagenetic reasons (Rieber 1968, 1973a, 1973b).

The bivalves belong mainly to the genus *Daonella*, which is a very thin-shelled form and has a worldwide distribution. Because these bivalves occur basinwide and predominantly as single valves, they were interpreted as having had a pseudoplanktonic mode of life (Rieber 1968). Yet there are several reasons why this lifestyle does not seem to be adequate to explain the paleontologic facts. Pseudoplanktonic species not only should show a very widespread facies distribution and be independent of benthic conditions, but also should be rare, compared with the benthic and/or pelagic fauna of the same deposit (Wignall and Simms 1990). *Daonella* specimens are much too abundant in the Grenzbitumenzone and have never been found associated with floating objects. One would also expect to find a highly diverse community of encrusters associated with *Daonella* occurrences, but this is not the case. Furthermore, different species of *Daonella* that occur in the southern Alps do show a facies dependence (Rieber, personal communication, 1996), which makes a benthic mode of life even more probable.

Twelve species of *Daonella* were described from Monte San Giorgio, and some of the short-lived species have proved to be very useful in biostratigraphy (Rieber 1968, 1969). Other bivalves do occur in the Grenzbitumenzone, mainly undescribed posidoniid species for which a pseudoplanktonic mode of life was assumed as well, and rare members of the genera *Gervilleia,* "*Modiolus,*" and "*Mytilus*" (Kuhn-Schnyder 1974).

Most of the cephalopods from Monte San Giorgio belong to the ceratitid ammonoids, many of them being endemic to this region (Rieber 1973b). Other species have allowed a very detailed subdivision of the Grenzbitumenzone and correlation with other Middle Triassic deposits of the southern Alps (Brack and Rieber 1993). Members of the nautiloids (*Michelinoceras*) and coleoids are much rarer (Rieber 1973b).

Additional invertebrates and protozoans include radiolarians (observed in many horizons), brachiopods known from only four doubtful specimens, very rare large and quite common small gastropods, a decapod crustacean known from only one individual, a few undescribed thylacocephalan arthropods, one echinoid spine, and rare conodonts (Rieber 1973a; Kuhn-Schnyder 1974; Etter 1994). A few plant remains are known and include shallow-water dascycladacean algae and terrestrial conifers (*Voltzia?*) (Kuhn-Schnyder 1974). The most common groups of invertebrates are, according to the new model, nektonic (cephalopods) and opportunistic benthic animals (bivalves, small gastropods). It must be emphasized that other benthic organisms like the large gastropods, brachiopods, arthropods, and echinoderms are indeed very rare and known from just one or a few specimens each (Rieber 1973a; Kuhn-Schnyder 1974; Etter 1994), which is striking considering the large volume of rock that has been quarried at Monte San Giorgio. It seems most likely that the latter animals are allochthonous. The same mechanism must account for the plants found in the Grenzbitumenzone and the heterogeneous assemblage of vertebrates, which includes openwater forms as well as nearshore durophagous reef dwellers and even terrestrial species.

CONCLUSIONS

Monte San Giorgio has yielded the world's most spectacular marine vertebrate fauna of Triassic times, which was a crucial time in the evolution of ichthyosaurs, nothosaurs, and ray-finned fishes (Bürgin et al. 1989). The large quantity of recovered vertebrate remains has allowed very detailed functional studies and even investigation of the life history of certain groups (Sander 1989). Among the invertebrates, nektonic taxa (ammonoids) and opportunistic epibenthic colonizers very tolerant to severe dysoxia (the bivalve *Daonella*) predominate. This deposit is characterized by perfectly laminated sediments, the complete absence of bioturbation and trace fossils, and a monotypic but abundant epibenthic fauna. According to the new depositional model proposed in this chapter, the

Grenzbitumenzone of Monte San Giorgio formed in a small intraplatform basin under severely dysoxic but not anoxic bottom-waters.

Although the understanding of the Monte San Giorgio deposit has greatly benefited as a result of recent geochemical investigations (Bernasconi 1991, 1994; Bernasconi and Riva 1993), many unanswered questions remain. A detailed sedimentological or basinwide analysis has never been done, and a comprehensive study integrating all the paleontologic, taphonomic, and sedimentologic data has yet to be published.

ACKNOWLEDGMENTS

H. Rieber critically read the chapter and provided literature.

REFERENCES

Allison, P. A., and D. E. G. Briggs. 1991. Taphonomy of nonmineralized tissues. In P. A. Allison and D. E. G. Briggs, eds., *Taphonomy: Releasing the Data Locked in the Fossil Record*, pp. 25–70. New York: Plenum.

Allison, P. A., and D. E. G. Briggs. 1993. Exceptional fossil record: Distribution of soft-tissue preservation through the Phanerozoic. *Geology* 21:527–530.

Bassani, F. 1886. Sui fossile e sull'età degli schisti bituminosi triasici di Besano in Lombardia. *Atti della Società Italiana delle Sciencie Naturale* 29:15–72.

Bernasconi, S. M. 1991. Geochemical and microbial controls on dolomite formation and organic matter production/preservation in anoxic environments: A case study from the Middle Triassic Grenzbitumenzone, southern Alps (Ticino, Switzerland). Ph.D. diss., Eidgenössische Technische Hochschule, Zurich.

Bernasconi, S. M. 1994. *Geochemical and Microbial Controls on Dolomite Formation in Anoxic Environments: A Case Study from the Middle Triassic (Ticino, Switzerland)*. Contributions to Sedimentology, no. 19. Stuttgart: Schweizerbart'sche Verlagsbuchhandlung.

Bernasconi, S. M., and A. Riva. 1993. Organic geochemistry and depositional environment of a hydrocarbon source rock: The Middle Triassic Grenzbitumenzone Formation, southern Alps, Italy/Switzerland. In A. M. Spencer, ed., *Generation, Accumulation and Production of Europe's Hydrocarbons*, vol. 3, pp. 179–190. Special Publication of the European Association of Petroleum Geologists, no. 3. Berlin: Springer-Verlag.

Brack, P., and H. Rieber. 1993. Towards a better definition of the Anisian/Ladinian boundary: New biostratigraphic data and correlations of boundary sections from the Southern Alps. *Eclogae geologicae Helvetiae* 86:415–527.

Brett, C. E., and A. Seilacher. 1991. Fossil Lagerstätten: A taphonomic consequence of event sedimentation. In G. Einsele, W. Ricken, and A. Seilacher, eds., *Cycles and Events in Stratigraphy*, pp. 283–297. Berlin: Springer-Verlag.

Brinkmann, W. 1996. Ein Mixosaurier (Reptilia, Ichthyosauria) mit Embryonen aus der Grenzbitumenzone (Mitteltrias) des Monte San Giorgio (Schweiz, Kanton Tessin). *Eclogae geologicae Helvetiae* 89:1321–1344.

Brinkmann, W. 1998. Die Ichthyosaurier (Reptilia) aus der Grenzbitumenzone (Mitteltrias) des Monte San Giorgio (Tessin, Schweiz)—Neue Ergebnisse. *Vierteljahresschrift der naturforschenden Gesellschaft in Zürich* 143:165–177.

Bürgin, T. 1992. *Basal Ray-finned Fishes (Osteichthyes; Actinopterygii) from the Middle Triassic of Monte San Giorgio (Canton Tessin, Switzerland).* Schweizerische paläontologische Abhandlungen, no. 114. Basel: Birkhäuser Verlag.

Bürgin, T. 1996. Diversity in the feeding apparatus of perleidid fishes (Actinopterygii) from the Middle Triassic of Monte San Giorgio (Switzerland). In G. Arratia and G. Viohl, eds., *Mesozoic Fishes—Systematics and Paleoecology*, pp. 555–565. Munich: Verlag Pfeil.

Bürgin, T., O. Rieppel, P. M. Sander, and K. Tschanz. 1989. The fossils of Monte San Giorgio. *Scientific American* 260:74–81.

Etter, W. 1994. A new penaeid shrimp (*Antrimpos mirigiolensis* n. sp., Crustacea, Decapoda) from the Middle Triassic of the Monte San Giorgio (Ticino, Switzerland). *Neues Jahrbuch für Geologie und Paläontologie, Monatshefte* 1994: 223–230.

Frauenfelder, A. 1916. Beiträge zur Geologie der Tessiner Kalkalpen. *Eclogae geologicae Helvetiae* 14:247–371.

Furrer, H. 1995. The Kalkschieferzone (upper Meride Limestone; Ladinian) near Meride (Canton Ticino, Southern Switzerland) and the evolution of a Middle Triassic intraplatform basin. *Eclogae geologicae Helvetiae* 88:827–852.

Krebs, B. 1965. Ticinosuchus ferox *nov. gen. nov. sp.* Schweizerische paläontologische Abhandlungen, no. 81. Basel: Birkhäuser Verlag.

Kuhn-Schnyder, E. 1974. Die Triasfauna der Tessiner Kalkalpen. *Neujahrsblatt der naturforschenden Gesellschaft in Zürich* 176:1–119.

Müller, W. 1969. Beitrag zur Sedimentologie der Grenzbitumenzone vom Monte San Giorgio (Kt. Tessin) mit Rücksicht auf die Beziehung Fossil-Sediment. Ph.D. diss., University of Basel.

Müller, W., R. Schmid, and P. Vogt. 1964. Vulkanogene Lagen aus der Grenzbitumenzone (Mittlere Trias) des Monte San Giorgio in den Tessiner Kalkalpen. *Eclogae geologicae Helvetiae* 57:431–450.

Mundil, R., P. Brack, M. Meier, H. Rieber, and F. Oberli. 1996. High-resolution U-Pb dating of Middle Triassic volcanoclastics: Time-scale calibration and verification of tuning parameters for carbonate sedimentation. *Earth and Planetary Sciences Letters* 141:137–151.

Nopcsa, F. von. 1923. Neubeschreibung des Trias-Pterosauriers *Tribelesodon*. *Paläontologische Zeitschrift* 5:161–181.

Peyer, B. 1931. Placodontia (Die Triasfauna der Tessiner Kalkalpen III). *Abhandlungen der schweizerischen paläontologischen Gesellschaft* 50:7–110.

Peyer, B. 1944. Die Reptilien vom Monte San Giorgio. *Neujahrsblatt der naturforschenden Gesellschaft in Zürich* 146:1–95.

Premru, E. 1991. Beschreibung eines neuen Fundes von *Macrocnemus bassanii* Nopcsa (Reptilia, Squamata, Prolacertiliformes) aus der Grenzbitumenzone (Anis/Ladin) von Besano, Italien. Diploma thesis, University of Zurich.

Rieber, H. 1968. Die Artengruppe der *Daonella elongata* Mojs. aus der Grenzbitumenzone der Mittleren Trias des Monte San Giorgio (Kt. Tessin, Schweiz). *Paläontologische Zeitschrift* 42:33–61.

Rieber, H. 1969. Daonellen aus der Grenzbitumenzone der Mittleren Trias des Monte San Giorgio (Kt. Tessin, Schweiz). *Eclogae geologicae Helvetiae* 62:657–683.

Rieber, H. 1970. *Phragmoteuthis? ticinensis* n. sp., ein Coleoidea-Rest aus der Grenzbitumenzone (Mittlere Trias) des Monte San Giorgio (Kt. Tessin, Schweiz). *Paläontologische Zeitschrift* 44:32–40.

Rieber, H. 1973a. Ergebnisse paläontologisch-stratigraphischer Untersuchungen in der Grenzbitumenzone (Mittlere Trias) des Monte San Giorgio (Kanton Tessin, Schweiz). *Eclogae geologicae Helvetiae* 66:667–685.

Rieber, H. 1973b. *Cephalopoden aus der Grenzbitumenzone (Mittlere Trias) des Monte San Giorgio (Kanton Tessin, Schweiz).* Schweizerische paläontologische Abhandlungen, no. 93. Basel: Birkhäuser Verlag.

Rieber, H. 1975. Der Posidonienschiefer (oberer Lias) von Holzmaden und die Grenzbitumenzone (mittlere Trias) des Monte San Giorgio: Ein Vergleich zweier Lagerstätten fossiler Wirbeltiere. *Jahreshefte der Gesellschaft für Naturkunde Württemberg* 130:163–190.

Rieber, H. 1980. Ein Conodonten-Cluster aus der Grenzbitumenzone (Mittlere Trias) des Monte San Giorgio (Kanton Tessin/Schweiz). *Annalen des Naturhistorischen Museums* (Vienna) 83:265–274.

Rieber, H. 1982. The formation of the bituminous layers of the Middle Triassic of Ticino (Switzerland) [abstract]. In G. Einsele and A. Seilacher, eds., *Cyclic and Event Stratification*, p. 527. Berlin: Springer-Verlag.

Rieber, H., and L. Sorbini. 1983. Middle Triassic bituminous shales of Monte San Giorgio (Tessin, Switzerland). In H. Rieber and L. Sorbini, eds., *First International Congress on Paleoecology, Excursion 11A Guidebook*, pp. 1–17.

Rieppel, O. 1982. A new genus of shark from the Middle Triassic of Monte San Giorgio. *Palaeontology* 25:399–412.

Rieppel, O. 1985. *Die Gattung* Saurichthys *(Pisces, Actinopterygii) aus der Mittleren Trias des Monte San Giorgio, Kanton Tessin.* Schweizerische paläontologische Abhandlungen, no. 108. Basel: Birkhäuser Verlag.

Rieppel, O. 1987. *Clarazia* and *Hescheleria*: A re-investigation of two problematical reptiles from the Middle Triassic of Monte San Giorgio (Switzerland). *Palaeontographica, A* 195:101–129.

Rieppel, O. 1992. A new species of the genus *Saurichthys* (Pisces, Actinopterygii) from the Middle Triassic of Monte San Giorgio (Switzerland), with comments on the phylogenetic interrelationships of the genus. *Palaeontographica, A* 221:63–94.

Sander, P. M. 1988. A fossil reptile embryo from the Middle Triassic of the Alps. *Science* 239:780–783.

Sander, P. M. 1989. The pachypleurosaurids (Reptilia: Nothosauria) from the Middle Triassic of Monte San Giorgio (Switzerland) with the description of a new species. *Philosophical Transactions of the Royal Society of London, B* 325:561–670.

Schwarz, W. 1970. Birgeria stensiöi *Aldinger.* Schweizerische paläontologische Abhandlungen, no. 89. Basel: Birkhäuser Verlag.

Seilacher, A., W. E. Reif, and F. Westphal. 1985. Sedimentological, ecological and temporal patterns of fossil Lagerstätten. *Philosophical Transactions of the Royal Society of London, B* 311:5–23.

Tintori, A. 1992. Fish taphonomy and Triassic anoxic basins from the alps: A case history. *Rivista Italiana di Paleontologia e Stratigrafia* 97:393–408.

Tschanz, K. 1988. Allometry and heterochrony in the growth of the neck of Triassic prolacertiform reptiles. *Palaeontology* 31:997–1011.

Wignall, P. B., and M. J. Simms. 1990. Pseudoplankton. *Palaeontology* 33: 359–378.

Wild, R. 1974. Tanystropheus longobardicus *(Bassani)*. Schweizerische paläontologische Abhandlungen, no. 95. Basel: Birkhäuser Verlag.

Wirz, A. 1945. *Beiträge zur Kenntnis des Ladinikums im Gebiet des Monte San Giorgio.* Schweizerische paläontologische Abhandlungen, no. 65. Basel: Birkhäuser Verlag.

Ziegler, P. A. 1988. Post-Hercynian plate reorganization in the Tethys and Arctic-North Atlantic domains. In W. Manspeizer, ed., *Triassic-Jurassic Rifting: Continental Breakup and the Origin of the Atlantic Ocean and Passive Margins*, part B, pp. 711–755. Developments in Geotectonics, vol. 22. Amsterdam: Elsevier.

Zorn, H. 1971. *Paläontologische, stratigraphische und sedimentologische Untersuchungen des Salvatoredolomits (Mitteltrias) der Tessiner Kalkalpen.* Schweizerische paläontologische Abhandlungen, no. 91. Basel: Birkhäuser Verlag.

13

Berlin-Ichthyosaur: Preserving Some of the Earth's Largest Marine Vertebrates

David J. Bottjer

REMARKABLE FOSSIL DEPOSITS THAT PRESERVE ARTICU-lated marine reptiles are common among Mesozoic Lagerstätten. Although some of these Lagerstätten, such as Monte San Giorgio (Chapter 12), preserve a startling diversity of articulated marine reptiles, and some, such as the Posidonia Shale (Chapter 15), have extensive soft-tissue preservation, others lay their claim to fame on different factors. Complete skeletons of the ichthyosaur *Shonisaurus popularis* are found within the Upper Triassic Luning Formation in the Shoshone Mountains at Berlin-Ichthyosaur State Park, Nye County, Nevada (Camp 1981; Orndorff, Wieder, and Filkorn 2001) (Figures 13.1 and 13.2). These skeletons come from a deposit reputed to be the richest source of ichthyosaurs in North America (Hogler 1992). The park takes its name not only from the ichthyosaur fossils, but also from the nearby ghost town of Berlin, founded in the late nineteenth century by miners who are said to have used ichthyosaur vertebrae as dinner plates. They easily could have functioned as dinner plates, because the ichthyosaur skeletons show that some individuals apparently reached a length of 18 m, with skulls 3 m long (Figure 13.3) and a weight in life of approximately 40 tons (Camp 1981). Such large carnivores have been termed *superpredators* (Fischer and Arthur 1977), and these Late Triassic ichthyosaurs were apparently some of the largest marine animals known up until that time (Alexander 1998). Ichthyosaurs from this locality are interpreted to have been deposited as part of a conservation Lagerstätte that was formed as a stagnation deposit (Hogler 1992).

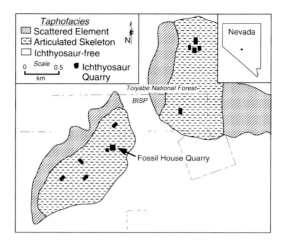

FIGURE 13.1 Location of ichthyosaur quarries at Berlin-Ichthyosaur State Park (BISP), Nevada, and map of ichthyosaur taphofacies within the Upper Triassic Luning Formation. (Modified from Hogler 1992)

Knowledge of these ichthyosaurs comes largely through the work of Charles Camp, a vertebrate paleontologist who taught at the University of California, Berkeley. Camp excavated many of the known skeletons in the area during the 1950s and was instrumental in developing the exhibits currently at the park (Camp 1981). Subsequently, Jennifer Hogler (1992) studied the sedimentology and paleoecology of the Luning Formation in the state park area, and much of the information presented here on paleoenvironments and taphonomy is based on her observations.

GEOLOGICAL CONTEXT

The Luning Formation in the Shoshone Mountains, Nevada, is over 1,000 m thick (Silberling 1959; Hogler 1992) (Figure 13.2). Luning strata in the Shoshone Mountains and surrounding region were deposited on an allocthonous terrane that was located during the Late Triassic in the Proto-Pacific Ocean (Panthalassia) at tropical latitudes offshore of North America (Tozer 1982). Such terranes represent ancient oceanic islands or microcontinents that were originally situated on top of ocean crust. Due to plate tectonic transport, the Luning terrane collided with and was sutured to North America, probably before the end of the Jurassic (Tozer 1982). These plate tectonic processes have continued to accrete material to North America so that the Luning terrane is now separated from the Pacific Ocean by several hundred kilometers of additional continental crust.

PALEOENVIRONMENTAL SETTING

Earlier studies interpreted the portion of the Luning Formation that contains articulated ichthyosaurs to have been deposited in an intertidal setting (Camp 1980, 1981). Subsequent studies, however, have shown that Luning sedimentary facies represent carbonate bank deposition as well as associated deeper shelf and basinal settings (Hogler 1992). Hogler (1992) has defined intervals of the Luning based on ichthyosaur taphofacies (Figure 13.2). The basal 250 m of the Luning Formation does not contain ichthyosaur remains, but disassociated elements are

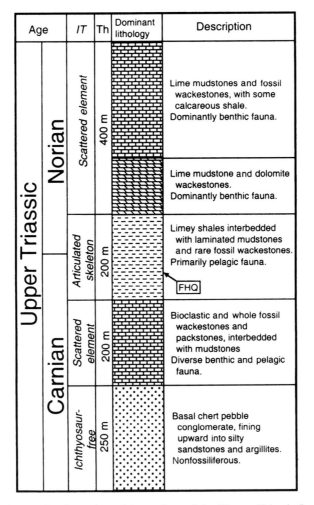

Age	IT	Th	Dominant lithology	Description
Upper Triassic — Norian	Scattered element	400 m		Lime mudstones and fossil wackestones, with some calcareous shale. Dominantly benthic fauna.
				Lime mudstone and dolomite wackestones. Dominantly benthic fauna.
	Articulated skeleton	200 m		Limey shales interbedded with laminated mudstones and rare fossil wackestones. Primarily pelagic fauna. FHQ
Upper Triassic — Carnian	Scattered element	200 m		Bioclastic and whole fossil wackestones and packstones, interbedded with mudstones Diverse benthic and pelagic fauna.
	Ichthyosaur-free	250 m		Basal chert pebble conglomerate, fining upward into silty sandstones and argillites. Nonfossiliferous.

FIGURE 13.2 Generalized stratigraphic section of the Upper Triassic Luning Formation (Shoshone Mountains, Nevada), showing stratigraphic location of the Fossil House Quarry (FHQ), as well as distribution of lithology and ichthyosaur taphofacies (IT). (Modified from Hogler 1992)

found from 250 to 450 m above the base of the formation and form the scattered element taphofacies. Articulated skeletons found at the park occur at 450 to 650 m, and their presence defines Hogler's (1992) articulated skeleton taphofacies. The scattered element taphofacies is also found from 650 to 1,050 m above the base of the Luning.

The portion of the Luning that contains the scattered element taphofacies occurs in rocks indicating carbonate platform conditions deposited near the platform edge (Hogler 1992). Evidence for these conditions includes limestone beds with diverse shallow-water benthic invertebrate faunas that are interbedded with beds of mudstone and shale more indicative of pelagic or basinal depositional conditions (Hogler 1992).

The central Luning interval, which contains articulated ichthyosaur skeletons and represents a conservation Lagerstatte, is composed primarily of laminated black mudstones and thinly bedded gray shales. Associated fossils include an abundant pelagic fauna, but fossils of benthic invertebrates are almost nonexistent (Hogler 1992). Some of the fossil deposits found in this portion of the unit appear to represent separate ichthyosaur and ammonite mass-mortality events (Hogler 1992). Hogler (1992) interprets this Luning interval as representing basinal deposition.

TAPHONOMY

Hogler (1992) has done a detailed study of the taphonomy of the ichthyosaur remains (Figure 13.2), and the following discussion is a summary of her results. The scattered element taphofacies includes dissociated elements and partial individuals. These ichthyosaur bones are commonly abraded and broken, and most likely experienced vigorous current and/or scavenging activity. Large elements such as pelvic and pectoral girdles are widely dispersed, indicating that bones may have originated from floating carcasses. Encrusters such as bivalves are commonly found attached to these bones, which is good evidence that many of them remained at the seafloor for extended periods before final burial.

The articulated skeleton taphofacies typically includes mostly complete individuals of *Shonisaurus popularis,* although some fragmented skeletons are found. Many skeletons show little postmortem disturbance, and while most are found alone, some occur in aggregations. In particular, the central fossil display site of the park, the Fossil House Quarry (Figures 13.1 and 13.2), contains a spectacular example of aggregation, with nine articulated skeletons of adult *Shonisaurus popularis* found along a bedding surface 8 m wide by 20 m long. Hogler (1992) interprets this aggregation to represent a mass mortality, and although evidence for its cause is not apparent, she suggests phytotoxin poisoning that occurred during a dinoflagellate red tide. Preservation of the bones in these skeletons is remarkably good, with no encrusters or borers, indicating that reduced levels of dissolved oxygen in bottom-waters inhibited scavengers and other benthic organisms, thus leading to the formation of a conservation Lagerstätte that is a stagnation deposit.

PALEOBIOLOGY AND PALEOECOLOGY

During the several decades that he worked on these ichthyosaurs, Camp excavated at least 40 skeletons of *Shonisaurus popularis,* four of which were removed from the state park area for detailed study (Camp 1980). The occurrence of such a large number of specimens makes this species the best known of all Late Triassic ichthyosaurs (Carroll 1988). *Shonisaurus popularis* is also the largest of all ichthyosaurs (Camp 1980; Carroll 1988). Two other species of *Shonisaurus* are found in the Luning Formation, *S. silberlingi* and *S. mulleri,* but they are represented only by largely incomplete, single skeletons (Camp 1976). Because the color and tone of the ichthyosaur bones at this site are similar to those of the surrounding sedimentary rock, informative photographs showing complete articulated skeletons (Figure 13.3) are not available. Perhaps the best photographs, showing partial skeletons, are found in Hogler (1992).

These ichthyosaurs occur both as isolated individuals and as groups of skeletons on bedding planes, so that *Shonisaurus popularis* is interpreted to have been subject to both attritional and catastrophic mortality (Hogler 1992). The great abundance of dissociated skeletal elements, as well as numerous individual skeletons, indicates that this ichthyosaur lived in fairly high-density populations. Camp (1980, 1981), who carried out his studies before modern sedimentological analysis of these rocks had been done, postulated that the mode of death for most of the articulated Luning ichthyosaur skeletons was by stranding in nearshore sediments during unusually low tides. Hogler (1992) has abandoned this interpretation because the sedimentology of the rocks in which the ichthyosaurs are found, as described earlier, indicates a deep-water depositional environment. Massare (1988) explains the great abundance of ichthyosaur remains in the Luning as the result of this region being an ichthyosaur breeding or birthing area in the Late Triassic. However, because juvenile ichthyosaurs, which should be common in birthing areas, are very rare, Hogler (1992) suggests that generally favorable environmental conditions caused the great abundance of ichthyosaurs. In particular, Hogler (1992) postulates that this may have been a significant upwelling area that could have supported high productivity, thus providing good feeding grounds for ichthyosaurs. The presence of clusters of skeletons indicates that at times adults engaged in group activities, but their exact nature is currently unknown (Hogler 1992).

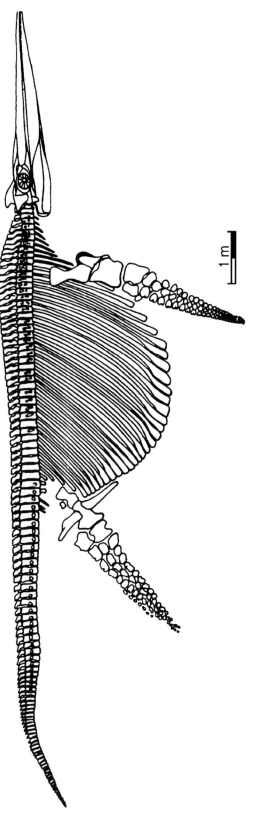

FIGURE 13.3 Complete skeleton of *Shonisaurus popularis*, which reached a length of 15 m. The heads of the ribs are not drawn, in order to show the articulating surfaces of the vertebrae. (Modified from Camp 1980)

CONCLUSIONS

Due to seafloor spreading and the continual destruction of ocean crust in subduction zones, no deep-sea sedimentary sequences currently exist in the oceans that are older than Jurassic in age. Thus, we must depend on allocthonous terranes, which once existed out in the ocean but are now "plastered" onto the continents, as sources of evidence on open-ocean life and environments for times older than the Jurassic, such as this Triassic example. The ichthyosaurs found in the conservation Lagerstätte at Berlin-Ichthyosaur State Park tell us that in the Late Triassic, the largest ichthyosaurs ever known roamed the Proto-Pacific Ocean, feeding along the offshore margins of carbonate banks. Intriguingly, this was during the initial evolutionary history of ichthyosaurs, which first appear in the fossil record in Lower Triassic rocks about 10 million years older than the Luning Formation, and reached their greatest diversity only 20 to 30 million years later in the Early Jurassic, whereupon they experienced a slow decline until final extinction 100 million years later in the early Late Cretaceous (Carroll 1988). Thus, although this ichthyosaur Lagerstätte does not preserve evidence of soft tissues, such as that found in the Posidonia Shale (Chapter 15), it is perhaps the most informative site currently known for information on the paleobiology of early ichthyosaurs.

ACKNOWLEDGMENTS

I thank J. Hogler for discussions that led to the improvement of this chapter.

REFERENCES

Alexander, R. M. 1998. All-time giants: The largest animals and their problems. *Palaeontology* 41:1231–1245.

Camp, C. L. 1976. Vorlaufige Mitteilung uber grosse Ichthyosaurier aus der oberen Trias von Nevada. *Osterreichische Akademie der Wissenschaften, Mathematische-Naturwissenschaftliche Klasse, Sitzungberichte* I,185:125–134.

Camp, C. L. 1980. Large ichthyosaurs from the Upper Triassic of Nevada. *Palaeontographica, A* 170:139–200.

Camp, C. L. 1981. *Child of the Rocks: The Story of Berlin-Icthyosaur State Park.* Nevada Bureau of Mines and Geology, Special Publication, no. 5. Reno: Nevada Bureau of Mines and Geology, in association with the Nevada Division of State Parks and the Nevada Natural History Association.

Carroll, R. L. 1988. *Vertebrate Paleontology and Evolution.* New York: Freeman.

Fischer, A. G., and M. A. Arthur. 1977. Secular variations in the pelagic realm. In H. E. Cook and P. Enos, eds., *Deep-Water Carbonate Environments,* pp. 19–50. Society of Economic Paleontologists and Mineralogists Special Publication, no. 25. Tulsa, Okla.: Society of Economic Paleontologists and Mineralogists.

Hogler, J. A. 1992. Taphonomy and paleoecology of *Shonisaurus popularis* (Reptilia: Icthyosauria). *Palaios* 7:108–117.

Massare, J. A. 1988. Live birth in ichthyosaurs: Evidence and implications. *Journal of Vertebrate Paleontology* 8:21A.

Orndorff, R. L., R. W. Wieder, and H. F. Filkorn. 2001. How the west was swum. *Natural History* 110:22–24.

Silberling, N. J. 1959. Pre-Tertiary stratigraphy and Upper Triassic paleontology of the Union District, Shoshone Mountains, Nevada. *United States Geological Survey Professional Paper* 322:1–67.

Tozer, E. T. 1982. Marine Triassic faunas of North America: Their significance for assessing plate and terrane movements. *Geologische Rundschau* 71:1077–1104.

14

Osteno: Jurassic Preservation to the Cellular Level

Carol M. Tang

JURASSIC DEPOSITS NEAR OSTENO IN NORTHERN ITALY HAVE yielded a beautifully preserved marine fauna that contains not only fish, sharks, and crustaceans, but also polychaetes, nematodes, and one of the few enteropneusts (acorn worms) in the fossil record. Although nonmarine fossils are not a dominant part of the assemblage, terrestrial plants and the first Italian dinosaur have also been recovered from this deposit. Fossils are fairly sparse in this stagnation Lagerstätte, but the yield is well worth the search as the spectacular preservation has allowed for soft-part preservation even at the cellular level. The presence of details—ranging from cephalopod musculature patterns to polychaete coloration to arthropod stomach contents—provides paleobiological and paleoecological information that is extremely rare in the fossil record. The Osteno deposit was discovered in 1964, and the first fossils were described in 1967 by Giovanni Pinna. The first information on these soft-body organisms was published by Arduini, Pinna, and Teruzzi in 1980, and these fossils have been systematically collected, beginning in 1980, by the Natural History Museum in Milan. Surprisingly, the extraordinary Osteno fauna and mode of soft-body preservation have not achieved the worldwide fame that many other Jurassic Lagerstätten have enjoyed.

GEOLOGICAL CONTEXT

The Osteno Lagerstätte is part of the Lombardische Kieselkalk Formation, also known as the Moltrasio Limestone Formation. The deposit is exposed in quarries along the northeastern bank of Lake Ceresio in the province of Como in northern Italy (Figure 14.1). The fossil-bearing unit commonly outcrops as vertical cliffs along the shore, making detailed study and collection difficult and sometimes dangerous. In addition, the many structural folds in the area further complicate the study of the geology of this formation (G. Teruzzi, personal communication, 1996).

The Lombardische Kieselkalk Formation is composed of mostly nonfossiliferous saccharoidal siliceous limestone with ammonites present in some layers. Fossils with soft-part preservation occur throughout the 4 m thick Osteno unit. This unit is both vertically and laterally restricted and is composed of a massive bed of gray micrite bounded both above and below by nonfossiliferous marl horizons that grade into the characteristically nonfossiliferous Lombardische Kieselkalk limestone facies (Pinna 1985).

Carbonate petrology of the Osteno unit indicates that it is composed of microcrystalline calcite, pyrite, hematite, and limonite (Pinna 1985). The unit contains large amounts of sponge and radiolarian spicules, which have been mostly replaced by calcite. Small geodes containing hydrocarbons are also found in some areas (Pinna 1985).

The unit is a massive micrite with no evidence of stratification, crossbedding, grading, bioturbation, or current activity (Pinna 1985). However, pseudostratification can be detected as a result of differential diagenetic concentration of limonitic material in some layers (Pinna 1985).

The age of the Osteno Lagerstätte unit is inferred to be Early Sinemurian (*Bucklandi* zone) based on ammonite biostratigraphy (Pinna

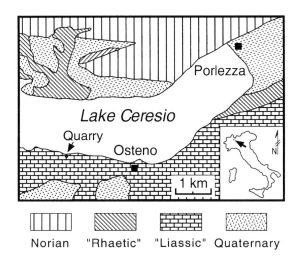

FIGURE 14.1 Geological setting of Osteno deposits in northern Italy. (Modified from Duffin and Patterson 1993)

1967). In addition, similarities between the Osteno fauna (e.g., crustaceans) and the Sinemurian fauna from Lyme Regis, England, support this age determination (Pinna 1985).

PALEOENVIRONMENTAL SETTING

The restricted nature of this unit—limited lateral and vertical extent—suggests that the Osteno Lagerstätte was the result of a brief sedimentary episode that did not affect a large area (Pinna 1985). The unit was probably deposited in a depression on the seafloor in a shelfal marine environment (G. Teruzzi, personal communication, 1996).

The absence of significant sedimentary features, size sorting, and fossil alignment suggests that there was little to no wave or current energy in this environment during deposition (Pinna 1985). The environment was probably a reducing one, as evidenced by the pyrite and high organic content in the sediment (Pinna 1985). In addition, the dissolution of aragonite and calcite suggests that in and/or near the seafloor, the seawater had a low pH and dissolved hydrogen sulfide. The lack of trace fossils, body fossils of infaunal organisms, and signs of decomposition indicates low oxygen conditions (Pinna 1985). However, the presence of a low-diversity benthic fauna suggests that some oxygen was available in the overlying bottom-waters and that oxygen deprivation probably occurred only in the sediment. The $H_2S–O_2$ boundary is hypothesized to have been only slightly deeper than the water–sediment interface (Pinna 1985).

Terrestrial plant fossils indicate that the regional climate was warm and relatively arid (Bonci and Vanucci 1986). There are large numbers of sponge spicules in the Osteno strata, and it has been hypothesized that the formation could represent an allochthonous deposit from a "siliceous sponge grassland" (Teruzzi 1990).

TAPHONOMY

Although skeletonized macrofauna—including molluscs, brachiopods, and rare ophiuroids—are present in this deposit, almost no calcareous skeletal parts are preserved (Pinna 1985). The ophiuroids have clearly undergone recrystallization (Arduini, Pinna, and Teruzzi 1983b). Bivalves and ammonites are found primarily as molds and imprints. However, noncalcareous skeletal parts, such as ammonite periostraca, siphuncles, and anaptychi, can be found as well as phosphatic and chitinous parts of aragonitic organisms. Fossilized hard parts composed of calcium phosphate, such as fish scales, coleoid hooks, and arthropod exoskeletons, are well preserved in Osteno (Pinna 1985).

Although the preservation of articulated fish and arthropods alone would classify Osteno as a conservation Lagerstätte, it is the preservation of soft-bodied organisms that makes this deposit an even more spectacular one. Among the fossils found at Osteno are coleoids, polychaetes, and nematodes—some preserved complete with digestive tracts (Arduini, Pinna, and Teruzzi 1982, 1983a). Other examples of soft-part preservation come from fish (Duffin 1987) and the muscles and branchia of crustaceans (Pinna et al. 1985).

Soft-part preservation occasionally occurred through replacement of the original organic matter by fine-grained minerals (Pinna 1985). Scanning electron microscope (SEM) work indicates that molecules of organic material have been replaced by tiny colloidal spherules of calcium phosphate less than 1 μm in diameter. The small size of the replacing colloidal material has allowed for the preservation of fine features, even capturing details at the cellular level. The organic matter appears to have been preserved without undergoing any decomposition on the seafloor (Pinna 1985), and the low level of decomposition, high concentrations of organic carbon, and abundance of authigenic pyrite all suggest seafloor reducing conditions (Pinna 1985).

Preservation of phosphatized soft tissues is usually represented as coatings of mineralized microbes (Wilby, Briggs, and Pinna 1995). Because arthropods, especially crustaceans, more commonly exhibit phosphatized soft-tissue preservation than do other organisms, it is hypothesized that the phosphate was derived mostly from the tissues of the decaying organisms, not from the seawater or sediments (Wilby, Briggs, and Pinna 1995). Some cells in nematodes are also preserved in phosphate, while soft parts of teuthids are preserved as kerogen films (G. Teruzzi, personal communication, 1996).

The benthic organisms are believed to be parautochthonous assemblages (G. Teruzzi, personal communication, 1996). The fine preservation of complete arthropod molts and detailed anatomical features, such as antennae and limbs, indicates that very limited transport must have occurred. Nektic organisms are thought to have been derived from the overlying water column and did not experience great amounts of lateral transport (Pinna 1985). However, the lack of bioturbation suggests that the many benthic organisms presumed to have been burrowers were transported and deposited here. The occurrence of terrestrial plant material also indicates that at least some fossil elements were transported to the area (Pinna 1985). The presence of many disarticulated fish—many of which lack skulls and vertebral elements—is attributed to predation and/or scavenging (Pinna 1985; Duffin and Patterson 1993) and possible regurgitation by predators (Pinna 1985).

PALEOBIOLOGY AND PALEOECOLOGY

Plants

Land plants are relatively common in Osteno, including horsetails (*Equisetites*), pteridosperms (*Pachypteris*), cycadeoids (*Zamites, Otozamites, Williamsonia*), and conifers (*Brachyphullum, Pagiophyllum*) (Bonci and Vannucci 1986). Floral analyses indicate that the fossils are derived from the tropical Tethyan phytogeographic realm and can be differentiated as representing several continental environments ranging from coastal to inland areas (Bonci and Vannucci 1986).

Invertebrates

There is a wide variety of taxa, including crustaceans, ammonites, squids, polycheates, brittle stars, brachiopods, bivalves, and nematodes. However, crustaceans dominate the fauna in terms of both species (over 60 percent of macroinvertebrate species) and individuals (92 percent of all collected specimens and 78 percent of all fossils in one bed) (Pinna 1985). Decapods alone make up about 50 percent of the macroinvertebrate individuals in the Osteno fauna (Garassino and Teruzzi 1990), and in the upper section of the unit, where a systematic paleontological survey was conducted, arthropods of the class Thylacocephala were found to be the most common (Pinna 1985).

Decapods include members of the infraorder Astacidea (crayfish, clawed lobsters) and the infraorder Palinura (spiny and slipper lobsters) (Garassino 1996). Osteno also contains one of the oldest known occurrences of mantis shrimp. Detailed systematic work has also been published on the penaeid shrimps (Garassino and Teruzzi 1990) and eryonoid lobsters (Pinna 1968, 1969; Teruzzi 1990).

Based on hundreds of very well preserved Osteno specimens and one described species, *Ostenocaris cypriformis* (Arduini, Pinna, and Teruzzi 1980), Pinna et al. (1982) established a new class of arthropods: the Thylacocephala. Arduini and Pinna (1989) have proposed that Thylacocephala includes about two dozen genera, spans from the Cambrian to the Upper Cretaceous, and includes specimens found in the Devonian of Australia, the Pennsylvanian Mazon Creek Lagerstätte (Chapter 10), and the Jurassic of France. Although most Thylacocephala specimens from Osteno remain undescribed, there appear to be three genera and numerous species (Teruzzi 1990). There has been much controversy over this class due to their unusual nature and lack of characteristics similar to those of other arthropods (Schram 1990).

As a result of extremely good preservation, the muscles, gills, stomach contents, and even possible remnants of the nervous system of Osteno Thylacocephala can be studied. *Ostenocaris cypriformis* is unique in that it does not have an optic notch and its cephalon has been modified into a large, protruding sac (Figure 14.2). This sac has been interpreted by some as eyes (Rolfe 1985; Secretan 1985), but other workers have asserted that it is highly sclerotized muscular organs used in the digestive process. The sac contains one or two layers of sclerites and strongly developed musculature (Alessandrello et al. 1991). The presence of what appear to be stomach contents—vertebrae of a common Osteno shark, teuthid hooks, and exoskeletons of decapods—found near the cephalic sac region has led Alessandrello et al. (1991) to propose that the sac was a blind extension of the Thylacocephala gut and used to break down large food pieces with hard parts. Pinna et al. (1985) hypothesized that the cephalic sac may have been used as a burrowing organ (Figure 14.2), but without more evidence, this idea remains a theory.

O. cypriformis is hypothesized to have had an unsegmented anterior cephalon followed by a segmented thorax that was covered by a large carapace attached to the body with muscles (Pinna et al. 1985). Many of the coprolites found in the deposit seem to match the stomach contents of Thylacocephala specimens. These include remains of cephalopod

FIGURE 14.2 Reconstruction of morphology and hypothetical life habitat of the thylacocephalan *Ostenocaris cypriformis*. Individuals could attain a length of up to 20 cm. (Modified from Pinna et al. 1985)

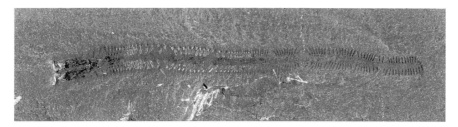

FIGURE 14.3 The polychaete *Melanoraphia maculata*. Specimen is about 83 mm in length. (Photo courtesy of Museo Civico di Storia Naturale di Milano)

hooks, vertebrae of small sharks, fish remains, and carapaces of crustaceans, as well as other thylacocephalans (Pinna et al. 1985). Thylacocephala were probably carnivorous scavengers (Alessandrello et al. 1991), as evidenced by these stomach contents and their spine-bearing limbs (Pinna et al. 1985).

Cephalopods include partially complete ammonites and coleoids, found either as detached hooks or as almost complete specimens (Pinna 1972; Garassino and Donovan 2000). Coleoid hooks have been found individually, in coprolites of marine reptiles, and in the stomach of Thylacocephala specimens. Soft-part preservation of coleoids shows eyes, muscles, jaws, and arms with hooks attached (Garassino and Donovan 2000).

Isolated parts of polychaete jaw apparatuses are found throughout several Osteno horizons. Fossils of the worms themselves are rare, since the deposit is massive and does not split along bedding planes (G. Teruzzi, personal communication, 1996). However, about a dozen specimens of polychaetes exhibiting soft-part preservation have been found (Figure 14.3). The largest polychaete specimens are about 15 cm long and 7 mm wide, although many are much smaller. Some of them have been assigned to the species *Melanoraphia maculata* (Arduini, Pinna, and Teruzzi 1982) (Figure 14.3). These polychaetes were mobile (errant) and are thought to be analogous to the living eunicid worms. Some specimens are nearly complete and have been preserved with their jaw apparatuses, acicula (rods that muscles attach to and move), an organic film where the body was, traces of the digestive tract, and, in some cases, marks that have been interpreted as coloration (Arduini, Pinna, and Teruzzi 1982). The pigmentation patterns are quite complex and consist of two types: (1) white spherical patches, which increase and decrease in size down the body, are situated in pairs on segments, and are hypothesized to correspond to gill placement; and (2) orange patches, which surround the aciculae in the front portion of the animal and are thought to correspond to the accumulation of pigment in specific glands (Arduini, Pinna, and Teruzzi 1982).

A new taxon of nematodes, *Eophasma jurasicum,* was erected based on Osteno specimens (Arduini, Pinna, and Teruzzi 1983a). These specimens are about 55 mm long and 2 mm wide with a tapering posterior end. They commonly exhibit traces of the esophagus and other digestive organs, and in some of the specimens, even the muscles of the pharynx and the esophagus have been preserved. Some possible denticles and setae are also found on these animals. The nematodes were probably free-living, but their relationship with modern taxa is undetermined.

One of the few descriptions of an enteropneust (acorn worm) in the fossil record was obtained from Osteno (Arduini, Pinna, and Teruzzi 1981). The specimen was first found in 1980 by an amateur paleontologist and was subsequently classified as *Megaderaion sinemuriense,* a new genus and species of the class Enteropneusta (Figures 14.4 and 14.5). The specimen is only about 20 mm long and 1.6 mm wide and is folded back on itself, which prevents a detailed description. However, the specimen clearly has three parts: a proboscis, collar, and trunk (Figure 14.5). Although most of the structures are not preserved in enough detail to study closely, two oval structures on the trunk have been interpreted as external gonads (Arduini, Pinna, and Teruzzi 1981) (Figure 14.5).

FIGURE 14.4 The first fossil acorn worm to be discovered, *Megaderaion sinemuriense.* Specimen is about 20.5 mm in length. (Photo courtesy of Museo Civico di Storia Naturale di Milano)

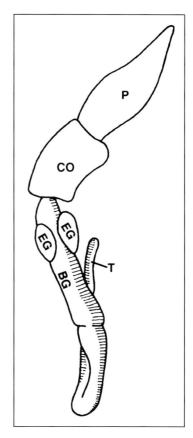

FIGURE 14.5 Reconstruction of the specimen of *Megaderaion sinemuriense*, where P refers to the proboscis, CO to the collar, EG to external gonads, T to the trunk, and BG to the branchio-genital region. (Modified from Arduini, Pinna, and Teruzzi 1981)

Vertebrates

A total of 12 fish and shark taxa have been identified in the Osteno deposit, accounting for about 10 percent of the total number of fossil specimens (Duffin and Patterson 1993). Among the fish fossils, only four taxa account for 90 percent of the specimens (Duffin and Patterson 1993). Some of these species are identified by isolated scales, some by partially articulated specimens, and some by beautifully preserved specimens, including those where the whole profile of the body is preserved (Arduini, Pinna, and Teruzzi 1983b). Of the 12 taxa, most either are present in slightly younger rocks at Lyme Regis in southern England or are related to forms found there (Duffin and Patterson 1993).

All sharks at Osteno are neoselachians, a group that includes all extant sharks and rays. The presence of modern shark taxa distinguishes the fauna at Osteno from that at Lyme Regis, which is dominated by an extinct group of hybodontid sharks (Duffin and Patterson 1993). A common taxon is an eel-shaped shark, *Ostenoselache stenosoma* (Duffin 1998), referred to by Osteno workers as a "skinny shark," based on its unusual morphology (Duffin and Patterson 1993). This organism is found in a range of preservation styles—as disarticulated vertebrae in the stomach contents of *Ostenocaris,* as parts of regurgitated ejecta, and as well-preserved individuals (Duffin and Patterson 1993). Based on soft-part preservation, it has been determined that this organism is unlike any shark, past or present, in its unusual body morphology. This includes a slender body up to 30 cm long, containing about 150 small vertebrae, and only one long median fin (Duffin and Patterson 1993; Duffin 1998). In one specimen, traces of the stomach and intestine show that the anus was also unusual in its forward placement (Duffin and Patterson 1993). The specimens have large orbital sockets, and the original organic pigment layers of the eye are commonly still preserved (Duffin and Patterson 1993). Because of the large size of the eye orbits, its unusual intestinal placement, and the similarity of this fish to modern knifefish and knife eels, Duffin and Patterson (1993) and Duffin (1998) suggest that the "skinny shark" may have been able to generate an electric field in its tail region and use electrolocation like modern knifefish and knife eels to navigate in turbid waters. In addition, they hypothesize that *Ostenoselache* may have used its anal fin for slow backward swimming in order to sense an area, as modern knife eels do, and detect predators such as *Ostenocaris* and other sharks.

Another shark, *Palaeospinax pinnai,* is believed to be one of the oldest and most primitive of the modern sharks (Duffin 1987). While one specimen consists of only one tooth (Duffin and Patterson 1993), the other is composed of teeth, scales from the head region, and complexly folded skin (Duffin 1987). Because *Paleospinax* has primitive neoselachian teeth that closely resemble hybodont teeth, the use of ultrastructural analysis of teeth was important in determining the systematic position of this taxon within the neoselachians (Duffin and Patterson 1993).

Other vertebrates include holocephalans, actinopterygians, and coelacanths (Schaeffer and Patterson 1984; Duffin 1992). One holocephalan, *Squaloraja polyspondyla,* is quite unusual in that it possesses a huge cartilaginous rostrum up to half of the entire length of the fish, which projected in front of the organism (Figure 14.6). Due to the excellent preservation at Osteno, this specimen represents the only known fossil shark or chimaeroid in which ichthyologists can reconstruct the entire pattern of sensory canals by examining the arrangement of crescent-shaped bones on the rostrum (Duffin and Patterson 1993). This rostrum may have been used for providing extrasensory information to the organism, but its function is not entirely clear (Duffin and Patterson 1993). In addition, the Osteno specimen is significant in that when analyzed in conjunction with *Squaloraja* specimens from Lyme Regis, it appears to suggest that sexual dimorphism does exist within the genus: males have frontal claspers, while females (such as the Osteno specimen) do not.

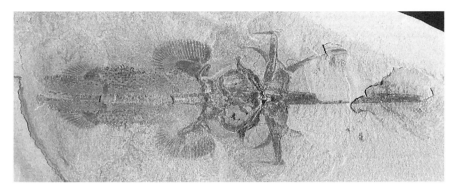

FIGURE 14.6 The holocephalan *Squaloraja polyspondyla*, which possesses a unique frontal rostra. Specimen is estimated to have had a total length of 29 cm. (Photo courtesy of Museo Civico di Storia Naturale di Milano)

Pholidophorus and *Pholidolepis*, the two genera of teleosts present in Osteno and interpreted as very primitive, are some of the most common vertebrate fossils found in the deposit. *Pholidophorus bechei* is a thin fish similar in shape to a herring but possessing thick cranial bones and enameled scales, unlike modern teleosts (Duffin and Patterson 1993). *Pholidolepis*—the most common bony fish, making up 23 percent of the total number of specimens—is usually found incomplete, and these specimens appear to have been preyed upon before burial (Duffin and Patterson 1993). Unlike the other genus of teleosts, this fish had thin modern-type scales, but had other primitive characteristics, suggesting that *Pholidoepis* is close to the base of the teleost clade (Duffin and Patterson 1993).

The first dinosaur collected in Italy was found within the Osteno deposits and is preserved with the skin and stomach contents (G. Teruzzi, personal communication, 1996).

Paleoecology

Due to the large number of invertebrate organisms and their excellent state of preservation, it is believed that the benthic fauna is parautochthonous and that the nektic organisms lived in the water column directly above and were not transported from afar. Pinna (1985) points out that many of the nektic organisms are represented as remains in coprolites and suggests that the nekton played an important ecological role in the benthic community, providing a food source for the scavenging crustaceans. In addition, judging from their dentition, at least some of the pelagic vertebrates, such as *Squaloraja, Palaeospinax,* and the bony fish, probably consumed benthic fauna and thus were important in the benthic ecosystem (Duffin and Patterson 1993).

The diversity of crustacean genera and species is very high, with 10 genera of decapods, five species of *Coleia,* and six morphological types of

Aeger (Garassino and Teruzzi 1990; Teruzzi 1990). The high degree of differentiation between species and genera has been attributed by Teruzzi (1990) to trophic niche specialization. He believes that this high crustacean diversity is analogous to that found in modern reefs, where crustaceans exploit a great diversity of available trophic niches and cryptic environments, and that the Osteno fauna is similar to modern demosponge communities, where crustaceans are the main accessory organisms. Teruzzi (1990) interprets the paleocommunity as a siliceous sponge grassland that supported a dominant and diverse crustacean community as well as nematodes, polychaetes, and ophiuroids, with associated nektic cephalopods and vertebrates.

CONCLUSIONS

With the extraordinary fossils and the small-scale details preserved in Osteno, it is surprising that this Lagerstätte has not received more scientific attention. Although there appears to be much more systematic, paleoecological, taphonomic, sedimentological, and geochemical work that could be attempted, only a small group of workers has studied these deposits in the past three decades. More detailed taphonomic and geochemical studies could contribute much to our understanding of the sedimentological and oceanographic conditions that led to this rare preservation style.

Although several taxonomic studies have already been conducted, including the erection of an entire class of crustaceans as well as numerous genera and species, there are still unidentified and unclassified organisms from this unit. For example, it has been estimated that among the fish fauna, one out of 10 specimens has represented a new taxon (Duffin and Patterson 1993). Specimens from this deposit have already influenced systematic interpretations of the crustacea, and more information about evolutionary relationships and life habits can be derived with additional study of the Osteno Lagerstätte.

ACKNOWLEDGMENTS

G. Teruzzi and G. Pinna provided very helpful reviews of this chapter, and G. Teruzzi and P. Arduini provided photographs. Special acknowledgment goes to G. Teruzzi, who provided much useful information during his visit to the United States.

REFERENCES

Alessandrello, A., P. Arduini, G. Pinna, and G. Teruzzi. 1991. New observations on the Thylacocephala (Arthropoda, Crustacea). In A. M. Simonetta and S. Conway Morris, eds., *The Early Evolution of Metazoa and the Significance of Problematic Taxa,* pp. 245–251. Cambridge: Cambridge University Press.

Arduini, P., and G. Pinna. 1989. *Tilacocefali: Una nuova classe di crostacei fossili.* Milan: Museo Civico di Storia Naturale di Milano.

Arduini, P., G. Pinna, and G. Teruzzi. 1980. A new and unusual Lower Jurassic cirriped from Osteno in Lombardy: *Ostenia cypriformis* n. g. n. sp. (preliminary note). *Atti della Società Italiana di Scienze Naturali e del Museo Civico di Storia Naturale de Milano* 121:360–370.

Arduini, P., G. Pinna, and G. Teruzzi. 1981. *Megaderaion sinemuriense* n. g. n. sp., a new fossil enteropneust of the Sinemurian of Osteno in Lombardy. *Atti della Società Italiana di Scienze Naturali e del Museo Civico di Storia Naturale de Milano* 122:104–108.

Arduini, P., G. Pinna, and G. Teruzzi. 1982. *Melanoraphia maculata* n. g. n. sp., a new fossil polychaete of the Sinemurian of Osteno in Lombardy. *Atti della Società Italiana di Scienze Naturali e del Museo Civico di Storia Naturale de Milano* 123:462–468.

Arduini, P., G. Pinna, and G. Teruzzi. 1983a. *Eophasma jurasicum* n. g. n. sp., a new fossil nematode of the Sinemurian of Osteno in Lombardy. *Atti della Società Italiana di Scienze Naturali e del Museo Civico di Storia Naturale de Milano* 124:61–64.

Arduini, P., G. Pinna, and G. Teruzzi. 1983b. The Sinemurian deposit of Osteno. In H. Rieber and L. Sorbini, eds., *First International Congress on Paleoecology, Excursion 11A Guidebook,* pp. 34–40.

Bonci, M. C., and G. Vannucci. 1986. I vegetali sinemuriani di Osteno (Lombardia). *Atti della Società Italiana di Scienze Naturali e del Museo Civico di Storia Naturale de Milano* 127:107–127.

Duffin, C. J. 1987. *Palaeospinax pinnai* n. sp., a new palaeospinacid shark from the Sinemurian (Lower Jurassic) of Osteno (Lombardy, Italy). *Atti della Società Italiana di Scienze Naturali e del Museo Civico di Storia Naturale de Milano* 128:185–202.

Duffin, C. J. 1992. A myriacanthid holocephalan from the Sinemurian (Lower Jurassic) of Osteno (Lombardy, Italy). *Atti della Società Italiana di Scienze Naturali e del Museo Civico di Storia Naturale de Milano* 132: 293–308.

Duffin, C. J. 1998. *Ostenoselache stenosoma* n. g. n. sp., a new neoselachian shark from the Sinemurian (Early Jurassic) of Osteno (Lombardy, Italy). *Paleontologia Lombarda* 9:1–27.

Duffin, C. J., and C. Patterson. 1993. I pesci fossili di Osteno: Una nuova finestra sulla vita del Giurassico Inferiore. *Paleocronache: Novità e Informazioni Paleontologiche* 2:18–38.

Garassino, A. 1996. The family Erymidae Van Straelen, 1924 and the superfamily Glypheoidea Zittel, 1885 in the Sinemurian of Osteno in Lombardy (Crustacea, Decapoda). *Atti della Società Italiana di Scienze Naturali e del Museo Civico di Storia Naturale de Milano* 135:333–373.

Garassino, A., and D. T. Donovan. 2000. A new family of coleoids from the Lower Jurassic of Osteno, Northern Italy. *Palaeontology* 43:1019–1038.

Garassino, A., and G. Teruzzi. 1990. The genus *Aeger* Münster, 1839 in the Sinemurian of Osteno in Lombardy (Crustacea, Decapoda). *Atti della Società Italiana di Scienze Naturali e del Museo Civico di Storia Naturale de Milano* 131:105–136.

Pinna, G. 1967. Découverte d'une nouvelle faune à crustacés du Sinémurien inferieur dans la région du lac Ceresio (Lombardie, Italie). *Atti della Società Italiana di Scienze Naturali e del Museo Civico di Storia Naturale de Milano* 106:183–185.

Pinna, G. 1968. Gli erioidei della nuova fauna sinemuriana a crostacei decapodi di Osteno in Lombardia. *Atti della Società Italiana di Scienze Naturali e del Museo Civico di Storia Naturale de Milano* 107:93–134.

Pinna, G. 1969. Due nuovi esemplari di Coleia viallii Pinna, del Sinemuriano inferiore di Osteno in Lombardia (Crustacea Decapoda). *Annali del Museo Civico di Storia Naturali* 77:626–632.

Pinna, G. 1972. Ritrovamento di un raro cefalopode coleoideo nel giacimento sinemuriano di Osteno in Lombardia. *Atti della Società Italiana di Scienze Naturali e del Museo Civico di Storia Naturale de Milano* 113:141–149.

Pinna, G. 1985. Exceptional preservation in the Jurassic of Osteno. *Philosophical Transactions of the Royal Society of London, B* 311:171–180.

Pinna, G., P. Arduini, C. Pesarini, and G. Teruzzi. 1982. Thylacocephala: Una nuova classe di crostacei fossili. *Atti della Società Italiana di Scienze Naturali e del Museo Civico di Storia Naturale de Milano* 123:469–482.

Pinna, G., P. Arduini, C. Pesarini, and G. Teruzzi. 1985. Some controversial aspects of the morphology and anatomy of *Ostenocaris cypriformis* (Crustacea, Thylacocephala). *Transactions of the Royal Society of Edinburgh: Earth Sciences* 76:373–379.

Rolfe, W. D. I. 1985. Form and function in Thylacocephala, Conchyliocarida and Concavicarida (?Crustacea): A problem of interpretation. *Transactions of the Royal Society of Edinburgh: Earth Sciences* 76:391–399.

Schaeffer, B., and C. Patterson. 1984. Jurassic fishes from the western United States, with comments on Jurassic fish distribution. *American Museum Novitates* 2796:1–86.

Schram, F. R. 1990. On Mazon Creek Thylacocephala. *Proceedings of the San Diego Society of Natural History* 3:1–16.

Secretan, S. 1985. Conchyliocarida, a class of fossil crustaceans: Relationships to Malacostraca and postulated behaviour. *Transactions of the Royal Society of Edinburgh: Earth Sciences* 76:381–389.

Teruzzi, G. 1990. The genus *Coleia* Broderip, 1835 (Crustacea, Decapoda) in the Sinemurian of Osteno in Lombardy. *Atti della Società Italiana di Scienze Naturali e del Museo Civico di Storia Naturale de Milano* 131:85–104.

Wilby, P. R., D. E. G. Briggs, and G. Pinna. 1995. Soft tissue preservation in the Sinemurian of Osteno, Italy—Comparisons with Solnhofen and with experimental results. *Extended Abstracts of the Second International Symposium on Lithographic Limestones*, 163–164.

15
Posidonia Shale:
Germany's Jurassic Marine Park

Walter Etter and Carol M. Tang

THE LOWER JURASSIC POSIDONIA SHALE (*POSIDONIEN-schiefer*) of southern Germany and, especially, the Holzmaden region is one of the most celebrated of all Lagerstätten. This deposit has, in places, been quarried for centuries, and large numbers of splendidly preserved fossils have found their way into museums and private collections around the world. Since quarrying in the Holzmaden region continues, even today spectacular fossils are being uncovered. Famous finds include not only fully articulated marine reptiles and fish, but also articulated crinoids attached to logs and belemnites with preserved soft parts (Hauff and Hauff 1981; Seilacher 1990; Urlichs, Wild, and Ziegler 1994). The most frequently found fossils include, however, a variety of ammonites and bivalves, which can be found on almost every bedding plane. The black and laminated bituminous Posidonia Shale has long been treated as the prototype stagnation deposit (Seilacher 1982b; Seilacher, Reif, and Westphal 1985). This classification still holds true, although over the years it has become clear that there were several episodes with oxygenated bottom-waters (Röhl 1998; Röhl et al. 2001).

Quarrying in the Holzmaden region dates back for centuries. In fact, floor tiles coming from the Posidonia Shale can be found in the Hohenstaufen castle, built in the twelfth century (Seilacher 1990). Early attempts to distill oil from the highly bituminous black shales date back to the late sixteenth century (Seilacher 1990). Fossils have been found since that time (historical review in Urlichs, Wild, and Ziegler 1994). The first fundamental treatments of the Posidonia Shale and its fossils were published in the early nineteenth century by Schlotheim, Zieten, Quensted, and Fraas (Riegraf, Werner, and Lörcher 1984; Seilacher 1990). The first comprehensive study of the Posidonia Shale of the Holzmaden region appeared in 1921 (Hauff 1921). Since then, many workers have studied various aspects of the paleontology, sedimentology, and geochemistry of this deposit. But even with such extensive examination, there is still some controversy about the exact nature of the paleoenvironmental conditions that existed during the deposition of this Lagerstätte and the mode of life of some fossil species.

GEOLOGICAL CONTEXT

The lower Toarcian Posidonia Shale of southwestern Germany crops out in a southwest–northeast-trending belt extending from Waldshut to the Nördlinger Ries in the foreland of the Swabian Alb (Urlichs 1977; Riegraf, Werner, and Lörcher 1984; Urlichs, Wild, and Ziegler 1994) (Figure 15.1). In somewhat reduced thickness, the Posidonia Shale extends in the southwest into northern Switzerland (Riegraf 1985; Kuhn and Etter 1994) and in the northeast into the Franconian Alb (Urlichs 1971; Bandel and Knitter 1986). However, the best exposures can be found in quarries around the villages of Ohmden, Zell, Bad Boll, and, especially, Holzmaden (Figure 15.1), a place that is inextricably linked to this Lagerstätte (Hauff and Hauff 1981).

The Posidonia Shale of the Swabian Alb is a 6 to 14 m thick succession of dark gray to black bituminous shales with intercalated bituminous limestones (Riegraf, Werner, and Lörcher 1984) (Figure 15.2). It has long been recognized that each of the individual beds of the Posidonia Shale shows a distinct sedimentological, paleontological, and taphonomic signature. Successful attempts to correlate marker horizons over large geographic distances date back to the nineteenth century, and the first detailed lithostratigraphic subdivision of the Posidonia Shale was made by Quenstedt (1858), who labeled the lithostratigraphic units of the Jurassic of southern Germany with Greek letters; Schwarzer Jura ε ("epsilon") designates the Posidonia Shale. An even more detailed subdivision was established by Hauff (1921), and this lithostratigraphy is still in use (Riegraf, Werner, and Lörcher 1984; Riegraf 1985; Urlichs, Wild, and Ziegler 1994) (Figure 15.2). According to this subdivision, a lower (ε I), a middle (ε II), and an upper part (ε III) of the Posidonia Shale can be recognized, with each of the individual marker beds specifically named and numbered. On a regional scale where individual laminae can be traced over a distance of several kilometers, an even finer correlation is possible (Seilacher 1990).

The biostratigraphy of the Posidonia Shale is well established, by both ammonites (Riegraf, Werner, and Lörcher 1984) and microfossils (ostracodes, foraminifers) (Riegraf 1985). The lower boundary of the Posido-

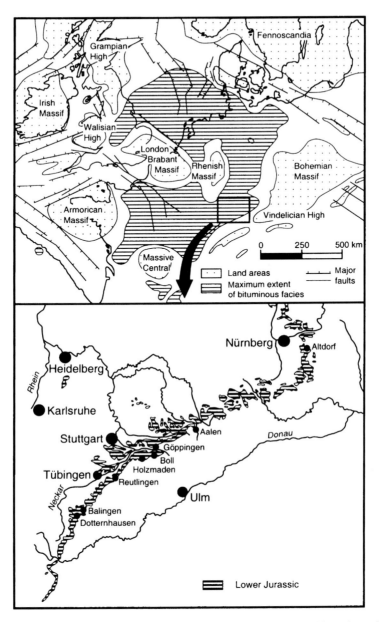

FIGURE 15.1 Paleogeography during the early Toarcian (*top*) and location of outcrops (*bottom*). (Modified from Urlichs 1977 and Riegraf 1985)

nia Shale is drawn with the onset of the first bituminous horizons (>2 percent organic matter content) (Riegraf, Werner, and Lörcher 1985). In the Holzmaden region, the bituminous facies starts in the early *tenuicostatum* zone (lowermost Toarcian) (Figure 15.2), but toward both the southwest (northern Switzerland) and the northeast (Franconian

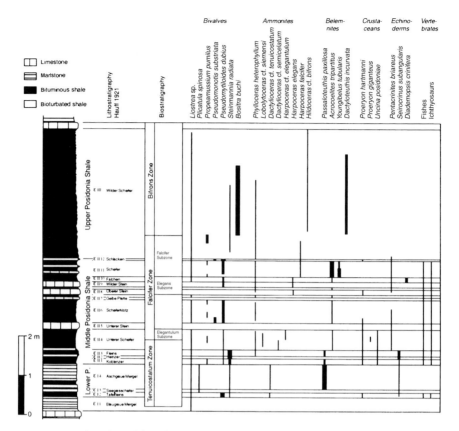

FIGURE 15.2 Stratigraphic column with range of common fossils indicated. (Modified from Hauff and Hauff 1981, Riegraf 1985, and Urlichs, Wild, and Ziegler 1994)

Alb), the basal layers of the Posidonia Shale are missing and the lowermost bituminous horizons are dated as late *tenuicostatum* zone or even early *falciferum* zone (Urlichs 1977; Riegraf 1985; Kuhn and Etter 1994).

The upper boundary of the Posidonia Shale normally coincides with an erosive surface that truncates the bituminous facies (Riegraf, Werner, and Lörcher 1984). This truncation can reach down to the lower *bifrons* zone or even the upper *falciferum* zone (Riegraf, Werner, and Lörcher 1984; Riegraf 1985). Where the transition from lower to upper Toarcian is continuous, the youngest bituminous layers have been dated as belonging to the middle *bifrons* zone (Riegraf, Werner, and Lörcher 1984). The deposition of the Posidonia Shale in southern Germany, where it is not secondarily reduced by erosion, lasted for 2.5 to 3 Ma (Riegraf 1985). Assuming a compaction rate of the sediment of 1:20, an average accumulation rate of 10 cm of uncompacted sediment per 1 Ka has been calculated for the *falciferum* zone of Gomaringen (Riegraf 1985). In the Holzmaden region, the thickness of the Posidonia Shale is more reduced, and accumulation rates are likely to be only half that value.

Apart from several bioturbated intervals, which are more prominently developed at the base and near the top of the sections, the whole sequence is finely laminated (Seilacher 1982b; Riegraf, Werner, and Lörcher 1984). The submillimetric laminations are an alternation of light-colored carbonate-rich laminae, consisting mainly of coccolith debris (especially *Schizosphaerella*), and darker laminae, consisting of clay minerals, organic material, and pyrite (Riegraf, Werner, and Lörcher 1984; Riegraf 1985). Besides these differences between individual laminae, there are major geochemical differences between calcareous beds and bituminous shales. As a general rule, the middle Posidonia Shale is more bituminous than the lower and upper parts of the sections (Urlichs, Wild, and Ziegler 1994).

The limestones consist of 60 to 80 weight percent (wt%) carbonate and 10 to 20 wt% clay minerals, mainly illite and muscovite, less commonly kaolinite, montmorillonite, and chlorite (Einsele and Mosebach 1955; Riegraf 1985). Quartz contributes around 5 wt%, pyrite adds 2 to 5 wt%, and the contents in organic material, mainly bitumen, have values of 3 to 7 wt% (exceptionally, more than 10 wt%) (Brockamp et al. 1944; Einsele and Mosebach 1955; Heller 1965; Riegraf 1985).

The bituminous shales, though, consist mainly of clay minerals (the same minerals found in the limestones), which account for up to 70 wt% (Einsele and Mosebach 1955). The carbonate content is usually less than 40 wt%, and quartz is present with up to 15 wt% (Einsele and Mosebach 1955; Riegraf 1985). Finely disseminated pyrite reaches values of 5 to 10 wt%, and organic carbon averages 4 to 10 wt%, although occasionally higher values (up to 12 wt%) have been observed (Einsele and Mosebach 1955). In certain layers, pyrite can occur abundantly in nodular or disk-shaped concretions.

PALEOENVIRONMENTAL SETTING

The Posidonia Shale can be correlated with other organic-rich shales in Europe. The most prominent examples are the Jet Rock and Alum Shale in England (Hallam 1967; Morris 1979, 1980) and the Schistes Cartons in France (Thomas 1977), but contemporaneous epicontinental black shales are also known from northern Germany (Loh et al. 1986; Littke et al. 1991) and from boreholes in the Netherlands and the North Sea (Ziegler 1990) (Figure 15.1). In fact, organic-rich shales from the lower Toarcian, which were deposited in deeper-water settings in the Tethys, are also found in Austria, Italy, Switzerland, Hungary, Greece, and Tunisia (Jenkyns 1988). Even outside Europe, strata of early Toarcian age are not uncommonly organic-rich, and this was the reason to coin the term *early Toarcian anoxic event* (Jenkyns 1988). The most widespread development of these black shales is in the *falciferum* zone (Jenkyns 1988), which corresponds to the middle Posidonia Shale.

The geological and paleontological documentation of the Posidonia Shale is probably more voluminous than that of any other Lagerstätten, but the environmental conditions during deposition of the Posidonia Shale have been a matter of debate for a long time. Early in the twentieth century, Pompeckj (1901) compared the sediments of the Posidonia Shale with those found today in the Black Sea. This was the first time that deposition of these laminated black shales was linked to anoxic bottom-water conditions. Ever since then, various workers have proposed different and sometimes conflicting scenarios for deposition of the Posidonia Shale.

The stagnant basin model introduced by Pompeckj (1901) has found many supporters during the history of research on the Posidonia Shale (Fischer 1961), and it is, with some modifications, still the most popular paleoenvironmental model (Seilacher 1982b, 1990; Seilacher, Reif, and Westphal 1985). There are, indeed, many indications that during deposition of the Posidonia Shale, the near-bottom environment was either anoxic or severely oxygen depleted. This evidence includes the preservation of laminations and the near absence of bioturbation, as well as the predominance of pelagic life forms among the fossils (Seilacher 1982b, 1990).

Another feature that is usually linked to severely oxygen-depleted environments is the preservation of high amounts of organic carbon. Indeed, apart from localized occurrences of Sinemurian oil-shales, the Posidonia Shale is the only sedimentary rock in southern Germany that shows values of total organic carbon (TOC) higher than 4 wt% (Riegraf, Werner, and Lörcher 1984). The high content of pyrite, mainly in the form of disseminated framboids, also supports a low-oxygen paleoenvironment (Brett and Baird 1986). In addition, the concentrations of iron and manganese in the Posidonia Shale are similar to those expected in anoxic basins (Veizer 1977).

A strict application of the stagnant basin model has been questioned by many paleontologists. The first to cite evidence against the Black Sea hypothesis was Beurlen (1925), who mentioned the occurrence of benthic crustaceans from the Posidonia Shale of Holzmaden. Since then, many workers have observed benthic life-forms, including bivalves, gastropods, crustaceans, and echinoderms (review in Riegraf, Werner, and Lörcher 1984). Additional evidence for benthic life comes from the presence of trace fossils (Brenner and Seilacher 1978) and autochthonous microfossils (Riegraf 1985). However, most of the authors reporting benthic life still maintain that during deposition of the Posidonia Shale, the bottom-waters were, as a rule, severely oxygen depleted or even anoxic.

Only brief oxygenation events allowed metazoan life-forms to colonize the seafloor (Brenner and Seilacher 1978; Röhl 1998).

A radically different view has been presented by Kauffman (1978, 1981), according to whom the water column was almost normally oxygenated throughout the early Toarcian. At or a few centimeters above the seafloor, however, an algal–fungal mat periodically developed, trapping anoxic water beneath it and allowing only benthic colonization of elevated shells that projected above it: the so-called ammonite shell surface community (Kauffman 1981). During times of stronger bottom currents, these mats may have been destroyed, allowing the colonization of the sediment surface and the bioturbation of sediment (Kauffman 1981). While many of Kauffman's observations are certainly valuable, they are not supported by sufficient data. In particular, his ammonite shell surface community probably represents ammonite shells that were overgrown while the shells were still afloat (Seilacher 1982a).

The current consensus is that the bottom-waters of the Posidonia Shale sea were severely oxygen depleted throughout the deposition of this formation, allowing the colonization of the seafloor by specialized or opportunistic taxa only at certain times. However, the mechanism that led to this unique and widespread deposition of black shales is still disputed. During early Toarcian times, southern Germany (as well as northern Switzerland, eastern France, northern Germany, and parts of England) was a basinal setting in an epicontinental seaway during a time of high sea-level stand (Riegraf 1985; Jenkyns 1988; Ziegler 1990). A highly structured bottom relief is evidenced by marked lateral variability in thickness of the Posidonia Shale (Riegraf, Werner, and Lörcher 1984; Riegraf 1985). Detrital material was shed mainly by the Bohemian Massif (Ziegler 1990). Bottom currents, which are not thought to be related to oxygenation events, followed the contours of the basins in a mainly south–north direction (Urlichs 1971; Brenner and Seilacher 1978).

Several models have been developed to explain the widespread occurrence of black shales and the exceptional preservation of fossils in this setting. The first is the classic stagnant basin model, involving salinity stratification. This explanation, which was favored by earlier authors, has recently been supported by palynologists who have observed that during deposition of the lowermost Posidonia Shale, a normal-marine phytoplankton association (consisting mainly of dinoflagellates, coccoliths, and acritarchs) was rather suddenly replaced by an assemblage dominated by prasinophytes (Loh et al. 1986; Prauss and Riegel 1989). This has been interpreted as the result of lowered salinity of surface water, produced by increasing river runoff. The same conclusion has been drawn by Brumsack (1988), who has documented high contents in manganese, lead, and cobalt in the Posidonia Shale, which, he believes, can have been derived only from enhanced river influx.

However, the salinity stratification model seems an unlikely explanation for the deposition of the Posidonia Shale for several reasons. First, apparently stenohaline pelagic organisms (radiolarians, cephalopods) occur throughout the sequence (Riegraf, Werner, and Lörcher 1984; Riegraf 1985). Second, geochemical tracers other than those cited by Brumsack (1988) seem to indicate normal-marine salinity (Küspert 1982). And third, the paleogeographic situation, which is known in detail (Ziegler 1990), does not seem to be appropriate for large-scale salinity stratification (Tyson and Pearson 1991).

A variation of the salinity stratification model has been proposed by Jordan (1974), according to whom a halocline developed not by high freshwater input (and hence a brackish surface layer), but by hypersaline bottom-waters produced by salt diapirism in northern Germany. However, corresponding diapirs are not known to have ascended during the early Toarcian in southern Germany, France, and England, and it seems very unlikely that enough hypersaline water could have been produced by this mechanism to cause such a large stratified body of water (Riegraf 1985).

A second model involving thermal stratification has been applied to the Posidonia Shale by Hallam (1987). While it is generally true that a thermally stratified water mass is less stable than a salinity stratified one, and is likely to experience seasonal mixing, a thermocline could be quite stable under equable climate (Hallam 1987). As has been pointed out, the irregular bottom topography of extensive shelf areas limited the amount of horizontal advection (Hallam and Bradshaw 1979; Hallam 1987). During the early stages of a transgression, the basinal settings of an epicontinental sea would have become deep enough to be situated beneath the thermocline. Black shales could thus have been formed in these stagnant water masses (Hallam and Bradshaw 1979). With further transgression, the expanded thermocline would have become less stable, and better circulation in the deeper, more extensive seas would have allowed more oxygenated water to enter the basins, leading to a complete mixing of the water column. This would have caused cessation of black shale formation (Hallam and Bradshaw 1979).

In its original formulation, this model also predicted the deposition of black shales during the early stages of regression, when water circulation was again restricted, but water depth of the depositional environment was still deep enough to be below the thermocline (Wignall and Hallam 1991). In a new variation of the theme, the expanding puddle model has been proposed, which incorporates additional sedimentological observations (Wignall and Hallam 1991). According to this model, black shales formed during the initial stages of transgressions in basinal settings, while land-derived sediment was trapped in newly formed estuaries. Indeed, the marginal areas of Posidonia Shale deposition (Franconian Alb, northern Switzerland) show condensed deposits and hiatal surfaces at the base of the Posidonia Shale, as predicted by the model (Bandel and Knitter 1986; Kuhn and Etter 1994; Urlichs, Wild, and Ziegler 1994). With ongoing transgression, sediment influx would have been reestablished and the basin (through sediment input) would have become shallower, ending black shale formation (Wignall and Hallam 1991). This was, indeed, the case in southern Germany, with the termination of the bituminous facies occurring in the *bifrons* zone, which was the time of the highest sea-level stand (Wignall and Hallam 1991).

While the expanding puddle model explains many aspects of the deposition of the Posidonia Shale, it does not account for the synchronous occurrence of organic-rich shales in deep-marine settings (Jenkyns 1988). Therefore, a third model must be considered: the oxygen minimum zone model (OMZ). There is, indeed, good evidence that Tethyan mid-waters were oxygen depleted during the early Toarcian (Jenkyns 1988). This is supported not only by the widespread occurrence of organic-rich shales (Dercourt et al. 2000), but also by the analysis of carbon isotopes. Sediments of early Toarcian age in all parts of Europe show a pronounced positive carbon isotope excursion (Jenkyns 1988). This event can be dated as occurring in the *exaratum* subzone of the *falciferum* zone, and is interpreted as reflecting a time interval during which anomally high rates of production, sedimentation, and burial of organic carbon took place (Jenkyns 1988). Since organic matter is isotopically much lighter (much more enriched in ^{12}C) than carbonate, removal of large amounts of organic material from the world's oceans would have left this carbon reservoir enriched in the heavier isotope ^{13}C. Carbonate produced from this reservoir (e.g., coccoliths) would thus show a shift to very positive $\delta^{13}C$ values (Jenkyns 1988). In contrast, the very low $\delta^{13}C$ values of the organic carbon at the base of the Posidonia Shale are thought to reflect upwelling of isotopically light organic carbon during times of anoxic bottom-waters (Küspert 1982).

Various authors have proposed that during the early Toarcian transgression, waters from the Tethyan oxygen minimum zone entered the epicontinental seas of central and western Europe (Knitter 1983; Jenkyns 1988; consistent with paleocurrent measurements: Brenner and Seilacher 1978). Together with the presumably highly fertile surface waters on the shelf, the accumulation of black shales and the preservation of high amounts of organic material would, therefore, largely reflect abnormally high plankton productivity (Jenkyns 1988). While the OMZ model is supported by geochemical evidence, many sedimentological and stratigraphic features of the Posidonia Shale are better explained with a model involving a thermally stratified basin during the early stages of a transgression. At present, it seems best to allow for both mechanisms to operate and to combine stratification (e.g., expanding puddle) with the OMZ model.

TAPHONOMY

The Posidonia Shale of southern Germany is most famous for its rich yield of fully articulated vertebrates. Indeed, thousands of spectacularly preserved marine reptiles and fish have been collected, with occasional soft-part preservation. Although the number of vertebrate specimens now on display in museums around the world is indeed impressive, it must be noted that this large number of fossils is the result of intense sampling efforts over more than 200 years (Hauff and Hauff 1981; Seilacher 1990). While the density of vertebrate occurrences in the Holzmaden region may actually be higher than in the Posidonia Shale of adjacent regions (Hauff 1921; Hauff and Hauff 1981; but see Seilacher 1990, who considers a sampling bias), spectacular finds are rare. Based on reported numbers of recovered specimens, Hauff and Hauff (1981) estimated that, on average, 12 quarry workers would have to dig for a year to recover one fully articulated ichthyosaur. Partially disarticulated and incomplete specimens are more common and account for about 90 percent of the finds (Hauff and Hauff 1981).

For most people, the ichthyosaurs are most readily associated with the Posidonia Shale of Holzmaden. Since most of the information about soft-part morphology of ichthyosaurs comes from specimens of this Lagerstätte, they indeed deserve their own paragraph. More than 500 fully articulated specimens have been found (Hauff and Hauff 1981), and some of them show remarkable soft-part preservation. Most spectacular is the preservation of the skin, which allowed for the first detailed reconstruction of the body outline (Hauff and Hauff 1981) (Figure 15.3), with a large mid-dorsal fin and a shark-like caudal fin (with the vertebral column, however, projecting into the lower lobe). The authenticity of this skin preservation has been questioned, and the presence of a distinct body outline was attributed to forgery by clever preparators (Martill 1987a, 1987b). However, a detailed investigation of the microfacies and geochemistry of ichthyosaurs showing skin preservation has confirmed their authenticity (Keller 1992) and subsequent study of the skin has provided important paleobiological information (Lingham-Soliar 2001). Other remarkable features associated with ichthyosaurs are the preservation of embryos in the body cavity of female specimens (Hauff and Hauff 1981; Böttcher 1990; Urlichs, Wild, and Ziegler 1994) and of stomach contents, mainly belemnite hooks (Keller 1976; Böttcher 1989).

Other reptiles from the Posidonia Shale are also known from fully articulated specimens, but they are much rarer and do not show excep-

FIGURE 15.3 Ichthyosaur *Stenopterygius* sp. with skin preservation. Length of specimen is 2.85 m. (Photo by H. Lumpe, courtesy of Staatliches Museum für Naturkunde, Stuttgart)

tional features such as skin preservation. Among the fish, however, which outnumber the reptiles by far, some specimens do show soft-part preservation. Preservation of the digestive tract (possibly phosphatized) is known from *Leptolepis* and *Pachycormus* (Hauff and Hauff 1981). Some sharks (*Hybodus*) and chimaeres (*Acanthorhina*) show preservation of the cartilaginous skeleton, parts of the skin, and stomach contents (Keller 1977; Urlichs, Wild, and Ziegler 1994). Again, incomplete specimens are much more common than fully articulated ones (Hauff and Hauff 1981).

Spectacular examples of exceptional preservation are also found among many invertebrates. Most notable is the crinoid *Seirocrinus subangularis,* which attained a column length of up to 20 m (Seilacher, Drozdzewski, and Haude 1968). Many perfectly preserved specimens have been found, and they are, unless broken, always attached to logs (Seilacher, Drozdzewski, and Haude 1968; Seilacher 1990) (Figure 15.4). Articulated specimens of other echinoderms are known as well (echinoids, asteroids, ophiuroids), but they are exceedingly rare (Seilacher 1990). The same is true for decapod crustaceans (Hauff and Hauff 1981; Seilacher 1990).

The Posidonia Shale of southern Germany is, furthermore, the only Lagerstätte that has yielded belemnites with preserved soft parts. The first of these specimens were described in 1976 (Wiesenauer 1976), but they turned out to be forgeries (Riegraf and Reitner 1979). Only a few years later, however, true soft-bodied belemnites were discovered, showing the presence of an ink sac and 10 arms with double rows of hooks (Reitner and Urlichs 1983; Riegraf and Hauff 1983) (Figure 15.5). All these specimens have a fractured alveolar region. This feature seems to be an integral part of soft-body preservation in belemnites and was the reason to formulate the cherry stone hypothesis (Wiesenauer 1976; Seilacher 1990). According to this theory, the major predators on belemnites (mainly ichthyosaurs) were selectively biting off the calcareous belemite rostra and spitting them out before swallowing the animal. Occasionally, an ichthyosaur would injure a belemnite without consuming the animal. After such an unsuccessful attempt, gas would be released from the phragmocone of a bitten specimen, which would then sink to the seafloor and be buried with attached soft parts. Specimens that died naturally would have floated long enough to disintegrate in the water column (Seilacher 1990).

Preservation of soft parts is also known in teuthoid cephalopods from the Posidonia Shale (Hauff and Hauff 1981; Riegraf, Werner, and Lörcher 1984), and in a few specimens of ammonites, preserved crop and stomach contents have been observed (Riegraf, Werner, and Lörcher 1984). However, the frequent preservation of aptychi in their body chambers indicates that many ammonites were embedded while the soft parts still held the jaws in position (Riegraf, Werner, and Lörcher 1984; Seilacher 1990). This is also indicated by the fact that the sediment

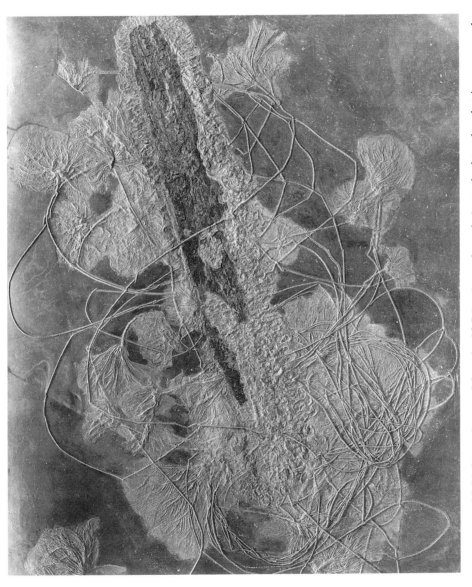

FIGURE 15-4 Multiple specimens of the crinoid *Seirocrinus subangularis* attached to a log that is overgrown by the bivalve *Pseudomytiloides dubius*. Length of log is 2.4 m. (Photo by H. Lumpe, courtesy of Staatliches Museum für Naturkunde, Stuttgart)

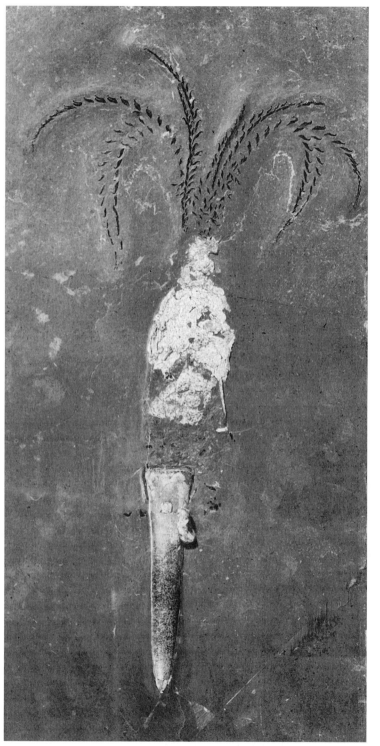

FIGURE 15-5 Belemnite *Passaloteuthis bisulcata* with soft parts. Length of rostrum is 11 cm. (Photo by H. Lumpe, courtesy of Staatliches Museum für Naturkunde, Stuttgart)

in the vicinity of ammonite body chambers contains organic matter contents, especially amino acids (Heller 1965). Among the other mollusc groups, soft-part preservation is not known, but at least one specimen of a bivalve (*Plagiostoma*) showing color preservation has been described (Riegraf, Werner, and Lörcher 1984). Worth mentioning also is the preservation of trace fossils in the Posidonia Shale. The rare bioturbation intervals show such a remarkable preservation of the lower tiers (frozen tiers) that this locality has even been called an Ichnofossil-Lagerstätte (Savrda and Bottjer 1989; for a discussion of trace fossils, see "Paleobiology and Paleoecology").

The exceptional preservation of fully articulated specimens, showing preserved soft parts in many cases, has long been explained by minimal decay under anoxic bottom-waters (Seilacher 1970; Seilacher, Reif, and Westphal 1985). Nektonic and pseudoplanktonic organisms would have, after death, sunk to the seafloor, where they were partially embedded in the very soft, soupy sediment. Under anoxic bottom-water conditions, decay was slow, and bioturbating and scavenging animals, which would normally disrupt multielement skeletons, were absent (Seilacher, Reif, and Westphal 1985). However, stagnation alone cannot explain the occurrence of articulated crinoids, because their skeletons would have disintegrated at the sediment surface within several hours to several days, even under anoxic conditions (Seilacher 1990). It is therefore believed that occasional sediment blanketing played an important role in the preservational history of such specimens (Brett and Seilacher 1991). That these sediment layers were thin is testified to by the fact that larger carcasses, which projected above the sediment–water interface, are better preserved on the underside than on their upper side, which is commonly partially disarticulated (Kauffman 1981).

Sediment was not invariably soupy throughout deposition of the Posidonia Shale. In some layers, drag and roll marks have been observed (Riegraf, Werner, and Lörcher 1984), and there is good evidence that microbial mats periodically covered the sediment surface, especially during deposition of the lower Posidonia Shale (Riegraf 1985). Whether these microbial mats played a crucial role in the preservation of fossils is not known at present.

The early diagenesis of Posidonia Shale fossils is characterized by strong compaction and dissolution of aragonite, but preservation of calcite, phosphate, and scleroproteins (and occasionally other tissues). The best clues for deciphering the early diagenetic history of the Posidonia Shale come from analysis of the ammonites (Seilacher et al. 1976). The Posidonia Shale ammonites show, unless they had a very rigid shell (*Dactylioceras*), a two-phase collapse (Seilacher et al. 1976). In the first stage, the body chamber collapsed, resulting in a fracture pattern. Later, the aragonitic shell started to dissolve, weakening the septal support of the phragmocone. This part of the shell collapsed without fractures because only the organic periostracum was left, resulting in a so-called leaf-preservation (Seilacher et al. 1976). The calcitic aptychi, however, were not dissolved. This diagenetic sequence is modified in concretions and some bituminous limestone beds where ammonites (and other fossils) can be preserved uncompacted, and sometimes with their shell recrystallized to calcite (Riegraf, Werner, and Lörcher 1984).

PALEOBIOLOGY AND PALEOECOLOGY

Fossils have long been considered the most useful tool for deciphering environmental conditions during deposition of the Posidonia Shale. At the same time, however, controversial interpretations of some species were the main reason for the different views that have been expressed on the origin of this Lagerstätte. While many macrofaunal species are clearly benthic and even more clearly pelagic, the life habits of some species are still debated. In addition to the large number of marine organisms, a few terrestrial species are also known from the Posidonia Shale. They include a few plant remains (cycads, ginkgos, conifers) (Hauff and Hauff 1981; Urlichs, Wild, and Ziegler 1994), rare pterosaurs (Wild 1975; Hauff and Hauff 1981; Seilacher 1990; Urlichs, Wild, and Ziegler 1994), and even rarer sphenodontid lizards and sauropod dinosaur remains (Wild 1978; Hauff and Hauff 1981; Urlichs, Wild, and Ziegler 1994). Although preserved in deposits at localities from the Franconian Alb and central eastern Germany, insects have not been found in the Posidonia Shale of the Swabian Alb (Urlichs, Wild, and Ziegler 1994).

Pelagic Organisms

Most of the fossils that made the Posidonia Shale of Holzmaden famous are nektonic animals (Seilacher 1990; Urlichs, Wild, and Ziegler 1994). They include the well-known ichthyosaurs, of which around a dozen species have been described (Riegraf, Werner, and Lörcher 1984; Maisch 1998a; but see Maisch 1998b, who accepts only eight valid species), and which range in adult size from about 2 to 10 m (Hauff and Hauff 1981). Other pelagic reptiles include plesiosaurs (four species with only 12 specimens) and marine crocodiles (three species, including *Platysuchus multiscrobiculatus*) (Hauff and Hauff 1981; Riegraf, Werner, and Lörcher 1984) (Figure 15.6). The ichthyosaurs spent their whole life in the sea, as is obvious from their morphology and from the female specimens carrying embryos at the near-hatching stage and even giving birth (Hauff and Hauff 1981; Urlichs, Wild, and Ziegler 1994). The plesiosaurs were less perfectly adapted to a permanent life in the sea, and the crocodiles most probably spent some time on land (Seilacher 1990; Urlichs, Wild, and Ziegler 1994).

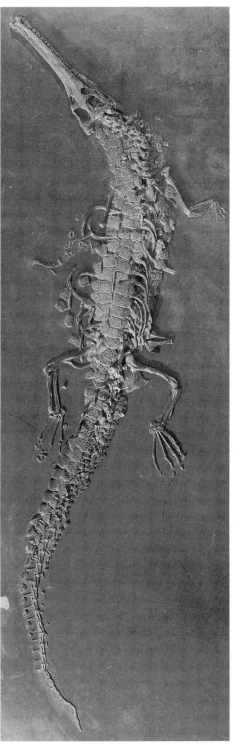

FIGURE 15.6 The marine crocodile *Platysuchus multiscrobiculatus*. Length of specimen is approximately 3 m. (Photo by H. Lumpe, courtesy of Staatliches Museum für Naturkunde, Stuttgart)

More than 20 species of bony fish have been described from the Posidonia Shale of southern Germany (Hauff and Hauff 1981; Riegraf, Werner, and Lörcher 1984). They range in size from only a few centimeters (*Leptolepis, Tetragonolepis*) to almost 3 m (*Chondrosteus, Ohmdenia*) (Hauff and Hauff 1981). Various taxonomic groups are present (Coelacanthidae: *Trachymetopon;* Teleostei: *Leptolepis, Pholidophorus;* Holostei: *Hypsocormus, Caturus, Pachycormus, Saurostomus, Euthynotus;* Subholostei: *Dapedium, Lepidotes, Tetragonolepis;* Chondrostei: *Ptycholepis, Saurorhynchus, Chondrosteus, Ohmdenia*), but all the species show morphologies of open- or shallow-water forms. Bottom-dwelling species seem to be absent (Aldinger 1965; Seilacher 1990; Urlichs, Wild, and Ziegler 1994). The same is true for the cartilaginous fishes (sharks and holocephalans), of which eight species are known (Riegraf, Werner, and Lörcher 1984).

Among the invertebrates, ammonites, squids, and belemnites are open-water swimmers (Seilacher 1990; Röhl 1998; but see Jäger and Fraaye 1997, who describe some ammonites of the genus *Harpoceras* with crustacean remains in their living chamber). However, some differences seem to exist in the mode of life of the various cephalopods. While the ammonites occur throughout the sections, belemnites are restricted to layers where ammonites are rare. Because these layers usually also contain some benthic bivalves, a nektobenthic mode of life was suggested for the belemnites (Röhl 1998). Besides the cephalopods, only rare specimens of a gooseneck barnacle (pseudoplanktonic scalpellid crustaceans) can be assigned, without doubt, to the pelagic realm (Seilacher 1990).

Benthic Organisms (Body Fossils)

Although no bottom-dwelling vertebrates are known from the Posidonia Shale, some invertebrate species are indisputably benthic. Most of them are quite rare, but their articulated state of preservation indicates that they are autochthonous. These unquestionably benthic organisms are mostly restricted to certain horizons and are found mainly in the lowermost and in the upper part of the Posidonia Shale (Brenner and Seilacher 1978; Urlichs, Wild, and Ziegler 1994) (Figure 15.2). Members of the benthos include rare asteroids and ophiuroids (Riegraf, Werner, and Lörcher 1984; Seilacher 1990), diademoid echinoids (Urlichs, Wild, and Ziegler 1994), and holothurians (only isolated ossicles in microfossil samples) (Riegraf 1985). Although the rare decapod crustaceans (*Uncina, Proeryon, Coleia*) (Figure 15.7) have been treated by some authors as possibly allochthonous (Hauff and Hauff 1981; Seilacher, Reif, and Westphal 1985), they are now widely regarded as autochthonous benthic organisms of the Posidonia Shale basin (Seilacher 1990).

Unquestionable members of the benthos also include some rare brachiopods (*Lingula,* "*Rhynchonella,*" *Spiriferina*), which occur only in beds

FIGURE 15.7 Decapod crustacean *Proeryon macrophthalmus.* Length of specimen is 15 cm. (Photo by H. Lumpe, courtesy of Staatliches Museum für Naturkunde, Stuttgart)

with low organic carbon content (Hauff and Hauff 1981; Riegraf, Werner, and Lörcher 1984; Röhl 1998). Larger gastropods are exceedingly rare in the Posidonia Shale, and most of the eight species are known from only a few specimens (Riegraf, Werner, and Lörcher 1984). Among the bivalves, several genera were endobenthic (*Solemya, Mesomiltha, Goniomya*) or semi-infaunal (*Cucullaea, Pinna*). Although these bivalves (which occur only in beds with low organic carbon content) are rare and were considered by some to be transported (Riegraf 1977), their preservation suggests that they are autochthonous (Bandel and Knitter 1986).

The microfossils of the Posidonia Shale, which are difficult to isolate from the bituminous shales and limestones, have recently been documented comprehensively (Riegraf 1985). In the strongly bituminous layers, foraminifers are either absent or represented by only an assemblage of small, mostly unsculptured species. Ostracodes are exceedingly rare in bituminous shales (Riegraf 1985). However, only a few horizons have not yielded foraminifers (Riegraf 1985). In bioturbated layers, the foraminiferan assemblage is more diverse and contains larger species, and ostracodes, although always a minor component, are present (Riegraf 1985).

Trace Fossils

Whereas the majority of the Posidonia Shale beds are finely laminated, as many as 14 bioturbated intervals containing a low-diversity assemblage of trace fossils were found throughout the sequence (Savrda and Bottjer 1989). The most common trace fossils are *Chondrites, Phymatoderma,* and

Thalassinoides (Savrda and Bottjer 1989), but in addition *Spongeliomorpha, Planolites,* and *Rhizocorallium* have been observed (Brenner and Seilacher 1978; Kauffman 1978, 1981). *Chondrites,* which usually occupies the lowermost tier, is interpreted to be produced by a worm-shaped animal living in symbiosis with sulfide-reducing bacteria (Fu 1991). *Phymatoderma* (formerly also known as *Chondrites*), on the contrary, was produced by a deposit feeder (Fu 1991). The bioturbated beds have been viewed as very short, one-phase benthic events (Brenner and Seilacher 1978). Whereas this seems to be true for thin horizons in the middle Posidonia Shale, thicker beds (*Seegrasschiefer*) found in the lower and uppermost part of the Posidonia Shale probably represented extended periods (several years) of bottom-water oxygenation (Savrda and Bottjer 1989).

Controversial Fossils (Pseudoplanktonic or Benthic)

Apart from the vertebrate fauna, the Posidonia Shale of southern Germany is most famous for the occurrence of large articulated crinoids (Hauff and Hauff 1981; Seilacher 1990; Urlichs, Wild, and Ziegler 1994). Specimens of the crinoid *Seirocrinus subangularis* may reach a length of up to 20 m and are, unless broken free, always attached to fossil logs (Seilacher, Drozdzewski, and Haude 1968) (Figure 15.4). Detailed investigation of a single occurrence revealed that these crinoids were attached to the underside of the log and were therefore pseudoplanktonic (Seilacher, Drozdzewski, and Haude 1968). This view has been challenged by Rasmussen (1977) and Kauffman (1978, 1981), who argued that these crinoids colonized the upper side of the logs at the seafloor and thus were truly benthic. However, analysis of the functional morphology of the stem of *Seirocrinus* has shown that this species had adapted to a pendent lifestyle and was indeed pseudoplanktonic (Simms 1986).

Associated with crinoids and also frequently found attached to logs is the inoceramid bivalve *Pseudomytiloides dubius* (Hauff and Hauff 1981; Urlichs, Wild, and Ziegler 1994). This attachment, as well as their scattered occurrence throughout the sequence, points to a pseudoplanktonic lifestyle (Seilacher 1982b, 1990; Seilacher, Reif, and Westphal 1985). However, although *Pseudomytiloides* is one of the most common bivalves of the Posidonia Shale, wood fragments are much rarer (Kauffman 1981). Furthermore, wood remains do not always seem to be overgrown on their underside, as would be expected when colonized by a byssally attached pseudoplanktonic bivalve (Kauffman 1981). It thus seems possible that *Pseudomytiloides* was facultatively pseudoplanktonic and that many specimens lived benthically (Wignall and Simms 1990; Etter 1996; Röhl 1998).

Other byssally attached bivalve genera of the Posidonia Shale include *Gervillia, Pteria, Plagiostoma, Antiquilima, Oxytoma,* and *Pseudomonotis*

(Riegraf, Werner, and Lörcher 1984). Earlier authors have regarded all these species as pseudoplanktonic (Fischer 1961). Again, Kauffman (1978, 1981) challenged this view and assigned all those bivalves to an epibenthic lifestyle. Articulated shells of these bivalves always occur in association with ammonite shells (Kauffman 1981; Seilacher 1982a). While Kauffman believed that colonization took place on dead ammonite shells at the seafloor, it can be shown that these ammonite shells were colonized while still afloat and the bivalves were thus pseudoplanktonic (Seilacher 1982a). Ammonite shells are overgrown on both flanks, which becomes impossible once an ammonite shell is lying with one flank on the seafloor (Seilacher 1982a). A possible exception to the pseudoplanktonic lifestyle is *Pseudomonotis substriata*, which occurs at certain localities in large numbers in a bed of the uppermost middle Posidonia Shale (Hauff and Hauff 1981); this species may have been benthic.

A pseudoplanktonic lifestyle is even more obvious for the cemented bivalves (*Liostrea, Exogyra, Plicatula*), attached inarticulate brachiopods (*Discinisca*), and serpulid worms. Again, they are associated with ammonite shells or found as isolated upper valves (Hauff and Hauff 1981; Seilacher 1982a). The fact that available benthic substrates like vertebrate bones and belemnite rostra show no encrustation at all (Seilacher 1982a; only one oyster attached to an ichthyosaur bone, which was recently found; A. Seilacher, personal communication, 1996) strongly supports the view that these organisms were truly pseudoplanktonic.

The most common fossils of the Posidonia Shale are the bivalves, which gave this formation its name: the posidoniids (Seilacher 1990; Urlichs, Wild, and Ziegler 1994). Two species can be distinguished: *Steinmannia bronni* (formerly *Posidonia bronni* var. *magna*) and *Bositra buchi* (= *Posidonia bronni* var. *parva;* but see Röhl 1998, who argues that there is only one species, *Bositra buchi,* occurring in different sizes). These species have gained some celebrity because of their widespread occurrence in black shales and the highly controversial interpretation of their mode of life. Earlier authors regarded them as epibenthic, but with the advent of the idea that black shales formed under anoxic bottom-waters, they were mostly interpreted as pseudoplanktonic (historical review in Jefferies and Minton 1965). However, there is no trace of a byssal notch, and it is doubtful that these species could byssally attach to a hard substrate (Jefferies and Minton 1965). As a radical solution, a nektonic mode of life was proposed to explain their occurrence in black shales (Jefferies and Minton 1965).

The interpretation of these bivalves as pelagic organisms has recently been supported, but now as fully planktonic, huge larvae (Oschmann 1993, 1994). However, this seems unlikely for three reasons. First, the original shell was probably thicker than reported by Jefferies and Minton

(1965; Kauffman 1981), which would preclude a floating lifestyle. Second, these bivalves do show a clear facies distribution in the Posidonia Shale, with *Steinmannia* occurring only in the lower part, and *Bositra* only in particular laminae in the upper part of the sequence (Riegraf, Werner, and Lörcher 1984; Urlichs, Wild, and Ziegler 1994; Röhl 1998). Third, well-preserved bedding-plane associations of articulated *Bositra* shells do show a highly regular spatial distribution, which would be impossible to explain if these shells had rained from the upper part of the water column (Etter 1996). A benthic mode of life thus seems the most likely explanation (Kauffman 1981; Seilacher 1990; Urlichs, Wild, and Ziegler 1994). It has been suggested that *Bositra* was chemosymbiotic (Kauffman 1988; Seilacher 1990), but this is difficult to imagine for an epibenthic species that is also known from sediments deposited under fully oxygenated conditions (Jefferies and Minton 1965).

A last genus that needs to be mentioned is the tiny gastropod *Coelodiscus* (maximum shell diameter, 3 mm). Because this gastropod was found mainly in carbonate-rich layers and in concretions around ichthyosaurs (Urlichs, Wild, and Ziegler 1994), it was concluded that it was a benthic scavenger (Einsele and Mosebach 1955; Fischer 1961). However, this genus occurs in microfossil samples throughout the Posidonia Shale (Riegraf 1985), and because of the close resemblance of its shell to those of certain modern heteropods (planktonic carnivores), it was concluded that *Coelodiscus* was planktonic (Bandel and Hemleben 1986).

In sum, the large majority of Posidonia Shale fossils were pelagic organisms. Even if some of the taxa formerly assigned to a pseudoplanktonic lifestyle actually belonged to the benthos, benthic organisms are still confined to certain horizons in the Posidonia Shale and are almost absent in the middle Posidonia Shale (*falciferum* zone). The paleoecological analysis, therefore, strongly supports a modified stagnant basin model.

CONCLUSIONS

The Posidonia Shale of southern Germany has probably received more scientific attention than any other Lagerstätte. This deposit is especially famous for its reptiles and fish and has yielded the best Jurassic marine vertebrate fauna of the world. It is from this Lagerstätte that we know about the actual body form of ichthyosaurs. The Posidonia Shale has also provided spectacularly preserved invertebrates, including fully articulated crinoids attached to logs, and the only belemnites showing soft-part preservation.

Although many features of this Lagerstätte have been studied thoroughly, there are still many controversies surrounding it. Strongly oxygen-depleted bottom-waters are indicated throughout the deposition of the Posidonia Shale, but there is still dispute about the amount of oxygen present and the duration of anoxic intervals. Although there is also some controversy about the exact nature of the depositional environment, a modified stagnant basin model is supported by the paleontological evidence. Whereas some of the fossils formerly considered as pseudoplanktonic or nektonic may actually represent benthic organisms, pelagic taxa are still predominant. Furthermore, benthic species preferentially occur in the lowermost and in the upper Posidonia Shale, and are considered to have been deposited under somewhat less severely oxygen-depleted bottom-waters.

ACKNOWLEDGMENTS

The chapter was thoroughly reviewed by A. Seilacher and M. Urlichs. Both offered many helpful thoughts and comments and provided some additional references.

REFERENCES

Aldinger, H. 1965. Zur Ökologie und Stratinomie der Fische des Posidonienschiefers (Lias epsilon). *Senckenbergiana lethaea* 46a:1–12.

Bandel, K., and C. Hemleben. 1986. Jurassic heteropods and their modern counterparts (planktonic Gastropoda, Mollusca). *Neues Jahrbuch für Geologie und Paläontologie, Abhandlungen* 174:1–22.

Bandel, K., and H. Knitter. 1986. On the origin and diagenesis of the bituminous Posidonia Shale (Toarcian) of southern Germany. *Mitteilungen des geologisch-paläontologischen Institutes der Universität Hamburg* 60:151–177.

Beurlen, K. 1925. Einige Bemerkungen zur Sedimentation in dem Posidonienschiefer Holzmadens. *Jahresberichte des oberrheinischen geologischen Vereins, neue Folge* 14:298–302.

Böttcher, R. 1989. Über die Nahrung eines *Leptopterygius* (Ichthyosauria, Reptilia) aus dem süddeutschen Posidonienschiefer (Unterer Jura) mit Bemerkungen über den Magen der Ichthyosaurier. *Stuttgarter Beiträge zur Naturkunde, B* 155:1–19.

Böttcher, R. 1990. Neue Erkenntnisse über die Fortpflanzungsbiologie der Ichthyosaurier. *Stuttgarter Beiträge zur Naturkunde, B* 164:1–51.

Brenner, K., and A. Seilacher. 1978. New aspects about the origin of the Toarcian Posidonia Shales. *Neues Jahrbuch für Geologie und Paläontologie, Abhandlungen* 157:11–18.

Brett, C. E., and G. C. Baird. 1986. Comparative taphonomy: A key to paleoenvironmental interpretation based on fossil preservation. *Palaios* 1:207–227.

Brett, C. E., and A. Seilacher. 1991. Fossil Lagerstätten: A taphonomic consequence of event sedimentation. In G. Einsele, W. Ricken, and A. Seilacher, eds., *Cycles and Events in Stratigraphy*, pp. 283–297. Berlin: Springer-Verlag.

Brockamp, B., G. Berg, H. Mojen, K. Staesche, R. Teichmüller, and F. Thiergart. 1944. Zur Paläogeographie und Bitumenführung des Posidonienschiefers im deutschen Lias. *Archive für Lagerstättenforschung* 77:1–59.

Brumsack, H.-J. 1988. *Rezente C_{org}-reiche Sedimente als Schlüssel zum Verständnis fossiler Schwarzschiefer.* Göttingen: Universität Göttingen.

Dercourt, J., M. Gaetani, B. Vrielynck, E. Barrier, B. Biju-Duval, M. F. Brunet, J. P. Cadet, S. Crasquin, and M. Sandulescu. 2000. *Peri-Tethys Atlas: Palaeogeographical Maps, Explanatory Notes.* Paris: Commission de la carte géologique du monde.

Einsele, G., and R. Mosebach. 1955. Zur Petrographie, Fossilerhaltung und Entstehung der Gesteine des Posidonienschiefers im Schwäbischen Jura. *Neues Jahrbuch für Geologie und Paläontologie, Abhandlungen* 101:319–430.

Etter, W. 1996. Pseudoplanktonic and benthic invertebrates in the Middle Jurassic Opalinum Clay, northern Switzerland. *Palaeogeography, Palaeoclimatology, Palaeoecology* 126:325–341.

Fischer, W. 1961. Über die Bildungsbedingungen der Posidonienschiefer in Süddeutschland. *Neues Jahrbuch für Geologie und Paläontologie, Abhandlungen* 111:326–340.

Fu, S. 1991. *Funktion, Verhalten und Einteilung fucoider und lophocteniider Lebensspuren.* Courier Forschungsinstitut Senckenberg, no. 135. Frankfurt: Senckenbergisches Naturforschende Gesellschaft.

Hallam, A. 1967. An environmental study of the upper Domerian and lower Toarcian in Great Britain. *Philosophical Transactions of the Royal Society of London, B* 252:393–445.

Hallam, A. 1975. *Jurassic Environments.* Cambridge: Cambridge University Press.

Hallam, A. 1987. Mesozoic marine organic-rich shales. In J. Brooks and A. J. Fleet, eds., *Marine Petroleum Source Rocks,* pp. 251–261. Special Publication, no. 26. London: Geological Society.

Hallam, A., and M. J. Bradshaw. 1979. Bituminous shales and oolithic ironstones as indicators of transgressions and regressions. *Journal of the Geological Society of London* 136:157–164.

Hauff, B., Sr. 1921. Untersuchung der Fossilfundstätten von Holzmaden im Posidonienschiefer des Oberen Lias Württembergs. *Palaeontographica* 64:1–42.

Hauff, B. and R. Hauff. 1981. *Das Holzmadenbuch.* 2nd ed. Holzmaden: Selbstverlag.

Heller, W. 1965. Organisch-chemische Untersuchungen im Posidonienschiefer Schwabens. *Neues Jahrbuch für Geologie und Paläontologie, Monatshefte* 1965:65–68.

Jäger, M., and R. Fraaye. 1997. The diet of the early Toarcian ammonite *Harpoceras falciferum. Palaeontology* 40:557–574.

Jefferies, R. P. S., and P. Minton. 1965. The mode of life of two Jurassic species of *"Posidonia." Palaeontology* 8:156–185.

Jenkyns, H. C. 1988. The early Toarcian (Jurassic) anoxic event: Stratigraphic, sedimentary, and geochemical evidence. *American Journal of Science* 288:101–151.

Jordan, R. 1974. Salz- und Erdöl/Erdgas-Austritt als Fazies bestimmende Faktoren im Mesozoikum Nordwest-Deutschlands. *Geologisches Jahrbuch, A* 13:1–64.

Kauffman, E. G. 1978. Benthic environments and paleoecology of the Posidonienschiefer (Toarcian). *Neues Jahrbuch für Geologie und Paläontologie, Abhandlungen* 157:18–36.

Kauffman, E. G. 1981. Ecological reappraisal of the German Posidonienschiefer (Toarcian) and the stagnant basin model. In J. Gray, A. J. Boucot, and W. B.

N. Berry, eds., *Communities of the Past*, pp. 311–381. Stroudsburg, Pa.: Hutchinson Ross.

Kauffman, E. G. 1988. The case of the missing community: Low-oxygen adapted Paleozoic and Mesozoic bivalves ("flat clams") and bacterial symbiosis in typical Phanerozoic seas. *Geological Society of America, Abstracts with Program* 20:A48.

Keller, T. 1976. Magen- und Darminhalte von Ichthyosauriern des süddeutschen Posidonienschiefers. *Neues Jahrbuch für Geologie und Paläontologie, Monatshefte* 1976:266–283.

Keller, T. 1977. Frassreste im süddeutschen Posidonienschiefer. *Jahreshefte der Gesellschaft für Naturkunde Württemberg* 132:117–134.

Keller, T. 1992. "Weichteil-Erhaltung" bei grossen Vertebraten (Ichthyosauriern) des Posidonienschiefers Holzmadens (Oberer Lias, Mesozoikum Süddeutschlands). *Kaupia (Darmstädter Beiträge zur Naturgeschichte)* 1:23–62.

Knitter, H. 1983. Biostratigraphische Untersuchungen mit Ostracoden im Toarcien Süddeutschlands. *Facies* 8:213–262.

Kuhn, O., and W. Etter. 1994. Der Posidonienschiefer der Nordschweiz: Lithostratigraphie, Biostratigraphie, Fazies. *Eclogae geologicae Helvetiae* 87:113–138.

Küspert, W. 1982. Environmental changes during oil shale deposition as deduced from stable isotope ratios. In G. Einsele and A. Seilacher, eds., *Cyclic and Event Stratification*, pp. 482–501. Berlin: Springer-Verlag.

Lingham-Soliar, T. 2001. The ichthyosaur integument: Skin fibers, a means for strong, flexible and smooth skin. *Lethaia* 34:287–302.

Littke, R., D. R. Baker, D. Leythaeuser, and J. Rullkötter. 1991. Keys to the depositional history of the Posidonia Shale (Toarcian) in the Hils Syncline, northern Germany. In R. V. Tyson and T. H. Pearson, eds., *Modern and Ancient Continental Shelf Anoxia*, pp. 311–333. Special Publication, no. 58. London: Geological Society.

Loh, H., B. Maul, M. Prauss, and W. Riegel. 1986. Primary production, maceral formation and carbonate species in the Posidonia Shale of NW Germany. *Mitteilungen des geologisch-paläontologischen Institutes der Universität Hamburg* 60:397–421.

Maisch, M. W. 1998a. A new ichthyosaur genus from the Posidonia Shale (lower Toarcian, Jurassic) of Holzmaden, SW Germany with comments on the phylogeny of post-Triassic ichthyosaurs. *Neues Jahrbuch für Geologie und Paläontologie, Abhandlungen* 209:47–78.

Maisch, M. W. 1998b. Kurze Übersicht der Ichthyosaurier des Posidonienschiefers mit Bemerkungen zur Taxionomie der Stenopterygiidae und Temnodontosauridae. *Neues Jahrbuch für Geologie und Paläontologie, Abhandlungen* 209: 401–431.

Martill, D. M. 1987a. Prokaryote mats replacing soft tissues in Mesozoic marine reptiles. *Modern Geology* 11:265–269.

Martill, D. M. 1987b. Ichthyosaurs with soft tissues: Additional comments. *Transactions of the Leicester Literature and Philosophical Society* 81:35–45.

Morris, K. A. 1979. A classification of Jurassic marine shale sequences: An example from the Toarcian (Lower Jurassic) of Great Britain. *Palaeogeography, Palaeoclimatology, Palaeoecology* 26:117–126.

Morris, K. A. 1980. Comparison of major sequences of organic-rich mud deposition in the British Jurassic. *Journal of the Geological Society of London* 137:157–170.

Oschmann, W. 1993. Environmental oxygen fluctuations and the adaptive response of marine benthic organisms. *Journal of the Geological Society of London* 150:187–191.

Oschmann, W. 1994. Adaptive pathways of marine benthic organisms in oxygen-controlled environments. *Neues Jahrbuch für Geologie und Paläontologie, Abhandlungen* 191:393–444.

Pompeckj, J. F. 1901. Die Jura-Ablagerungen zwischen Regensburg und Regenstrauf. *Geognostische Jahreshefte* (Munich) 14:139–220.

Prauss, M., and W. Riegel. 1989. Evidence from phytoplankton associations for causes of black shale formation in epicontinental seas. *Neues Jahrbuch für Geologie und Paläontologie, Monatshefte* 1989:671–682.

Quenstedt, F. A. 1858. *Der Jura.* Tübingen: Laupp.

Rasmussen, H. W. 1977. Function and attachment of the stem of Isocrinidae and Pentacrinitidae: Review and interpretation. *Lethaia* 10:51–57.

Reitner, J., and M. Urlichs. 1983. Echte Weichteilbelemniten aus dem Untertoarcium (Posidonienschiefer) Südwestdeutschlands. *Neues Jahrbuch für Geologie und Paläontologie, Abhandlungen* 153:129–146.

Riegraf, W. 1977. *Goniomya rhombifera* (Goldfuss) in the Posidonia Shales (Lias epsilon). *Neues Jahrbuch für Geologie und Paläontologie, Monatshefte* 977:446–448.

Riegraf, W. 1985. *Mikrofauna, Biostratigraphie und Fazies im Unteren Toarcium Südwestdeutschlands und Vergleiche mit benachbarten Gebieten.* Tübinger mikropaläontologische Mitteilungen, no. 3. Tübingen: Institut und Museum für Geologie und Paläontologie der Universität Tübingen.

Riegraf, W., and R. Hauff. 1983. Belemnitenfunde mit Weichkörper, Fangarmen und Gladius aus dem Untertoarcium (Posidonienschiefer) und Unteraalenium (Opalinuston) Südwestdeutschlands. *Neues Jahrbuch für Geologie und Paläontologie, Abhandlungen* 165:466–483.

Riegraf, W., and J. Reitner. 1979. Die "Weichteilbelemniten" des Posidonienschiefers (Untertoarcium) von Holzmaden (Baden-Württemberg) sind Fälschungen. *Neues Jahrbuch für Geologie und Paläontologie, Monatshefte* 1979:291–304.

Riegraf, W., G. Werner, and F. Lörcher. 1984. *Der Posidonienschiefer: Biostratigraphie, Fauna und Fazies des südwestdeutschen Untertoarciums (Lias ε).* Stuttgart: Ferdinand Enke Verlag.

Röhl, H.-J. 1998. *Hochauflösende palökologische und sedimentologische Untersuchungen im Posidonienschiefer (Lias ε) von SW-Deutschland.* Tübinger Geowissenschaftliche Arbeiten, A, no. 47. Tübingen: Institut und Museum für Geologie und Paläontologie der Universität Tübingen.

Röhl, H.-J., A. Schmid-Röhl, W. Oschmann, A. Frimmel, and L. Schwark. 2001. The Posidonia Shale of SW Germany: An oxygen-depleted ecosystem controlled by sea level and palaeoclimate. *Palaeogeography, Palaeoclimatology, Palaeoecology* 165:27–52.

Savrda, C. E., and D. J. Bottjer. 1989. Anatomy and implications of bioturbated beds in "black shale" sequences: Examples from the Jurassic Posidonienschiefer (southern Germany). *Palaios* 4:330–342.

Seilacher, A. 1970. Begriff und Bedeutung der Fossil-Lagerstätten. *Neues Jahrbuch für Geologie und Paläontologie, Monatshefte* 1970:34–39.

Seilacher, A. 1982a. Ammonite shells as habitats in the Posidonia Shales of Holzmaden—Floats or benthic islands? *Neues Jahrbuch für Geologie und Paläontologie, Monatshefte* 1982:98–114.

Seilacher, A. 1982b. Posidonia Shales (Toarcian, S. Germany)—Stagnant basin model revalidated. In E. M. Gallitelli, ed., *Paleontology, Essential of Historical Geology: Proceedings of the First International Meeting on "Paleontology, Essential of Historical Geology,"* pp. 279–298. Modena: STEM Mucchi.

Seilacher, A. 1990. Die Holzmadener Posidonienschiefer—Entstehung der Fossillagerstätte und eines Erdölmuttergesteins. In. K. W. Weidert, ed., *Klassische Fundstellen der Paläontologie*, vol. 2, pp. 107–131. Korb: Goldschneck-Verlag.

Seilacher, A., F. Andalib, G. Dietl, and H. Gocht. 1976. Preservational history of compressed Jurassic ammonites from southern Germany. *Neues Jahrbuch für Geologie und Paläontologie, Abhandlungen* 152:307–356.

Seilacher, A., G. Drozdzewski, and R. Haude. 1968. Form and function of the stem in a pseudoplanktonic crinoid (*Seirocrinus*). *Palaeontology* 11:275–282.

Seilacher, A., W. E. Reif, and F. Westphal. 1985. Sedimentological, ecological and temporal patterns of fossil Lagerstätten. *Philosophical Transactions of the Royal Society of London, B* 311:5–23.

Simms, M. J. 1986. Contrasting lifestyles in Lower Jurassic crinoids: A comparison of benthic and pseudopelagic Isocrinidae. *Palaeontology* 29:475–493.

Thomas, M. 1977. Les schistes bitumineux du Toarcien de Franche-Comté. *Annales Scientifiques de l'Université de Besançon*, 3rd ser., 28:73–76.

Tyson, R. V., and T. H. Pearson. 1991. Modern and ancient continental shelf anoxia: An overview. In R. V. Tyson and T. H. Pearson, eds., *Modern and Ancient Continental Shelf Anoxia*, pp. 1–24. Special Publication, no. 58. London: Geological Society.

Urlichs, M. 1971. Alter und Genese des Belemnitenschlachtfeldes im Toarcien von Franken. *Geologische Blätter für Nordost-Bayern* 21:29–38.

Urlichs, M. 1977. The Lower Jurassic in southwestern Germany. *Stuttgarter Beiträge zur Naturkunde, B* 24:1–41.

Urlichs, M., R. Wild, and B. Ziegler. 1994. Der Posidonienschiefer des unteren Juras und seine Fossilien. *Stuttgarter Beiträge zur Naturkunde, C* 36:1–95.

Veizer, J. 1977. Geochemistry of lithographic limestones and dark marls from the Jurassic of southern Germany. *Neues Jahrbuch für Geologie und Paläontologie, Abhandlungen* 153:129–146.

Wiesenauer, E. 1976. Vollständige Belemnitentiere aus dem Holzmadener Posidonienschiefer. *Neues Jahrbuch für Geologie und Paläontologie, Monatshefte* 1976:603–608.

Wignall, P. B., and A. Hallam. 1991. Biofacies, stratigraphic distribution and depositional models of British onshore Jurassic black shales. In R. V. Tyson and T. H. Pearson, eds., *Modern and Ancient Continental Shelf Anoxia*, pp. 291–309. Special Publication, no. 58. London: Geological Society.

Wignall, P. B., and M. J. Simms. 1990. Pseudoplankton. *Palaeontology* 33:359–378.

Wild, R. 1975. Ein Flugsaurier-Rest aus dem Lias Epsilon (Toarcium) von Erzingen (Schwäbischer Jura). *Stuttgarter Beiträge zur Naturkunde, B* 17:1–16.

Wild, R. 1978. Ein Sauropoden-Rest (Reptilia, Saurischia) aus dem Posidonienschiefer (Lias, Toarcium) von Holzmaden. *Stuttgarter Beiträge zur Naturkunde, B* 41:1–15.

Ziegler, P. A. 1990. *Geological Atlas of Western and Central Europe.* 2nd ed. The Hague: Shell International Petroleum Maatschappij.

16
La Voulte-sur-Rhône:
Exquisite Cephalopod Preservation

Walter Etter

Tᴴᴇ ᴍᴀʀɪɴᴇ ʟᴀɢᴇʀsᴛäᴛᴛᴇ ᴏꜰ ʟᴀ ᴠᴏᴜʟᴛᴇ-sᴜʀ-ʀʜôɴᴇ ɪs
poorly known outside France, even though generations of paleontologists and geologists have worked at this locality. The Callovian (Middle Jurassic) of La Voulte, a series of dark platy claystones and
marls with intercalated ferruginous limestone layers and carbonate concretions, is especially notable for the excellent preservation of ophiuroids, crustaceans, and fish, as well as perhaps the finest examples
known of fossil cephalopods exhibiting soft-part preservation. Apart
from Beecher's Trilobite Bed (Chapter 7) and the Hunsrück Slate
(Chapter 8), La Voulte-sur-Rhône is the only other locality where extensive pyritization of soft parts occurs (Bartels, Briggs, and Brassel 1998).
However, soft-part preservation in La Voulte-sur-Rhône is not confined
to pyritization, but also occurs associated with phosphatized and calcified tissues (Wilby, Briggs, and Riou 1995, 1996). The exceptional preservation in this locality in southeastern France is most likely the result of
both oxygen-depleted bottom-waters and repeated sediment blanketing.
This Lagerstätte thus qualifies as both a stagnation and an obrution deposit (*sensu* Seilacher, Reif, and Westphal 1985).

Interest in the fossiliferous layers of La Voulte-sur-Rhône dates back
to the early nineteenth century. At that time, the ferruginous beds were
exploited as iron ore, and as a result, various geologists tried to establish
the stratigraphic position of these layers (Sayn and Roman 1928). By the
middle of the nineteenth century, the first remarkable fossils were discovered, including the bedding-plane accumulations of fully articulated
ophiuroids (Lory 1854). Although the stratigraphy of the Callovian of

La Voulte remained the main interest, enough excellently preserved fossils had been collected to be described in monographs in the early twentieth century (van Straelen 1923, 1925). In 1928 a comprehensive paleontological review of the site appeared (Sayn and Roman 1928), and it is still a standard reference for this Lagerstätte.

In more recent years, the paleoecology and taphonomy of the ophiuroids have received particular attention (Hess 1960; Dietl and Mundlos 1972), and a detailed stratigraphic framework for this locality was established (Elmi 1967). Currently, the Callovian of La Voulte-sur-Rhône is being reinvestigated. Intense sampling efforts in the 1980s yielded many new specimens, and in the past few years hitherto unknown fossils have been described (Fischer and Riou 1982a, 1982b; Secretan 1983; Secretan and Riou 1986; Carriol and Riou 1991). Many of these specimens are now on display in a newly opened small museum in La Voulte (curated by Bernard Riou). Although the taphonomy of this Lagerstätte was the subject of recent work (Wilby, Briggs, and Riou 1995, 1996), a comprehensive analysis of the sedimentology and paleoecology of this remarkable locality still awaits publication.

GEOLOGICAL CONTEXT

La Voulte-sur-Rhône is a small town on the western border of the Rhône River in southeastern France (department of Ardèche) (Figure 16.1). The remarkably fossiliferous layers, which are of lower Callovian age, are exposed in small creeks a few hundred meters west of La Voulte (Cayeux 1922; Sayn and Roman 1928; Elmi 1967). They are part of the eastern sediment cover of the Massif Central, a hercynian crystalline complex that was uplifted in the Cenozoic (Ziegler 1990).

At the contact between the metamorphic rocks of the Massif Central and the Mesozoic sediment cover, there is a 1 m thick tectonic breccia composed of angular fragments of micaschists and sand (Sayn and Roman 1928; Elmi 1967). This breccia is sometimes overlain by Triassic sandstones, Bajocian and Bathonian marls, and crinoid limestones (Elmi 1967). However, the Callovian of La Voulte typically rests directly on the breccia, because older Mesozoic units are commonly missing (Elmi 1967) (Figure 16.2).

The Callovian sequence begins with up to 25 m of dark gray to black clayey marls with some intercalated hematitic limestones (Elmi 1967) (Figure 16.2). Biostratigraphic studies have placed this unit in the upper *macrocephalus* to lower *koenigi* (= *calloviense*) zone of the lower Callovian (Elmi 1967), and the fauna consists primarily of the small bivalve *Bositra buchi*, some ammonites, and belemnites (Sayn and Roman 1928; Elmi 1967).

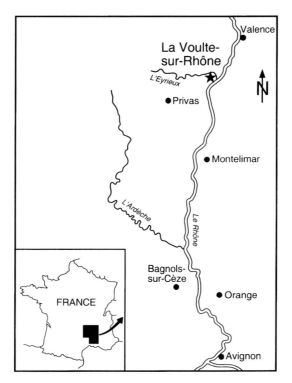

FIGURE 16.1 Geographic setting of the La Voulte-sur-Rhône locality.

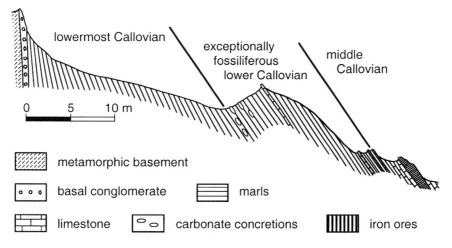

FIGURE 16.2 Stratigraphic section at "Ravin de Gramades." Only fossils in the layers of early Callovian age show exceptional preservation. (Modified from Elmi 1967)

The overlying 10 m contain the exceptionally preserved fossils (Figure 16.2). In the lower part, 3 to 7 m of black clayey marls contain calcareous nodules, which are most abundant near the base (Elmi 1967). Although the dark marls have yielded mainly a sparse fauna dominated by *Bositra buchi* and *Ophiopinna elegans,* some of the nodules contain cephalopods, crustaceans, and fish (Sayn and Roman 1928; Elmi 1967; Fischer and Riou 1982a). Many of these specimens exhibit soft-part preservation (Fischer and Riou 1982a; Secretan 1985). This nodule-bearing unit belongs to the upper *koenigi* to lower *ardescicum* zone of the lower Callovian (Elmi 1967).

The following unit is 5.5 m at its thickest parts and consists of an alternation of dark gray clayey marls, beds of calcareous marls, and limonitic limestones, sometimes developed as iron oolites (Sayn and Roman 1928; Elmi 1967). This uppermost lower Callovian sequence, the *ardescicum* zone (Elmi 1967), contains several horizons of bedding-plane accumulations with fully articulated ophiuroids, some crustaceans and cephalopods, and many *Bositra buchi* (Hess 1960; Elmi 1967).

Overlying these two units are calcareous marls and hematitic limestones of middle Callovian age (Figure 16.2), which in places contain a rich fauna of siliceous sponges (Sayn and Roman 1928). They are, in turn, overlain by sparsely fossiliferous marls of the upper Callovian and the Oxfordian (Sayn and Roman 1928).

Only the nodule-bearing unit and the layers with the articulated ophiuroids can be regarded as a conservation Lagerstätte, and the following discussion concentrates on these two units of early Callovian age. There is very high lateral variability within the lower Callovian sequence (Sayn and Roman 1928; Elmi 1967), which includes both marked changes in thickness of many units and changes in lithology. Most of the ferruginous limestone beds and the horizons with carbonate concretions do not appear to be traceable for more than several decimeters (Sayn and Roman 1928; Elmi 1967). However, the detailed reconstruction of lateral variation is hampered because many small-scale faults disrupt the sequence (Hess 1960).

The lower Callovian sediments of La Voulte are not laminated, although the clayey marls are quite fissile (Elmi 1967). In thin sections, both marly and calcareous beds appear homogeneous on the millimeter scale (Hess 1960), but thin individual layers, especially those covering bedding-plane accumulations of articulated fossils, show characteristics of mudflow deposits (Dietl and Mundlos 1972). The calcareous concretions are of early diagenetic origin, as is obvious from the largely uncompacted state of preservation of the embedded fossils (Sayn and Roman 1928). Some of the ferruginous limestone beds might also be of early diagenetic origin rather than the result of sedimentary carbonate supply (Dietl and Mundlos 1972). The geochemistry of the deposit is virtually unknown. Data exist for only the iron-ore beds, which have a content in iron oxides of up to 80 percent and are slightly phosphatic (Cayeux 1922). Values for organic carbon of the platy marls are not available, although they have been reported to be high (Fröhlich et al. 1992).

PALEOENVIRONMENTAL SETTING

The lower Callovian strata of La Voulte-sur-Rhône were deposited in an epicontinental sea bordering the eastern margin of the Massif Central. This crystalline complex was a small landmass at this time (Ziegler 1990) and was most likely the major source of terrigeneous clastics, which were shed to the region of La Voulte. The bottom relief was quite developed, as is evident from the high lateral variability of the lower Callovian. The water depth is difficult to constrain, but the absence of storm layers and shell beds indicates that deposition took place below storm wave base. Some reworked fossils occur, however, in the middle Callovian of La Voulte (Elmi 1967), and this might indicate that during the early Callovian, La Voulte was situated only slightly below storm wave base.

There is sedimentological evidence that many thin beds represent accumulation during single episodes of sediment blanketing. Thin mud-flow layers (>1 mm) seem to constitute the majority of the ophiuroid-bearing unit (Hess 1960) and suggest that most layers of the lower Callovian of La Voulte are event beds.

The bottom-waters were probably oxygen depleted during deposition of the lower Callovian (Dietl and Mundlos 1972; Fischer and Riou 1982a; Wilby, Briggs, and Riou 1995, 1996; Wilby 2001). The posidoniid bivalve *Bositra buchi,* also known from Jurassic black shales (Chapter 15), is the only abundant benthic mollusc species. Other benthic groups include some crustaceans and other arthropods, ophiuroids, and rare echinoids. In addition, there is a large proportion of nektonic animals, including cephalopods and fish (Dietl and Mundlos 1972). It is also notable that foraminifers are very sparse (Hess 1960) and ostracodes seem to be absent.

TAPHONOMY

Soft-part preservation in the lower Callovian of La Voulte is known from a variety of taxonomic groups. Most common is the ophiuroid *Ophiopinna elegans,* which forms dense bedding-plane accumulations of fully articulated specimens (Hess 1960) (Figure 16.3). The skeletons of these ophiuroids are pyritized in varying degrees, and sometimes even the muscular fibers of the arms are pyritized (Hess 1960).

More spectacular are the occurrences of crustaceans (Figure 16.4). They can be preserved in carbonate concretions, in the platy marls, or

FIGURE 16.3 Bedding plane covered by specimens of the ophiuroid *Ophiopinna elegans*. Field of view is approximately 39 cm. (Photo courtesy of B. Riou, Musée de Paléontologie, La Voulte-sur-Rhône, Ardèche, France)

FIGURE 16.4 The decapod crustacean *Eryma* sp. Length of specimen is 11 cm. (Photo courtesy of B. Riou, Musée de Paléontologie, La Voulte-sur-Rhône, Ardèche, France)

on top of limonitic limestones. Although they are preserved in three dimensions within concretions, they are heavily flattened in the marls (Fröhlich et al. 1992). Not only are many of these crustaceans preserved fully articulated, but sometimes the finest details of their appendages are visible (Secretan and Riou 1986; Carriol and Riou 1991). In exceptional instances, even soft parts at the cellular level have been phosphatized, like the retinal structures in the eyes of conchyliocarid crustaceans (Fröhlich et al. 1992; Wilby, Briggs, and Riou 1996).

Equally remarkable is the preservation of dibranchiate cephalopods. Again, they occur flattened in the platy marls and more or less undeformed in calcareous concretions (Fischer and Riou 1982a). The preserved soft parts of teuthoids include not only the massive body, but also the head with traces of the eyes, fins, and tentacles in remarkable detail (Fischer and Riou 1982a). Even a small octopus, probably the best fossil representative of this group, has been recovered (Fischer and Riou 1982b) (Figure 16.5). The soft parts of the cephalopods are preserved in calcite, limonite, or apatite, which can be lined by organic matter or pyrite (Fischer and Riou 1982a; Wilby, Briggs, and Riou 1996). Other examples of soft-part preservation include occurrences of pycnogonids and annelids, but these findings have not yet been described (Fischer and Riou 1982a, 1982b).

The lower Callovian of La Voulte-sur-Rhône is very remarkable in the sense that different styles of early diagenetic mineralization (pyritization, phosphatization, and calcite replacement) have led to the preservation of soft parts (Wilby, Briggs, and Riou 1995, 1996; Wilby 2001). These different preservational styles vary among layers (Fröhlich et al. 1992). In several specimens, however, these different minerals occur in association (Wilby, Briggs, and Riou 1996). Apatite is restricted largely to muscular tissues and sometimes replicates cellular details. During the subsequent steps in diagenesis, apatite acted as a template for later calcification and pyritization (Wilby, Briggs, and Riou 1996).

There is broad agreement that periodic rapid sediment blanketing was a prerequisite for the excellent preservation of these fossils (Hess 1960; Dietl and Mundlos 1972; Fischer and Riou 1982a; Secretan and Riou 1983; Fröhlich et al. 1992; Wilby, Briggs, and Riou 1996). Such obrution events are indicated not only by thin mudflow layers, but also by taphonomic evidence. It is known that modern ophiuroids disarticulate within less than one day after death (Schäfer 1972), so that articulated preservation requires sudden burial (Dietl and Mundlos 1972). Some of these blanketing events were obviously accompanied by currents; ophiuroids in these layers are reoriented, and the shells of *Bositra buchi* are oriented convex up (Dietl and Mundlos 1972). In many other layers, however, there are no indications of current activity (Hess 1960).

The early diagenetic mineralization of soft parts was most likely facilitated by severe oxygen depletion of the bottom-water (Dietl and

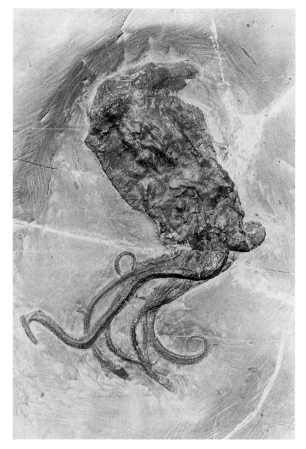

FIGURE 16.5 *Proteroctopus ribeti,* possibly the best-known example of a fossil octopus. Body length is 7.5 cm (without tentacles). (Photo courtesy of B. Riou, Musée de Paléontologie, La Voulte-sur-Rhône, Ardèche, France)

Mundlos 1972; Fischer and Riou 1982a), which may also have promoted slower decay of organic tissue (Fischer and Riou 1982a). It can be shown that at least the phosphatization of soft parts was mediated by bacteria (Fröhlich et al. 1992; Wilby, Briggs, and Riou 1996). In addition, the rapid mineralization of soft tissues required high concentrations of dissolved metal phosphates, carbonates, sulfates, and sulfides. These minerals became available through early diagenetic mobilization of sedimented metals (Wilby, Briggs, and Riou 1996).

PALEOBIOLOGY AND PALEOECOLOGY

The most obvious fossils in the lower Callovian of La Voulte are the ophiuroids (Sayn and Roman 1928; Hess 1960; Dietl and Mundlos 1972). Whereas earlier authors have described several species (Valette 1928), a detailed reinvestigation has shown that all the specimens belong to only one species: *Ophiopinna elegans* (Hess 1960). This rather small ophiuroid species has been documented in all postlarval ontogenetic stages, and can occur in sheer abundances of up to 3,000 specimens per m^2 (Hess 1960) (Figure 16.3).

This species has traditionally been considered to be benthic, as are all other ophiuroids (Hess 1960). However, *Ophiopinna elegans* shows some unusual morphological features, most notably feather- or paddle-like spines in the middle part of the arms (Hess 1960; Dietl and Mundlos 1972). This has been seen as evidence for swimming ability in *Ophiupinna elegans,* and in analogy to modern comatulid crinoids, a nekto-benthic mode of life has been proposed (Dietl and Mundlos 1972). Other echinoderms are rare and include only regular echinoids and a comatulid crinoid species (Fischer and Riou 1982a; Manni, Nicosia, and Riou 1985). The latter was certainly able to swim but probably had a benthic mode of life, like modern comatulids (Hess 1960).

The most diverse group of fossils in the lower Callovian of La Voulte are the crustaceans. At least 10 species of decapods are known (van Straelen 1925; Sayn and Roman 1928; Carriol and Riou 1991). Most common are nekto-benthic species with good swimming abilities (e.g., *Aeger, Antrimpos, Archaesolenocera, Udora*), whereas the truly benthic reptantian decapods (e.g., *Eryon, Eryma, Coleia, Glyphea*) are known from only a few specimens (Sayn and Roman 1928; Carriol and Riou 1991) (Figure 16.4). Mysidaceans, also a nekto-benthic group, are quite well represented, and six species have been described (Secretan and Riou 1986). A cumacean species is known as well, but only very few of the up to 4 mm long specimens have been collected (Bachmayer 1960).

Four species of the enigmatic Conchyliocarida have been described (Secretan and Riou 1983; Secretan 1985), but only the large (up to 30 cm long) *Dollocaris ingens* is well known (Secretan 1985). This species has been described as a member of the Mysidacea (van Straelen 1923), but new findings show the presence of some very unusual morphological features. *Dollocaris* has a large carapace covering the whole body, at least 12 abdominal segments, large raptorial thoracopods, large lamellar gills, and huge compound eyes (Secretan and Riou 1983; Secretan 1985). These traits exclude *Dollocaris* and related forms from any known crustacean groups, and therefore the class Conchiliocarida has been established (Secretan 1983).

The systematic position of this group as well as its relationship to the Concavicarida and Thylacocephala (Chapters 11 and 14) are still a matter of debate (Rolfe 1985; Schram 1990; Schram, Hof, and Steeman 1999). The huge compound eyes (up to 4 cm in diameter) are reminiscent of those in modern hyperiid amphipods, and, indeed, in analogy with that group, a mesopelagic lifestyle has been suggested for the Conchyliocarida

(Rolfe 1985). However, the heavily calcified carapace and thoracic limbs clearly exclude a pelagic mode of life, and the Conchyliocarida were most likely benthic predators (Secretan and Riou 1983; Secretan 1985). Pycnogonids (sea spiders) and annelid worms are very rare, and are yet to be described (Fischer and Riou 1982a).

The most common fossil in the lower Callovian of La Voulte is the bivalve *Bositra buchi*, which covers entire bedding planes (Hess 1960; Elmi 1967). Although the mode of life of this tiny bivalve has been a matter of debate for decades, it seems appropriate to assume a benthic lifestyle for this species (Chapter 15). Other bivalves have not been reported from this locality, with the possible exception of another posidoniid species (Sayn and Roman 1928). Cephalopods are quite rare in the fossiliferous layers. This applies not only to the exceptionally preserved teuthoids (four species) (Figure 16.6) and the octopus (Fischer and Riou 1982a, 1982b) (Figure 16.5), but also to ammonites and belemnites (Sayn and Roman 1928; Elmi 1967). All the cephalopods, including the octopus, are considered to have been pelagic forms (Fischer and Riou 1982b). Gastropods seem to be absent.

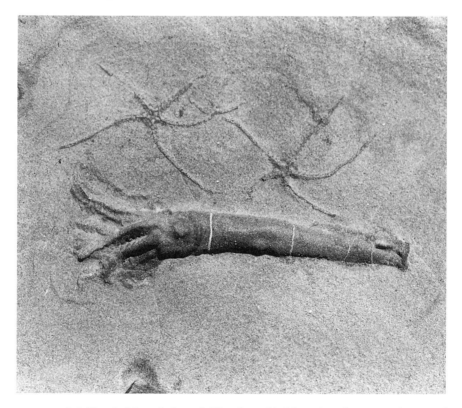

FIGURE 16.6 Teuthoid cephalopod *Rhomboteuthis lehmani* and two specimens of *Ophiopinna elegans*. Body length of *Rhomboteuthis* is 8 cm. (Photo courtesy of B. Riou, Musée de Paléontologie, La Voulte-sur-Rhône, Ardèche, France)

Fish are known from at least three species (Sayn and Roman 1928; Hess 1960), but the diversity of this group might be much higher (Dietl and Mundlos 1972). The reported species (*Notagogus* sp., *Pholidophorus* sp., shark remains) include open-water forms (Aldinger 1965); bottom-dwelling taxa seem to be absent. Other vertebrates are known from only a single skull of a marine crocodile (Kuhn-Schnyder 1960). Terrestrial elements, including both plants and animals, seem to be absent altogether.

The microfauna has not been studied in detail, but foraminifers are reportedly rare and of low diversity (genera include *Ammodiscus, Dentalina, Ophthalmidium,* and *Trochammina*) (Hess 1960). Ostracodes seem to be absent (Hess 1960). Distinct trace fossils are present but are also rare, and only fecal strings (possibly *Planolites*) and unnamed trace fossils have been mentioned in the literature (Hess 1960).

CONCLUSIONS

The lower Callovian of La Voulte-sur-Rhône has yielded a highly diverse fauna, including ophiuroids, crustaceans, fish, and cephalopods. Among the last are probably the best fossil examples of dibranchiate cephalopods in the world. Preservation of soft parts is known from many specimens from different taxonomic groups, and it is notable that different modes of early diagenetic mineralization contributed to soft-part preservation (Wilby, Briggs, and Riou 1995, 1996). The processes responsible for the exceptional preservation of these fossils are periodic sediment blanketing and oxygen depletion of the bottom-water. The large proportion of pelagic and nekto-benthic species supports the hypothesis of oxygen depletion, but the consistent presence of benthic animals also indicates that the bottom-water was never anoxic for a long time.

Although most of the fossils found at this locality are now described and the stratigraphy is known in detail, the understanding of this Lagerstätte could certainly be improved by a detailed sedimentological and geochemical investigation.

ACKNOWLEDGMENTS

J. C. Fischer carefully reviewed the chapter and provided some additional references. B. Riou provided information on the local museum at La Voulte-sur-Rhône. Access to older literature was made possible by H. Rieber.

REFERENCES

Aldinger, H. 1965. Zur Ökologie und Stratinomie der Fische des Posidonienschiefers (Lias epsilon). *Senckenbergiana lethaea* 46a:1–12.

Bachmayer, F. 1960. Eine fossile Cumaceenart (Crustacea, Malacostraca) aus dem Callovien von La Voulte-sur-Rhône (Ardèche). *Eclogae geologicae Helvetiae* 53:422–426.

Bartels, C., D. E. G. Briggs, and G. Brassel. 1998. *The Fossils of the Hunsrück Slate.* Cambridge: Cambridge University Press.

Carriol, R.-P., and B. Riou. 1991. Les Dendrobranchiata (Crustacea, Decapoda) du Callovien de La Voulte-sur-Rhône. *Annales de Paléontologie* 77:143–160.

Cayeux, L. 1922. *Les minerais de fer oolithiques de France.* Fascicule II, *Minerais de fer secondaires.* Mémoir Explicative de la Carte Géologique de France, Etudes des Gîtes Mineraux de France. Paris: Ministère des Traveux Publics.

Dietl, G., and R. Mundlos. 1972. Ökologie und Biostratinomie von *Ophiopinna elegans* (Ophiuroidea) aus dem Untercallovium von La Voulte (Südfrankreich). *Neues Jahrbuch für Geologie und Paläontologie, Monatshefte* 1972:449–464.

Elmi, S. 1967. *Le lias supérieur et le Jurassique moyen de l'Ardèche.* Documents des Laboratoires de Géologie de la Faculté des Sciences de Lyon, vol. 19. Lyon: Laboratoires de Géologie de l'Université de Lyon, Faculté de Sciences.

Fischer, J.-C., and B. Riou. 1982a. Les teuthoides (Cephalopoda, Dibranchiata) du Callovien inférieur de La Voulte-sur-Rhône (Ardèche, France). *Annales de Paléontologie* 68:295–325.

Fischer, J.-C., and B. Riou. 1982b. Le plus ancien octopode connu (Cephalopoda, Dibranchiata): *Proteroctopus ribeti* nov. gen., nov. sp., du Callovien de l'Ardèche (France). *Comptes Rendus de l'Académie des Sciences,* 2nd ser., 295:277–280.

Fröhlich, F., A. Mayrat, B. Riou, and S. Secretan. 1992. Structures rétiniennes phosphatées dans l'oeil géant de *Dollocaris,* un crustacé fossile. *Annales de Paléontologie* 78:193–204.

Hess, H. 1960. Neubeschreibung von *Geocoma elegans* (Ophiuroidea) aus dem unteren Callovien von La Voulte-sur-Rhône (Ardèche). *Eclogae geologicae Helvetiae* 53:335–385.

Kuhn-Schnyder, E. 1960. Ein Schädelfragment von *Metriorhynchus* aus dem unteren Callovien von La Voulte-sur-Rhône (Ardèche, France). *Eclogae geologicae Helvetiae* 53:793–804.

Lory, Ch. 1854. Compte rendu de l'excursion de la Voulte: Réunion extraordinaire de la Société Géologique à Valence. *Bulletin de la Société Géologique de la France,* 2nd ser., 11:737.

Manni, R., U. Nicosia, and B. Riou. 1985. *Rhodanometra lorioli* n. gen. n. spec. and other Callovian crinoids from La Voulte-sur-Rhône (Ardèche, France). *Geologica Romana* 24:87–100.

Rolfe, W. D. I. 1985. Form and function in Thylacocephala, Conchyliocarida and Concavicarida (?Crustacea): A problem of interpretation. *Transactions of the Royal Society of Edinburgh* 76:391–399.

Sayn, G., and F. Roman. 1928. *Etudes sur le Callovien de la Vallée du Rhône. II. Monographie stratigraphique et paléontologique du Jurassique moyen de La Voulte-sur-Rhône.* Travaux du Laboratoire de Géologie de la Faculté des Sciences de Lyon, vol. 13 (Mémoire II). Lyon: Laboratoire de Géologie de l'Université de Lyon, Faculté de Sciences.

Schäfer, W. 1972. *Ecology and Palaeoecology of Marine Environments.* Chicago: University of Chicago Press.

Schram, F. 1990. On Mazon Creek Thylacocephala. *Proceedings of the San Diego Society of Natural History* 1990:1–15.

Schram, F. R., C. H. J. Hof, and F. Steeman. 1999. Thylacocephala (Arthropoda: ?Crustacea) from the Cretaceous of Lebanon. *Palaeontology* 42:769–797.

Secretan, S. 1983. Une nouvelle classe fossile dans la super-classe des crustacés: Conchyliocarida. *Comptes Rendus de l'Académie des Sciences,* 2nd ser., 296:741–743.

Secretan, S. 1985. Conchyliocarida, a class of fossil crustaceans: Relationships to Malacostraca and postulated behaviour. *Transactions of the Royal Society of Edinburgh* 76:381–389.

Secretan, S., and B. Riou. 1983. Un group énigmatique de crustacés: Ses représentants du Callovien de La Voulte-sur-Rhône. *Annales de Paléontologie* 69:59–97.

Secretan, S., and B. Riou. 1986. Les mysidacés (Crustacea, Peracarida) du Callovien de La Voulte-sur-Rhône. *Annales de Paléontologie* 72:295–323.

Seilacher, A., W. E. Reif, and F. Westphal. 1985. Sedimentological, ecological and temporal patterns of fossil Lagerstätten. *Philosophical Transactions of the Royal Society of London, B* 311:5–23.

Valette, D. A. 1928. Note sur des Ophiurides du Callovien inférieur de La Voulte. In G. Sayn and F. Roman, *Etudes sur le Callovien de la Vallée du Rhône. II. Monographie tratigraphique et paléontologique du Jurassique moyen de La Voulte-sur-Rhône,* pp. 67–79. Travaux du Laboratoire de Géologie de la Faculté des Sciences de Lyon, vol. 13 (Mémoire II). Lyon: Laboratoire de Géologie de l' Université de Lyon, Faculté de Sciences.

van Straelen, V. 1923. Les mysidacés du Callovien de La Voulte-sur-Rhône (Ardèche). *Bulletin de la Société Géologique de la France,* 4th ser., 23:431–439.

van Straelen, V. 1925. *Contribution à l'étude des crustacés décapodes de la période Jurassique.* Mémoir de l'Académie Royale Belgique, Classe Scientifique, 2nd ser., vol. 7. Brussels: Hayez.

Wilby, P. R. 2001. La Voulte-sur-Rhône. In D. E. G. Briggs and P. R. Crowther, eds., *Palaeobiology II,* pp. 349–351. Oxford: Blackwell Science.

Wilby, P. R., D. E. G. Briggs, and B. Riou. 1995. La Voulte-sur-Rhône (Callovian): Fossilized soft tissues from a "toxic sludge." *Palaeontology Newsletter* 28:25.

Wilby, P. R., D. E. G. Briggs, and B. Riou. 1996. Mineralization of soft-bodied invertebrates in a Jurassic metalliferous deposit. *Geology* 24:847–850.

Ziegler, P. A. 1990. *Geological Atlas of Western and Central Europe.* 2nd ed. The Hague: Shell International Petroleum Maatschappij.

17
Oxford Clay:
England's Jurassic Marine Park

Carol M. Tang

OR OVER A CENTURY, JURASSIC MARINE STRATA IN GREAT
Britain have yielded an amazing array of well-preserved fossils, ranging from articulated marine reptiles to squid with preserved soft
parts. In the early ninteenth century, great finds of articulated "sea mon-
sters" from Jurassic strata exposed along the southern Dorset coast
started a fossil-hunting craze that put the town of Lyme Regis on the map
as a paleontological mecca. Mary Anning, the world's first well-known
female paleontologist, found the first articulated ichthyosaur in the cliffs
around Lyme Regis and later discovered several large marine vertebrates
from the cliff exposures of the Lias and Kimmeridge Clay.

The fossils from the British Jurassic Lagerstätten not only have revealed much about the morphology and paleobiology of extinct organisms, but also have played a significant role in the development of modern ideas in paleontology and biostratigraphy. Large numbers of both
professional and amateur paleontologists have devoted their lives to dis-
covering and studying these sometimes extremely well preserved and
beautiful fossils. Surprisingly, however, much still can be learned about
the paleontological, geological, and taphonomic aspects of these Lagerstätten. This chapter examines in detail the Oxford Clay, one of these
spectacular British deposits.

Fossils from the Oxford Clay were collected in the mid-nineteenth
century from rail cuts and borrow pits dug for rail construction, and
since the late nineteenth century from brick quarries in central England.
Because the Oxford Clay is particularly suitable for making bricks, quarries provided good exposures and thus ample collecting was possible. In

the nineteenth century, the Leeds brothers, who were amateur paleon-tologists, amassed the world's largest collection of marine reptiles by re-warding quarrymen who worked Oxford Clay brick pits by hand for mak-ing fossil discoveries. Today, mechanized work in brick pit quarries still allows for the discovery of spectacular fossils at good exposures. Al-though many outcrops are no longer accessible, new exposures can be created during the construction of roads. They can often provide much geological and paleontological information even during the short times they are accessible. For example, excavation in a worked-out gravel pit in Ashton Keynes yielded 11 squid with exquisite soft-tissue preservation, two fish, and over 100 ammonites before the pit became flooded under 4.5 m of rainwater ("Jurassic squid" 1996).

The Oxford Clay Formation can be considered as both a conserva-tion and a concentration Lagerstätte because of the presence of (1) an abundance of exquisitely preserved ammonites throughout the forma-tion, (2) articulated marine reptiles in the Peterborough Member, and (3) rare soft-part preservation in squid.

The Oxford Clay is scientifically significant in many ways. First, the organic-rich nature of these strata provides clues to the formation of petroleum-producing source rocks. Second, the abundance and variety of preservational modes exhibited by ammonites and large marine ver-tebrates, for which the British Jurassic is well known, have provided pa-leontologists with a much better understanding of the morphology and paleobiology of these extinct organisms. Finally, the soft-body preserva-tion found within the Oxford Clay provides an opportunity to study de-tails of many organisms not usually preserved in the fossil record.

GEOLOGICAL CONTEXT

The Oxford Clay Formation ranges from middle Callovian to lower Ox-fordian (Cope et al. 1980) (Figure 17.1). It is exposed along the Dorset and Yorkshire coasts as well as in brick pits, railroad exposures, and bor-row pits across southern and central England. There are two main facies in the Oxford Clay, with several subfacies: (1) the organic-rich mudstone of the lowest unit, the Peterborough Member (informally known as the lower Oxford Clay), and (2) the calcareous, organic-poor Stewartby and Weymouth Members (informally known as the middle and upper Ox-ford Clays, respectively) (Hudson and Palframan 1969; Duff 1975). Al-though the Oxford Clay had long been informally divided into lower, middle, and upper units, it was only in 1993 that the members were for-mally described according to modern stratigraphic nomenclature (Cox, Hudson, and Martill 1993). Because the lowest member of the Oxford Clay is more organic-rich and better for making inexpensive bricks, there

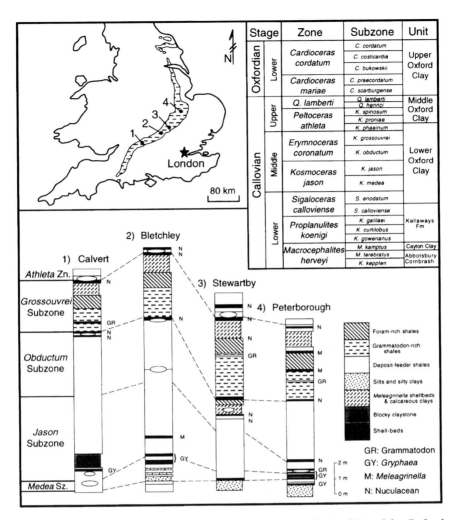

FIGURE 17.1 Geographic distribution and stratigraphic relationships of the Oxford Clay, including stratigraphic sections of the lower Oxford Clay (Peterborough unit), in central and southern England. The numbered sections represent localities. (Modified from Duff 1978 and Martill and Hudson 1991)

are more exposures of the Peterborough Member than of the Stewartby and Weymouth Members.

Organic-rich shales with high organic carbon content dominate the Peterborough Member (Figure 17.1). This member is composed primarily of quartz and illite, but also contains chlorite, kaolinite, K-feldspar, plagioclase, anatase, apatite, calcite, and siderite (Norry, Dunham, and Hudson 1994). Organic concentrations from the Peterborough Member range from 0.5 percent to a very high value of 16.6 percent (Kenig et al. 1994). This unit contains a number of shell beds commonly located at the tops of parasequences (MacQuaker and Howell 1999). Other lithologies include carbonate concretion layers, silts, blocky claystones, and calcareous shales dominated by distinct faunas (Duff 1975). Petrological studies of concretions that formed early in diagenesis and have not been greatly deformed suggest that much of the sediment of the lower Oxford Clay was composed of fecal pellets that were pelleted either in the water column or on the seafloor (Hudson and Martill 1994). The middle and upper members are more massive calcareous mudstones containing fewer fossils and having a lower organic carbon content. Shell beds, especially those dominated by large *Gryphaea* valves, are found throughout the formation, although they are most common in the basal Peterborough Member, disappear in the upper Peterborough Member, and reappear sporadically in the Stewartby and Weymouth Members (Hudson and Martill 1991).

The organic matter in the Oxford Clay is predominantly marine (Belin and Kenig 1994; Kenig et al. 1994). Variations in organic matter can be correlated with stratigraphic breaks as well as different environmental conditions (Kenig et al. 1994). The highest concentrations of organic matter are found in Peterborough Member shales, which are dominated by deposit-feeding bivalves, while lower levels of organic carbon are found in Peterborough Member shell beds and the Stewartby and Weymouth Members (Kenig et al. 1994). The organic carbon within the Stewartby may have had a longer residence time in a more oxygenated environment (Belin and Kenig 1994).

The Oxford Clay was not deposited continuously (MacQuaker and Howell 1999), and many diastems, condensed sections, and episodes of high sedimentation occur throughout this interval (Hudson and Martill 1991). It is estimated that only 13 m of sediment in the Peterborough region accumulated in about 1.67 Ma, with an average rate of about 0.008 mm per year (Cope 1993). This low accumulation rate in the Oxford Clay is believed to be a result of (1) winnowing and sediment bypass into offshore environments during times of low sea levels, and (2) sediment starvation during high sea levels (Macquaker 1994).

The Oxford Clay did not experience much burial or thermal alteration (Hudson and Martill 1991; Kenig et al. 1994), thus making it ideal for studying the sedimentary origins of petroleum-producing rocks.

PALEOENVIRONMENTAL SETTING

The Oxford Clay was deposited in a shallow epicontinental seaway in water depths of about 10 to 50 m (Martill, Taylor, and Duff 1994). Overall, there is a deepening upward trend through the Oxford Clay Formation (Duff 1975). The presence of autochthonous shell beds—interpreted as tempestites—indicates the passage of large storms or hurricanes across the seafloor (D. Martill, personal communication, 1985).

The land source was nearby and probably toward the north and east, as evidenced by sandy facies in Yorkshire and the southern North Sea, as well as the presence of terrestrial plant and vertebrate fossils in the Peterborough area (Hudson and Martill 1991). Judging from the high levels of organic matter in the sediments, nutrient supply was quite high in this seaway and was probably derived from terrestrial runoff (Hudson and Martill 1991). Restricted circulation in the basin may have acted to concentrate nutrients in this region (Martill, Taylor, and Duff 1994).

Oxygen isotopic analyses of well-preserved carbonate mollusc shells (ammonites, belemnites, bivalves) and phosphatic vertebrate bones (fish, marine reptiles) produced bottom-water temperature estimates ranging from 12° to 20°C, while the surface water temperature estimates range from 16° to 29°C (Williams 1988; Anderson et al. 1994). There do not appear to be time-dependent trends in paleoceanographic conditions through deposition of the Peterborough Member, although short-term fluctuations, possibly seasonal changes, were detected amid generally stable conditions (Anderson et al. 1994).

Based on the lack of deep-burrowing organisms and the level of preservation of marine reptile bones, it has been suggested that the seafloor was soupy during deposition of the Peterborough Member; the soupiness was possibly maintained by extensive bioturbation and input of water from an underlying aquifer (Martill, Taylor, and Duff 1994). The substrate is hypothesized to have been firmer during deposition of the Stewartby and Weymouth Members.

Unlike some Jurassic Lagerstätten, interpreted as having been deposited under anoxic conditions, the Oxford Clay contains an abundance of benthic organisms, including infaunal deposit feeders, which indicates that neither the bottom-waters nor the substrate was continuously anoxic (Duff 1975). Organic geochemical studies suggest that there were occasional, short-lived periods of water-column anoxia during deposition of the Peterborough Member, but not during Stewartby or Weymouth Member times (Kenig et al. 1994). Numerous microburrows—even in sediments with the highest concentrations of organic matter—suggest that the seafloor was rarely anoxic, but could have been dysoxic at times (Belin and Kenig 1994). Paleobiological studies indicate that oxygen levels may have been lower during the deposition of the Peterborough Member (especially during accumulation of the most organic-rich beds in the Peterborough) than at other times (Hudson and Martill 1991).

Some workers have proposed that high levels of organic matter preservation would require low oxygen conditions on the seafloor, possibly resulting from high productivity and stratification of the water column (Tyson and Pearson 1991). Stable isotopic results from both benthic and pelagic organisms of the Peterborough Member are compatible with the idea of seasonal alternation between stratified and mixed states in the Oxford Clay sea, but are not unequivocal (Williams 1988; Anderson et al. 1994). The abundance of large fecal pellets in the beds with highest organic content does suggest high productivity (Beling and Kenig 1994). Similarly, lithofacies studies indicate that relatively high levels of productivity existed during deposition of most of the Oxford Clay and that not much variation in productivity occurred through time (Macquaker 1994). Variations in organic matter do not appear to be related to changes in productivity, although they can be correlated with stratigraphic breaks, changes in environmental conditions, differences in food-web structure, and variations in reworking of the sediments (Kenig et al. 1994). Macquaker (1994) proposes that the organic matter was preserved when the sediments below the mixed layer were anoxic—even if bottom-waters were oxygenated—and that the units deposited quickly are most organic-rich.

TAPHONOMY

In the Peterborough Member, ammonites, bivalves, and gastropods are present on almost every bedding plane (Hudson and Martill 1991). The organisms are usually preserved in only two dimensions but are generally aragonitic, especially in the organic-rich shales (Hudson and Martill 1991). In organic-rich horizons, aragonite, with its original chemical and microstructural details, frequently shows beautiful iridescence (Hudson and Martill 1991). Fewer fossils are found in the Stewartby and Weymouth Members, although pyritized internal ammonite molds are common in the upper part of the Stewartby Member and in the Weymouth Member. Some ammonites preserved three-dimensionally in calcareous concretions occur in the *Lamberti* zone. In the younger two members, aragonite is usually dissolved, although in some localities like Ashton Keynes, aragonite is preserved (Hudson and Martill 1991).

Limestone concretions throughout the Oxford Clay are formed diagenetically and contain the same fossils as the surrounding shales. Both pre- and post-compaction concretions are found, with the former being septarian and containing fossils preserved in three dimensions. The late-forming concretions contain crushed fossils composed of secondary calcite (Duff 1975).

Organic biomarkers are still preserved in the Oxford Clay, and biomarker molecules of chlorophyll and bacterial proteins are abundant (Kenig et al. 1994). Organic cysts of some microplankton (including dinoflagellates) are also preserved (Woolam 1980).

Pyritized ammonites and bivalves are fairly common fossils, with pyrite preferentially precipitated in voids, producing internal molds. In the Peterborough Member, pyrite may also precipitate on the outside of shells and even replace aragonite or bone phosphate (Hudson and Martill 1991). In the Stewartby and Weymouth Members, the burrows can be pyritized (Martill 1991a). Pyrite in shale horizons occurs as patchy overgrowths on aragonitic shells (Duff 1975). In the more porous lithologies such as the shell beds, calcite and/or pyrite has frequently replaced the aragonite. It is hypothesized that the shell beds acted as aquifers for sulfide-rich waters and thus pyrite was preferentially deposited in the center of shell beds (Duff 1975).

Coprolites are quite abundant in the Peterborough Member (Martill 1985). These trace fossils may contain fish remains or cephalopod hooks that were easily preserved due to the large amount of fish-bone phosphate, which lithified the fecal material quickly. Fecal pellets are also very abundant in the most organic-rich sediments (Belin and Kenig 1994).

In general, the benthic organisms are considered to be autochthonous, as evidenced by the lack of abrasion (even on fragile fossils), the abundance of articulated valves, and the presence of complete, articulated ophiuroids and crustaceans (Duff 1975). However, few individuals are preserved in life position. It appears that even the shell beds are autochthonous, albeit winnowed, and are concentrated as a result of extremely slow sediment accumulation or storm reworking (Duff 1975; Hudson and Martill 1991).

The nektic organisms are thought to have been derived from the overlying water column and have not undergone significant lateral transport. Ammonites are typically found as complete tests, and, commonly, even peristomes and delicate lappets are preserved (Martill, Taylor, and Duff 1994) (Figure 17.2). These well-preserved examples can be found with broken pieces of ammonite shells that are interpreted to have been broken by predators, not abrasion (Martill, Taylor, and Duff 1994).

Similarly, articulated marine reptiles can also be found along with fragmented remains, which are believed to have resulted from scavenging. Marine vertebrates can be preserved in an articulated manner, but exhibit a great range of preservational styles (Martill 1985). Isolated teeth were most likely shed by the reptiles during life, and the isolated bones may have been derived from floating carcasses (Martill 1985). Commonly, disarticulated skeletons are associated with large shark teeth, suggesting scavenging or predation. There is a positive correlation between the degree of articulation and the organic content of sediment, which may be due to increased productivity in the water column or to low oxygenation levels at depth, preventing scavenging (Martill 1985).

Although, as discussed earlier, dysoxic conditions were present occasionally, Martill (1985) has suggested that the preservation of articulated vertebrate remains is not necessarily a result of low oxygen conditions, but occurred when dead organisms sank through the soupy sediment and were buried in mud immediately. This burial in soupy sediment is proposed to explain the preservation of soft parts of ichthyosaurs in the

FIGURE 17.2 Complete microconch of the ammonite *Indosphinctes,* with lappet. Mature microconchs are approximately 90 mm in size. (Photo courtesy of D. Martill, University of Portsmouth, United Kingdom)

Peterborough Member (Martill 1985). Commonly, a thin layer of pyrite precipitated on vertebrate bones and inside bone cavities, followed by precipitation of secondary calcite (Martill 1991e). If the cavities were filled entirely, the bones were usually able to withstand compaction. Otherwise, bones could be flattened (Martill 1991e). Typically, the upper surfaces of vertebrate skeletons are encrusted and are hypothesized to have formed "benthic islands" within the soft substrate, providing a hard substrate for boring or encrusting organisms (Martill 1987).

The soft-body preservation at Christian Malford is thought to be different from the preservational processes that operated during deposition of the rest of the Oxford Clay (Allison 1988). This horizon of soft-part preservation is in the upper part of the Peterborough Member in the *Athleta* zone (Figure 17.1). The soft parts of preserved cephalopods (Figure 17.3) are present here as a film of calcium phosphate (Allison 1988). The absence of bioturbation and the presence of laminations suggest that oxygenation levels were low. The carcasses likely became entrained in the sediment, which was probably quite soupy, before decay gases developed (Allison 1988). While some compaction and decomposition of the arms occurred, most of the squid found here were miner-

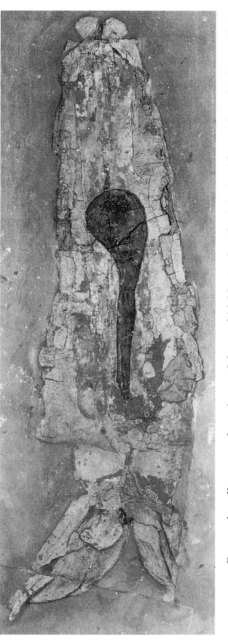

FIGURE 17.3 Extremely well preserved specimen of the teuthid *Mastigophora brevipinnis*, with mantle and ink sac still visible, collected from the lower Oxford Clay (*Athleta* zone) at Christian Malford. Length of specimen is approximately 52 cm. (Copyright The Natural History Museum, London, courtesy of Dr. Paul D. Taylor)

alized. Although beds that are laterally equivalent with this horizon at Christian Malford do contain a similar fauna, they do not exhibit soft-part preservation. Thus, it appears that only preservational conditions—not any paleobiological conditions—differed in this one horizon at Christian Malford (Page and Doyle 1991).

PALEOBIOLOGY AND PALEOECOLOGY

The Oxford Clay has contributed much to the paleobiological understanding of such extinct organisms as ammonites and belemnites. Buckman (1909–1930) described many ammonite taxa from the Oxford Clay, and although much of his work has proved to be invalid, it served as a starting point for subsequent paleontological work. Brinkmann's (1929) landmark study using statistical paleontology to document evolutionary trends in ammonites was based on *Kosmoceras* specimens from the Oxford Clay. This classic paper not only has contributed much to evolutionary research techniques, but has greatly influenced the field of biostratigraphy. Oxford Clay ammonites were also used as the basis for conclusive evidence of sexual dimorphism in this group, as documented by Callomon (1963) and followed up by Palframan (1966). Before this was recognized, micro- and macroconchs were often described as two species instead of two sexes of the same species. Soft-body preservation of belemnoids found at Christian Malford in the 1840s provides a more complete picture of the morphology of these organisms than that provided by the commonly found rostrum. A fairly detailed description of the fauna, written by many Oxford Clay experts, has been published as a popular book (Martill and Hudson 1991).

It appears that there was much interplay between the benthic and the pelagic food webs, possibly due to the shallow nature of the sea (Martill, Taylor, and Duff 1994). For example, some of the mobile organisms (e.g., sharks, fish, ammonites) may have been able to exist in both pelagic and nektobenthic settings, perhaps feeding on benthic organisms, and organic matter (in the form of fecal matter and carcasses) would have been transported through the pelagic realm onto the seafloor (Martill, Taylor, and Duff 1994).

Benthic Invertebrates

Among Oxford Clay benthic invertebrates, bivalves are the dominant group, with over 50 species described, and are the best documented (Duff 1978, 1991). Bivalves vary greatly in their density and distribution throughout the formation. Diversity can vary in horizons, with some shell beds dominated by a single species (Martill, Taylor, and Duff 1994). Although *Gryphaea* occurs throughout the Oxford Clay (Hudson and Martill 1991), it is most common in the lower part of the Peterborough Member, while *Grammatadon, Meleagrinella,* and nuculaceans dominate the shell beds higher in the member (Martill, Taylor, and Duff 1994). Deep-burrowing bivalves are uncommon in the Peterborough Member, whereas they are more common in the Stewartby and Weymouth Members (Hudson and Palframan 1969; Duff 1975). Other benthic molluscs include scaphopods and gastropods, which are locally common but are not diverse (Martill, Taylor, and Duff 1994).

Brachiopods may be locally abundant in the Weymouth Member, but overall are fairly rare, especially within shell beds of the Peterborough Member. Diversity was lowest in the Peterborough Member, which had a softer substrate than the two younger members (Prosser 1991). Taxa include lingulids, rhynchonellids, and terebratulids.

There are about a dozen crustacean species present in the Oxford Clay, and most of the trace fossils found in the unit, including *Ophiomorpha, Thalassinoides,* and *Favreina* fecal pellets, have been attributed to crustacean activities (Martill 1991a). Although few well-preserved specimens are found, malacostracan crustaceans are believed to have been quite abundant and diverse in the Peterborough Member (Martill 1991a). Colonial, encrusting, stalked cirripedes are reported from the Christian Malford locality (Martill 1991a). Ostracodes are rare in the Peterborough Member, but are quite common in the Stewartby and Weymouth Members (Martill 1991a).

Echinoderms can be locally abundant throughout the Oxford Clay (Martill 1991a). Holothurian remains are found in all members (Martill 1991a; Martill, Taylor, and Duff 1994), and ophiuroids can be found as single plates and, occasionally, as complete specimens (Martill 1991a). Crinoids are represented by individual plates, some articulated columns, and rare isocrinid cups (Martill 1991a). Rare regular and irregular echinoids have been found, mainly from the Stewartby and Weymouth Members, but their preservation is poor, whereas regular echinoid spines have been recovered from shell beds within the Peterborough Member (Martill 1991a).

Bryozoans are rare overall and are preserved as encrusters, borings, and bioimmured colonies (Martill 1991a). Cyclostomes and ctenostomes are most common in condensed sections where they are frequently found on grypheate oysters (Martill 1991a). Annelids are most commonly represented by *Serpula* encrusted on *Gryphaea* and by isolated *Genicularia* tubes (Martill, Taylor, and Duff 1994). Serpulids can also be found encrusting ammonites, wood, bones, and belemnites (Martill 1991a).

Foraminifera, including both encrusting and free-living forms, are very common, even dominating some facies (Duff 1975). As with the benthic macrofauna, foraminifera in the lower member of the Oxford Clay are less diverse than in the other two members, with 35 species from the Peterborough Member, 57 species from the Stewartby Member, and 54 species from the Weymouth Member (Shipp 1978). Other microfauna include calcareous nannofossils and dinoflagellate cysts (Martill, Taylor, and Duff 1994).

Pelagic Invertebrates

The other dominant molluscs are the ammonites (Figure 17.2). The abundance of ammonites, their beautiful preservation, and their usefulness in biostratigraphy have made them a focus for paleontological studies of the Oxford Clay. Even so, few detailed studies have been published, and only a few taxa have been monographed (Page 1991). Both Tethyan and Boreal ammonites are present (Callomon 1985). Generally, the Peterborough Member is dominated by Kosmoceratidae; the Stewartby Member, by Kosmoceratidae, Cardioceratidea, Perisphinctacea, and Oppeliidae; and the Weymouth Member, by Cardioceratidae (Page 1991). Ammonite aptychi are found throughout the unit, usually as isolated parts and rarely as pairs or in association with the body chamber (Page 1991).

Nautiloids are rare, but one genus, *Paracenoceras,* has been found (Page 1991). It grew up to 0.5 m in diameter, and one specimen has stomach contents consisting of phosphatic coprolites and teleost fish remains (Martill, Taylor, and Duff 1994). This suggests that it was a benthic opportunist that ingested crustaceans and small fish (Martill, Taylor, and Duff 1994).

Remains of belemnites are abundant throughout the Oxford Clay (Page and Doyle 1991). The two most common belemnites are *Cylindrotheuthis* and *Belemnopsis,* with *Cylindrotheuthis* being the more abundant of the two (Martill, Taylor, and Duff 1994). They are most frequently found as rostra, but phragmocones are also quite common, and rare examples of the pro-ostracum can also occur (Page and Doyle 1991). It has been suggested that many guards with damaged phragmocones are a result of predation by marine reptiles (Martill, Taylor, and Duff 1994).

Hooklets from other squid-like teuthids are common, suggesting that such animals were quite numerous even if they were rarely preserved (Page and Doyle 1991). More complete specimens of soft-bodied cephalopods have been recovered from the Christian Malford locality, where spectacular preservation of the soft-part morphology of coleoids, teuthids, and a cuttlefish occurred (Allison 1988). In some complete specimens, the arms, mantle, and ink sacs are clearly seen (Figure 17.3). Study of these Oxford Clay coleoids and their preserved arms, tentacles, and sucker-like structures has yielded insights into the evolutionary pattern of cephalopods (Vecchione et al. 1999). In one specimen from Christian Malford, a small fish is within the arms of a teuthid armed with hooklets (Boucot 1990).

Oxygen isotopic analyses of ammonite tests and belemnite rostra suggest that the *Kosmoceras* tended to live in the warmer near-surface waters, while the belemnites inhabited colder waters (Anderson et al. 1994). The belemnites could have been either nektobenthic or migrants from colder regions (Anderson et al. 1994).

Possible Pseudoplankton

Although there is a great diversity of both epifaunal and infaunal benthic bivalves, there are also some taxa that have been proposed to be pseudoplanktonic (attached to floating algae, logs, or other hard substrates) (Duff 1975). For example, bedding planes with dense accumulations of thin-shelled *Meleagrinella* and *Bositra* could be interpreted as mass kills of pseudoplanktonic forms attached to floating substrates (Martill, Taylor, and Duff 1994). However, these genera have also been interpreted as benthic opportunistic colonizers (Martill, Taylor, and Duff 1994). *Parainoceramus* is presumed to be a pseudoplanktonic byssate bivalve, which may have been attached to driftwood (Martill, Taylor, and Duff 1994).

Trace Fossils

Coprolites within the Oxford Clay are commonly between 2 and 10 mm in length (Martill 1985). Larger specimens sometimes contain fish bones and cephalopod hooks (Martill 1985). Fecal pellets, some of which have been identified as from *Favreina,* can be found dominating the sediment in some beds (Martill, Taylor, and Duff 1994). Other smaller pellets may have been made by zooplankton such as copepods (Martill, Taylor, and Duff 1994).

Levels of bioturbation can vary greatly. Some beds show almost complete bioturbation, while others contain only some discrete *Thalassinoides* (Martill, Taylor, and Duff 1994). Many of the burrows (e.g., *Thalassinoides, Ophiomorpha*) were probably made by crustaceans, wheras others were probably made by polychaete worms (Martill, Taylor, and Duff 1994). Large burrows are commonly found immediately below shell beds (Martill, Taylor, and Duff 1994).

Vertebrates

Fish are most common in the Peterborough Member, usually as fragments, although articulated specimens do occur (Martill 1991b). The diverse fish fauna is represented by 27 genera in the Peterborough Member alone (Martill, Taylor, and Duff 1994) and about 32 genera in the entire formation (Hudson and Martill 1991). Actinopterygians (ray-finned bony fishes), represented by otoliths, teeth, scales, and skeletons (usually fragmented), include common examples of *Lepidotes, Caturus,* and *Hypsocormus* (Martill 1991b). The fish filled a number of niches, feeding on many types of organisms, on different sizes of prey, and at different levels in the water column (Martill, Taylor, and Duff 1994). For

example, some fish are interpreted as fast-swimming, open-ocean predators that may have fed on fish and cephalopods, while others may have
fed on benthic invertebrates (Martill, Taylor, and Duff 1994). One of the
most interesting fish to be found is *Leedsichthys,* which may have been
over 20 m in length and is the largest bony fish of all time (Martill 1988).
Analogous to modern baleen whales, *Leedsichthys* was probably a filter
feeder that used giant gills with teeth to strain its food (Martill, Taylor,
and Duff 1994).

Chondrichthyes (cartilaginous fish such as sharks) are also commonly
found—usually as dental plates, teeth, and spines. The most common
taxa are large hybodont sharks and chimaeras (ratfish). Remains of
neoselachians (modern types of sharks) are also occasionally found
(Martill, Taylor, and Duff 1994). Although traditional paleoecological
reconstructions place the sharks and chimaeras as bottom-dwelling organisms, based on dentition seemingly adapted to preying on organisms
with shells (Martill, Taylor, and Duff 1994), stable isotopic analyses of
the bone materials suggest that many were surface dwellers along with
the marine reptiles (Anderson et al. 1994).

The marine reptiles that made the British Jurassic shales famous consist of 10 genera (Martill, Taylor, and Duff 1994). They include at least
one genus of ichthyosaurs, three genera of long-necked plesiosaurs, four
genera of pliosaurs, and two genera of marine crocodiles (Martill, Taylor, and Duff 1994; systematic review in Martill 1991c).

The ichthyosaurs are represented by the genus *Ophthalmosaurus,*
which probably grew up to 5 m in length (Martill 1991c). As its name
suggests, *Ophthalmosaurus* had very large eye orbits—perhaps the largest
of any known extinct or living vertebrate—and probably used its large
eyes to see under low light conditions (Martill, Taylor, and Duff 1994).
Based on its morphology, this genus is interpreted to have been a fast,
cruising swimmer that fed on small, fast prey (Martill, Taylor, and Duff
1994). Since the jaws are elongate and slender, and may have lacked
teeth in adults, *Opthalmosaurus* could have fed on cephalopods without
heavy skeletal parts and/or may have bitten or shaken off the calcareous
guards from belemnites before ingesting them (Martill, Taylor, and Duff
1994). Since few juveniles are represented in the Oxford Clay, these
ichthyosaurs may have bred in a different region and migrated into this
area (Martill, Taylor, and Duff 1994). There may also be a second
ichthyosaur genus, which is as yet undescribed (Martill 1991c).

Plesiosaurs are quite common in the Oxford Clay—especially *Cryp-
toclidus*—and growth series ranging from juvenile to advanced adult
stages have been obtained (Martill 1991c). The plesiosaur species seem
to have become specialized to their respective prey; some may have used
their teeth for sieving small prey, such as crustaceans, while others may
have eaten fish and cephalopods (Martill, Taylor, and Duff 1994). Their

long necks are not hydrodynamically efficient and thus must have been adaptive for other reasons—possibly for stirring up sediments to flush out fish and crustaceans or for hunting while their heads were elevated above the sea surface (Martill, Taylor, and Duff 1994).

The pliosaurs from the Oxford Clay represent some of the largest carnivorous reptiles of all time and may have been as large as *Tyrannosaurus* (Martill 1991c). For example, *Liopleurodon ferox* had a skull up to 3 m in length and probably grew to over 15 m long. It is one of the largest marine reptiles ever found (Martill 1991c) and may rank with the ichthyosaurs at Berlin-Ichthyosaur (Chapter 13) as one of the largest carnivorous reptiles to have roamed the seas. Stomach contents indicate that these pliosaurs preyed on fish, cephalopods, and other marine reptiles (Martill 1992). Some marine reptile bones have bite marks attributable to pliosaurs, and the partial skeletons common in the Oxford Clay may be the result of pliosaur predation (Martill, Taylor, and Duff 1994). Although the pliosaurs have wide heads that would have allowed them to swallow large prey, at least some of the smaller pliosaurs would have had trouble swallowing large cephalopods and other prey. However, pliosaur dentition was adapted to employ shake-feeding or twist-feeding techniques, which involved shaking or twisting off smaller pieces of prey for eating (Martill, Taylor, and Duff 1994).

Remains of the marine crocodiles *Metriorhynchus* and *Steneosaurus* are fairly common (Martill 1991b). *Steneosaurus* is interpreted to have been semiaquatic, feeding on heavily scaled fish and thin-shelled cephalopods in the water (Massare 1987) and resting on land (Martill, Taylor, and Duff 1994). *Metriorhynchus* was more marine and had a robust, wider snout, and both its snout and its body are similar to those of mosasaurs. Evidence for the diet of *Metriorhynchus* comes from skeletons with pebbles and cephalopod hooks in the stomach (Martill 1986) and a tooth found embedded in a *Leedsichthys* skull (Martill 1985). Both crocodiles are hypothesized to have been ambush predators (Massare 1988) that could sprint after their prey.

Terrestrial Organisms

Although most of the fossils found in the Oxford Clay are autochthonous or derived from the water column immediately above the substrate, some dinosaurs believed to have been derived from a landmass at least 80 km away are present as well (Martill 1991d). The dinosaurs are mostly incomplete, probably due to scavenging in the water column (Martill 1991d) and disarticulation during transport. Small pterosaurs, referred to as *Rhamphorhynchus,* have also been recovered from the Oxford Clay but are very rare (Martill, Taylor, and Duff 1994). These pterosaurs are assumed to have been local terrestrial inhabitants that hunted fish from the sea, but they could just as well have been migrants.

CONCLUSIONS

Although Oxford Clay fossils have been collected by both professional and amateur paleontologists for over 100 years, only recently has much modern sedimentological, geochemical, and paleoecological work been conducted. With the organization of the informal, multidisciplinary, internationally based Oxford Clay Working Group in 1990, new attention has been placed on understanding the geological, chemical, and biological aspects of the Oxford Clay, especially in context with other organic-rich shales.

While the application of new techniques will yield more information, new exposures would also greatly increase our knowledge about the Oxford Clay, especially the Stewartby and Weymouth Members. New specimens of marine reptiles are continually being excavated in brickyards. In the past, temporary exposures have yielded unique finds—for example, soft-part preservation at Christian Malford—but their short-lived nature prevented the collection of detailed stratigraphic, sedimentological, paleoecological, and geochemical data. New exposures or reexcavations of classic sites may yield new information on the fauna, flora, and geology of the Oxford Clay and may even provide new examples of soft-body preservation.

Oxford Clay studies have contributed in many areas for decades—petroleum geology, industrial applications, taphonomy, paleoceanography, diagenesis, and paleobiology—and will probably continue to do so for decades to come.

ACKNOWLEDGMENTS

Many thanks go to D. Martill and J. Hudson, who provided careful reviews, references, and encouragement for this chapter.

REFERENCES

Allison, P. A. 1988. Phosphatized soft-bodied squids from the Jurassic Oxford Clay. *Lethaia* 21:403–410.

Anderson, T. F., B. N. Popp, A. C. Williams, L. Z. Ho, and J. D. Hudson. 1994. The stable isotopic records of fossils from the Peterborough Member, Oxford Clay Formation (Jurassic), U.K.: Palaeoenvironmental implications. *Journal of the Geological Society of London* 151:125–138.

Belin, S., and F. Kenig. 1994. Petrographic analyses of organo-mineral relationships: Depositional conditions of the Oxford Clay Formation (Jurassic), U.K. *Journal of the Geological Societ of London* 151:153–160.

Boucot, A. J. 1990. *Evolutionary Paleobiology of Behavior and Coevolution.* Amsterdam: Elsevier.

Brinkmann, R. 1929. Statistisch-biostratigraphische Untersuchungen an mitteljurassichen Ammoniten über Artebegriff und Stammesentwicklung. *Ab-*

handlungen der Gesellschaft der Wissenschaft, Göttingen, mathematische und physische Klasse, n.s., 13:1–249.

Buckman, S. S. 1909–1930. *Type Ammonites.* London: Wesley.

Callomon, J. H. 1963. Sexual dimorphism in Jurassic ammonites. *Transactions of the Leicester Literary and Philosophical Society* 57:21–56.

Callomon, J. H. 1985. The evolution of the Jurassic ammonite family Cardioceratidae. *Special Papers in Palaeontology* 33:49–90.

Cope, J. C. W. 1993. High resolution biostratigraphy. In E. A. Hailwood and R. B. Kidd, eds., *High Resolution Stratigraphy,* pp. 257–265. Special Publication, no. 70. London: Geological Society.

Cope, J. C. W., K. L. Duff, C. F. Parsons, H. S. Torrens, W. A. Wimbledon, and J. K. Wright, eds. 1980. *A Correlation of Jurassic Rocks in the British Isles.* Part 2, *Middle and Upper Jurassic.* Special Report, no. 15. London: Geological Society.

Cox, B. M., J. D. Hudson, and D. M. Martill. 1993. Lithostratigraphic nomenclature of the Oxford Clay (Jurassic). *Proceedings of the Geologists' Association* 103:343–345.

Duff, K. L. 1975. Palaeoecology of a bituminous shale—The lower Oxford Clay of central England. *Palaeontology* 18:443–482.

Duff, K. L. 1978. *Bivalvia from the English Lower Oxford Clay (Middle Jurassic).* London: Palaeontographical Society.

Duff, K. L. 1991. Bivalves. In D. M. Martill and J. D. Hudson, eds., *Fossils of the Oxford Clay,* pp. 35–77. Palaeontological Association Field Guide to Fossils, no. 4. London: Palaeontological Association.

Hudson, J. D. 1994. Oxford Clay studies. *Journal of the Geological Society of London* 151:111–112.

Hudson, J. D., and D. M. Martill. 1991. Introduction. In D. M. Martill and J. D. Hudson, eds., *Fossils of the Oxford Clay,* pp. 11–34. Palaeontological Association Field Guide to Fossils, no. 4. London: Palaeontological Association.

Hudson, J. D., and D. M. Martill. 1994. The Peterborough Member (Callovian, Middle Jurassic) of the Oxford Clay Formation at Peterborough, U.K. *Journal of the Geological Society of London* 151:113–124.

Hudson, J. D., and D. F. B. Palframan. 1969. The ecology and preservation of the Oxford Clay fauna at Woodham, Buckinghamshire. *Journal of the Geological Society of London* 124:387–418.

Jurassic squid. 1996. *Science* 274:1473.

Kenig, F., J. M. Hayes, B. N. Popp, and R. E. Summons. 1994. Isotopic biogeochemistry of the Oxford Clay Formation (Jurassic), U.K. *Journal of the Geological Society of London* 151:139–152.

MacQuaker, J. H. S. 1994. A lithofacies study of the Peterborough Member, Oxford Clay Formation (Jurassic), U.K.: An example of sediment bypass in a mudstone succession. *Journal of the Geological Society of London* 151:161–172.

MacQuaker, J. H. S., and J. K. Howell. 1999. Small-scale (<5.0 m) vertical heterogeneity in mudstones: Implications for high-resolution stratigraphy in siliciclastic mudstone successions. *Journal of the Geological Society of London* 156:105–112.

Martill, D. M. 1985. The preservation of marine vertebrates in the lower Oxford Clay (Jurassic) of central England. *Philosophical Transactions of the Royal Society of London, B* 311:155–165.

Martill, D. M. 1986. The diet of *Metriorhynchus*, a Mesozoic marine crocodile. *Neues Jahrbuch für Geologie and Paläontologie, Monatshefte* 10:621–625.

Martill, D. M. 1987. Taphonomic and diagenetic case study of a partially articulated ichthyosaur. *Palaeontology* 30:543–555.

Martill, D. M. 1988. *Leedsichthys problematicus* Woodward, a giant plankton feeding teleost from the Oxford Clay. *Neues Jahrbuch für Geologie and Paläontologie, Monatshefte* 11:670–680.

Martill, D. M. 1991a. Other invertebrates. In D. M. Martill and J. D. Hudson, eds., *Fossils of the Oxford Clay*, pp. 167–191. Palaeontological Association Field Guide to Fossils, no. 4. London: Palaeontological Association.

Martill, D. M. 1991b. Fish. In D. M. Martill and J. D. Hudson, eds., *Fossils of the Oxford Clay*, pp. 197–225. Palaeontological Association Field Guide to Fossils, no. 4. London: Palaeontological Association.

Martill, D. M. 1991c. Marine reptiles. In D. M. Martill and J. D. Hudson, eds., *Fossils of the Oxford Clay*, pp. 226–243. Palaeontological Association Field Guide to Fossils, no. 4. London: Palaeontological Association.

Martill, D. M. 1991d. Terrestrial reptiles. In D. M. Martill and J. D. Hudson, eds., *Fossils of the Oxford Clay*, pp. 244–248. Palaeontological Association Field Guide to Fossils, no. 4. London: Palaeontological Association.

Martill, D. M. 1991e. Introduction of vertebrate fossils. In D. M. Martill and J. D. Hudson, eds., *Fossils of the Oxford Clay*, pp. 192–196. Palaeontological Association Field Guide to Fossils, no. 4. London: Palaeontological Association.

Martill, D. M. 1992. Pliosaur stomach contents from the Oxford Clay. *Mercian Geologist* 13:37–42.

Martill, D. M., and J. D. Hudson, eds. 1991. *Fossils of the Oxford Clay*. Palaeontological Association Field Guide to Fossils, no. 4. London: Palaeontological Association.

Martill, D. M., M. A. Taylor, and K. L. Duff. 1994. The trophic structure of the biota of the Peterborough Member, Oxford Clay Formation (Jurassic), U.K. *Journal of the Geological Society of London* 151:173–194.

Massare, J. A. 1987. Tooth morphology and prey preference of Mesozoic marine reptiles. *Journal of Vertebrate Paleontology* 7:121–137.

Massare, J. A. 1988. Swimming capabilities of Mesozoic marine reptiles: Implications for method of predation. *Paleobiology* 14:187–205.

Norry, M. J., A. C. Dunham, and J. D. Hudson. 1994. Mineralogy and geochemistry of the Peterborough Member, Oxford Clay, Jurassic, U.K.: Element fractionation during mudrock sedimentation. *Journal of the Geological Society of London* 151:95–207.

Page, K. N. 1991. Ammonites. In D. M. Martill and J. D. Hudson, eds., *Fossils of the Oxford Clay*, pp. 86–143. Palaeontological Association Field Guide to Fossils, no. 4. London: Palaeontological Association.

Page, K. N., and P. Doyle. 1991. Other cephalopods. In D. M. Martill and J. D. Hudson, eds., *Fossils of the Oxford Clay*, pp. 144–162. Palaeontological Association Field Guide to Fossils, no. 4. London: Palaeontological Association.

Palframan, D. F. B. 1966. Variation and ontogeny of some Oxfordian ammonites: *Taramelliceras richei* (de Loriol) and *Creniceras renggeri* (Oppel) from Woodham, Buckinghamshire. *Palaeontology* 9:290–311.

Prosser, C. D. 1991. Brachiopods. In D. M. Martill and J. D. Hudson, eds., *Fossils of the Oxford Clay*, pp. 163–166. Palaeontological Association Field Guide to Fossils, no. 4. London: Palaeontological Association.

Shipp, D. J. 1978. Foraminifera from the Oxford Clay and Corallian of England and the Kimmeridgian of the Boulannais, France. Ph.D. diss., University College, London.

Tyson, R. V., and T. H. Pearson. 1991. Modern and ancient continental shelf anoxia: An overview. In R. V. Tyson and T. H. Pearson, eds., *Modern and Ancient Continental Shelf Anoxia*, pp. 1–24. Special Publication, no. 58. London: Geological Society.

Vecchione, M., R. E. Young, D. T. Donovan, and P. G. Rodhouse. 1999. Reevaluation of coleoid cephalopod relationships based on modified arms in the Jurassic coleoid *Mastigophora*. *Lethaia* 32:113–118.

Williams, A. C. 1988. Palaeoecological and palaeoenvironmental variations in the Callovian, Oxfordian, and Kimmeridgian (Jurassic) of Britain. Ph.D. diss., Univ-ersity of Leicester.

Woolam, R. 1980. Jurassic dinocysts from shallow deposits of the Midlands, England. *Journal of the University of Sheffield Geological Society* 7:243–261.

18

Solnhofen: Plattenkalk Preservation with Archaeopteryx

Walter Etter

THE UPPER JURASSIC SOLNHOFEN PLATTENKALK (LITH-ographic limestone), named after a small village in southern Germany, is probably the world's most famous Lagerstätte. A wealth of exquisitely preserved fossils have been unearthed over the years, with many of them showing soft-part preservation, mainly in the form of imprints. This is also the place where the oldest known bird, *Archaeopteryx,* has been found (Hecht et al. 1985; Padian and Chiappe 1998). The most common fossils, however, include floating crinoids, ammonites, crustaceans, and fish, as well as occasional jellyfish, squid, insects, and fully articulated reptile skeletons.

The Solnhofen Plattenkalk, an up to 90 m thick succession of almost pure, thinly bedded limestone intercalated with shaley layers, was deposited in a restricted area in northwestern Bavaria, Germany, extending approximately 80 km in an east–west direction and only about 30 km in a north–south direction (Figure 18.1). Deposition of the Plattenkalk occurred in only a series of small basins surrounded by bioherms and reefs, probably under hypersaline and oxygen-depleted bottom-water conditions (Viohl 1985, 1996; Barthel, Swinburne, and Conway Morris 1990; Wellnhofer 1990). The Solnhofen Plattenkalk is, therefore, a classic conservation Lagerstätte that is a stagnation deposit, although a certain amount of obrution and bacterial sealing may have been crucial for the splendid preservation of the fossils (Seilacher, Reif, and Westphal 1985).

The Solnhofen Plattenkalk has a long history of exploitation. Some carvings on Solnhofen limestone date back to the late Stone Age, and since Roman times the Plattenkalk has been extensively used as building material. In 1793, Alois Senefelder invented lithography (Barthel 1978),

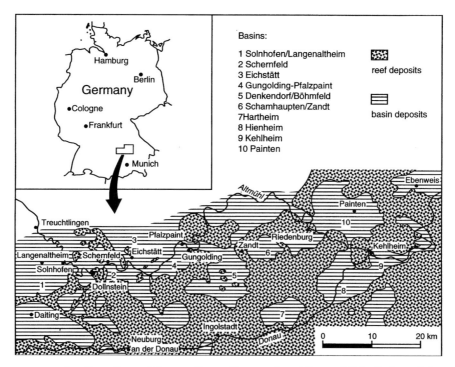

FIGURE 18.1 Location of the southern Franconian Alb, and lithofacies of the Solnhofen region in early Tithonian times. (Modified from Barthel, Swinburne, and Conway Morris 1990)

and only very fine-grained, pure limestone slabs could be used for this technique. The Solnhofen Plattenkalk proved to be by far the best material for lithographic stones and, as a consequence, was heavily quarried in the nineteenth century. With the advent of newer and cheaper technologies in the twentieth century, the Solnhofen Plattenkalk became less important as a source for lithographic limestone, but it is still quarried as a building stone (Barthel, Swinburne, and Conway Morris 1990).

Ever since the limestones of Solnhofen have been exploited, fossils have been collected from this deposit. Although fossils are actually quite rare in the Solnhofen Plattenkalk, a large number of exquisitely preserved remains have been found over the years. The first comprehensive accounts of the fauna of the Solnhofen limestone were published in the early nineteenth century by A. Hagen, Graf Georg von Münster, Albert Oppel, Baron Friedrich von Schlotheim, Andreas Wagner, and Hermann von Meyer, among others (historical review in Barthel 1978). In 1860, an isolated feather was found near Solnhofen, and in 1861, just two years after the publication of Darwin's *On the Origin of Species,* the discovery of the first fully articulated skeleton of *Archaeopteryx* was reported, making the Solnhofen Plattenkalk world famous (Barthel, Swinburne, and Conway Morris 1990).

In 1877, a second and even more complete *Archaeopteryx* was found and was sold to the Humboldt Museum of Berlin. It took almost 80 more years until the next discovery of an *Archaeopteryx* could be reported, and today seven specimens are known (Wellnhofer 1993). In actuality, one of them had already been discovered in 1855, but was then determined to be a small pterosaur and remained unnoticed for more than 100 years in a museum in Haarlem, the Netherlands. Sadly, one specimen, which was still in private hands, disappeared recently (Abbott 1992).

GEOLOGICAL CONTEXT

The area where the Solnhofen Plattenkalk occurs is part of northwestern Bavaria, Germany, known as the southern Franconian Alb (Figure 18.1). This region was covered in Jurassic times by a shallow epicontinental sea, which spread from the Paris Basin in the west to the Vienna region in the east. Marine connections were present in the south to the oceanic Tethys and in the north to the Lower Saxony Basin, in northern Germany (Meyer and Schmidt-Kaler 1984).

In the early portion of Late Jurassic times, mainly light-colored limestones and marls were deposited in the southern Franconian Alb (Meyer and Schmidt-Kaler 1984). Starting in the Oxfordian and proliferating especially in the Kimmeridgian, sponge–microbial mounds (reefs) developed (Figure 18.2), first in a few northwest–southeast-trending belts, but later becoming much more widespread (Meyer and Schmidt-Kaler 1984). The so-called Ries-Wisent sill, located where the Tertiary Ries meteorite impact later occurred, became a major submarine topographic high, separating the southern Franconian sea from the more westerly located Swabian Alb area (Meyer and Schmidt-Kaler 1984).

At the end of the middle Kimmeridgian, the shallowing of the sea reached its maximum. At this time, the topographic differences between mounds and basins were less pronounced (Meyer and Schmidt-Kaler 1989). A new subsidence in the late Kimmeridgian led to renewed growth of the sponge–microbial mounds and to the first Plattenkalk sedimentation in the basins (Meyer and Schmidt-Kaler 1989). By early Tithonian times, when deposition of the Solnhofen Plattenkalk started, a highly structured submarine relief was established (Meyer and Schmidt-Kaler 1984; Viohl 1985, 1996; Barthel, Swinburne, and Conway Morris 1990). Sponge–microbial mounds, rising to 50 m above the surrounding seafloor, formed a network that enclosed many small basins, locally known as *Wannen* (Meyer and Schmidt-Kaler 1984; Barthel, Swinburne, and Conway Morris 1990; Viohl 1996) (Figure 18.3). Because of the shallowing of the sea and the extensive growth of sponge–microbial reefs, the exchange of water masses with the sea to the north became very restricted. Likewise, the development of a shallow-water platform

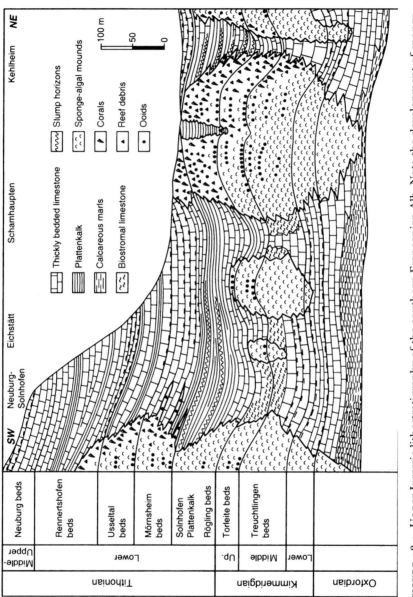

FIGURE 18.2 Upper Jurassic lithostratigraphy of the southern Franconian Alb. Note the development of sponge–microbial reefs through time. (Modified from Meyer and Schmidt-Kaler 1984)

with corals and ooid shoals farther to the south inhibited exchange with the oceanic waters of the Tethys (Meyer and Schmidt-Kaler 1984, 1989). These changing conditions were no longer optimal for sponge growth, and many of the reefs died in the early Tithonian or their tops were colonized by corals (Meyer and Schmidt-Kaler 1984; Barthel, Swinburne, and Conway Morris 1990). Nevertheless, localized basins still existed between submarine highs in early Tithonian times in the southern Franconian Alb, and it was in these small basins, extending from Langenaltheim in the west to Kehlheim and Ebenweis in the east, that the Solnhofen Plattenkalk was deposited (Röper and Rothgaenger 1996, 1998a, 1998b) (Figures 18.1 and 18.3).

The German word *Plattenkalk* is used here for this special kind of lithology (very fine-grained, pure, platy limestone) known from the Solnhofen region (and other deposits as well). "Platy limestone" is an imprecise English translation of *Plattenkalk* and does not imply the flat, tabular bedding implicit in the German word. Use of the alternative term "lithographic limestone" does not seem appropriate either because it is not descriptive and because the requirements for limestones suitable for lithography are currently known by only a few people (Hemleben and Swinburne 1991).

Locally, deposition of Plattenkalk facies had already started in the late Kimmeridgian. Sometimes the fossils enclosed in these layers show excellent preservation (Barthel, Swinburne, and Conway Morris 1990). However, only the lower Tithonian Solnhofen Plattenkalk has yielded the world-famous fauna that shows soft-part preservation.

The Solnhofen Plattenkalk is 30 to 90 m thick and is characterized by an alternation of very pure, micritic limestone, called Flinz beds, and more clayey limestone, called Fäule (Barthel 1978). The bedding surfaces are extremely flat, but they show some crenulations, which are thought to have formed during early diagenesis in the process of dewatering (Mayr 1967). Sometimes the bedding surfaces show fern-like iron and manganese oxide markings known as dendrites. The Flinz beds, which are originally light gray but weather to light yellow, contain 95 to 98 percent calcium carbonate and are finely laminated (Barthel, Swinburne, and Conway Morris 1990). The content in organic matter is very low and reaches values of only 0.2 percent C_{org} (Hückel 1974b). The thickness of individual Flinz beds varies considerably and may attain 20 to 30 cm in the Solnhofen area, whereas in the Eichstätt region, Flinz beds are on the order of a few millimeters to 5 cm thick (Wellnhofer 1990; Hemleben and Swinburne 1991). The gray to brown shaley Fäule intervals, a few millimeters to 2 cm thick and laminated as well, contain only 77 to 87 percent calcium carbonate, the remainder being mainly clay minerals (10 to 20 percent) and quartz (~3 percent) (Hückel 1974a; Barthel, Swinburne, and Conway Morris 1990). Values of organic carbon are somewhat higher than in the Flinz beds and may attain 1.0 percent (Hückel 1974b).

The Solnhofen Plattenkalk, locally known as litho-stratigraphic unit Malm ζ 2, is of early Tithonian age, and the various ammonites occurring in the beds indicate the *Hybonoticeras hybonotum* zone (Barthel 1978). Slump horizons are quite frequent in the Solnhofen Plattenkalk, but they are mainly concentrated in two layers (Figure 18.2): the so-called Trennende Krumme Lage (separating slump bed) and the Hangende Krumme Lage (overlying slump bed), both of which can be traced over long distances (Barthel 1978; Meyer and Schmidt-Kaler 1984; Barthel, Swinburne, and Conway Morris 1990). The Trennende Krumme Lage is actually the marker horizon for the subdivision of the Solnhofen Plattenkalk into a lower and an upper unit. The lower Solnhofen Plattenkalk, 10 to 35 m thick, is less regularly bedded than the upper Solnhofen Limestone and is slightly bioturbated (Meyer and Schmidt-Kaler 1984; Barthel, Swinburne, and Conway Morris 1990). Body fossils are rarely found. Only the up to 60 m thick upper Solnhofen Limestone (Malm ζ 2b) contains the world-famous fossiliferous deposit (Barthel, Swinburne, and Conway Morris 1990).

Based on the rough estimate that one ammonite zone in the Upper Jurassic spans approximately 1 Ma and that sedimentation of the Solnhofen Plattenkalk lasted for about half an ammonite zone, the deposition of this formation may have taken around 500 Ka (Barthel, Swinburne, and Conway Morris 1990). During this time interval, approximately 500 to 2,500 Flinz–Fäule couplets would have formed, giving an average time span of 200 to 1,000 years for each Flinz–Fäule cycle (Barthel, Swinburne, and Conway Morris 1990). However, these calculations are highly speculative. The time represented by the Fäulen beds is unknown, but each Flinz bed represents only one event (storm or monsoonal season). The time span of 0.5 Ma for the deposition of the Solnhofen Plattenkalk is therefore a maximum; probably it was much less (G. Viohl, personal communication, 1996).

PALEOENVIRONMENTAL SETTING

To reconstruct the paleoenvironmental setting and to develop a depositional model for the Solnhofen Plattenkalk, sedimentological, geochemical, paleogeographic, and paleontological patterns have been synthesized over several decades. Whereas the paleogeographic setting is now known in great detail, sedimentological and geochemical study of the limestone has encountered great difficulties, which have prevented the development of a generally accepted depositional model for the Solnhofen Plattenkalk.

Microfacies and Geochemistry
of the Solnhofen Plattenkalk

As mentioned earlier, the Flinz and Fäule beds are finely laminated, and in the Flinz beds the lamination is frequently enhanced by pressure solution (Hemleben 1977). Otherwise, the very fine-grained sediments show few or no sedimentological features that could be linked to a specific depositional environment. Furthermore, the Flinz beds were quite severely altered during diagenesis (Barthel, Swinburne, and Conway Morris 1990). This has hindered the clarification of the micrite origin, which is, in turn, crucial for the development of a depositional model. Much better results have been obtained by investigating the Fäule beds using scanning electron microscopy (Keupp 1977a, 1977b). Due to their higher clay content, the Fäule beds were much less altered by diagenesis than were the Flinz beds. Because Flinz and Fäule show almost identical size–frequency distributions of the individual particles, it can be assumed that their micrite components came from the same source (Flügel and Franz 1967; Keupp 1977a).

Coccoliths have been observed in Fäule and Flinz beds, but recognizable remains are common only in shaley laminae, especially of the Fäule (Keupp 1977a, 1977b). They occur mainly in isolated aggregates, which suggests accumulation as fecal pellets (Keupp 1977b) and, more rarely, as remains of entire coccospheres. Some of these layers have yielded a quite diverse association, and in addition to coccoliths, cysts of dinoflagellates (calcispheres) (Keupp 1977a), rare radiolarians, rare ostracodes (Gocht 1973), and some benthic foraminifers (Groiss 1967) have been described. Other more carbonaceous layers of the Fäule beds contain very small calcite-lined spherical cavities that, according to Keupp (1977a, 1977b), are the remains of coccoid cyanobacteria. Because the crystals that make up the spheroid wall correspond exactly to the observed grain size peak at 1 to 3 m, these coccoid cyanobacteria are believed to be major contributors to the Fäule carbonate even where no longer recognizable (Keupp 1977a, 1977b).

These observations have been applied to the Flinz beds as well. The small spheres are normally not recognizable in the Flinz, because the cavities have been filled with calcite cement (Keupp 1977a). According to Keupp (1977a), the bulk of the carbonate in the Flinz beds also comes from coccoid cyanobacteria, whereas coccoliths contribute only about 1 percent of Flinz carbonate. However, others still consider most of the Flinz carbonate to be diagenetically altered remains of coccoliths (Flügel and Franz 1967; Buisonjé 1972, 1985). It could be shown quite convincingly, however, that this is not the case because unaltered as well as diagenetically altered (through calcite dissolution and precipitation) coccoilth assemblages would not have produced the observed grain size–frequency distribution (Keupp 1977a).

In some Flinz layers, larger particles have been observed. They are aligned in thin bands and consist mainly of fragments of foraminifers, bivalves, gastropods, ostracodes, and echinoderms (Hemleben 1977). Occasionally, small-scale cross-lamination is visible in these bands, confirming that the larger particles actually represent reef debris (Hemleben 1977). Reef debris is much more common in the eastern basins (Kehlheim–Painten region), which were surrounded by rapidly growing coral reefs, than in the western basins (Solnhofen–Eichstätt region), which were surrounded by sills of dying sponge-microbial reefs that shed little detritus (Meyer and Schmidt-Kaler 1984).

In the past, mineralogical (Hückel 1974a) and geochemical investigations (Hückel 1974b; Veizer 1977) did not substantially increase the understanding of the Solnhofen Plattenkalk. The Sr content, together with oxygen isotope data, indicates that only about 33 percent of the carbonate was originally present as low-Mg calcite (e.g., coccoliths). The rest was originally aragonite and/or high-Mg calcite (Veizer 1977). The oxygen and carbon isotope values of the calcareous matrix ($\delta^{18}O$ of Flinz beds on average, 3.5; $\delta^{13}C = +2.0$) as well as the values for Fe and Mg seem to be rather typical of carbonate produced under fully oxic conditions (Veizer 1977). A new approach, measuring the mineral distribution bed by bed in various sections in Eichstätt, shows conclusively that both the carbonate and the clay fraction accumulated in the basins from suspension fallout (Bausch et al. 1994). This is, to date, the strongest support for Barthel's model of Plattenkalk deposition and contradicts Keupp's theory of carbonate production *in situ*, as discussed later.

Paleoenvironment and Climate

The sedimentological features of the Solnhofen Plattenkalk indicate deposition in very quiet waters, and the preservation style of the fossils clearly points to an environment hostile for metazoans (Barthel, Swinburne, and Conway Morris 1990). During Plattenkalk deposition, the Solnhofen area was bordered to the south by a broad shallow-water platform with coral reefs, and to the northwest and east by landmasses (Figure 18.3). It is this situation that has led to the term "Solnhofen lagoon." However, this is not a very appropriate term since this is a very broad, highly structured area (30+ km north–south by 80 km east–west), in which at least the surface waters of the region were somewhat connected to open oceanic waters (Viohl 1985; Barthel, Swinburne, and Conway Morris 1990) (Figure 18.3). Only in the very localized basins bordered by submarine highs (sponge–microbial mounds) did Plattenkalk accumulate, and to only these Wannen can the silled basin model be applied. Water depths are estimated at 50 to 60 m for the Solnhofen–

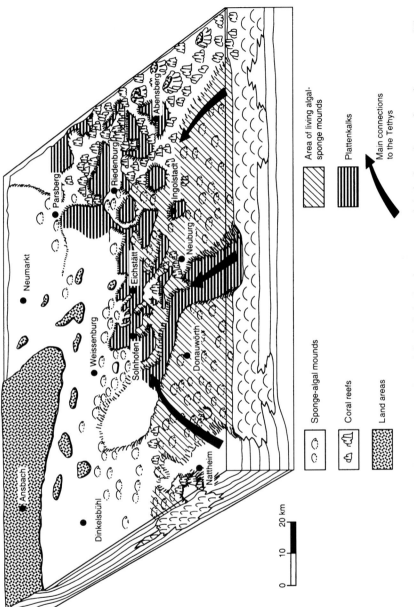

FIGURE 18.3 Paleogeographic reconstruction of the Solnhofen lagoon, overlain by locations of modern towns. View from the south. (Modified from Barthel, Swinburne, and Conway Morris 1990)

Langenaltheim basin, 30 m at most for the Eichstätt basin, 20 to 30 m for the Painten basin, and approximately 50 m for the Kehlheim basin (Barthel, Swinburne, and Conway Morris 1990).

In Tithonian times, Solnhofen was located at 25° to 30° north latitude in the semiarid subtropical belt. The best evidence for a generally dry climate comes from the terrestrial plant remains found in the Solnhofen Plattenkalk, which show leathery cuticules and stomata located only on the underside of the leaves in sunken pits (Meyer 1974; Barthel, Swinburne, and Conway Morris 1990). However, there must have been at least some seasonal freshwater runoff and small ponds on the nearby land areas, because most of the insect species found in the Solnhofen Plattenkalk develop through aquatic larval stages (Barthel 1978).

It is assumed that the special paleogeographic and climatic conditions led to a high evaporation rate and low freshwater input in the Solnhofen lagoon. Because the exchange with surrounding water masses was somewhat restricted, heavy, high-salinity waters developed in the basins and led to stagnant conditions of the bottom-water (Hemleben 1977; Keupp 1977a, 1977b; Barthel 1978; Barthel, Swinburne, and Conway Morris 1990). However, hypersalinity never reached values high enough for salt or gypsum to have precipitated. A major question is whether these hypersaline bottom-waters were oxygen deficient as well. In fact, hypersaline waters have less soluble oxygen, and can become more easily anoxic, than less saline waters (Barthel, Swinburne, and Conway Morris 1990). However, the low organic matter content of the Solnhofen Plattenkalk indicates low productivity, and in the absence of a substantial amount of degrading organic material, oxygen deficiency is unlikely to develop. Pyrite, which should have formed under anoxic conditions, is present in only very small amounts, but its formation may have been iron limited (Barthel, Swinburne, and Conway Morris 1990). Some workers maintain that the occurrence of foraminifers in some horizons indicates, at least temporarily, some oxygen in the bottom-waters (Groiss 1967), although these foraminifers could have been washed in from the adjacent topographic highs. At present, it is believed that oxygen concentrations varied from almost normal to almost anoxic (Barthel, Swinburne, and Conway Morris 1990).

Depositional Models

Ever since the fossils from the Solnhofen Plattenkalk were scientifically described, models have been developed to explain their preservation and the origin of the Plattenkalk (historical review in Barthel 1978). Many of the earlier authors (Gümbel 1891; Rothpletz 1909; Abel 1927) have seen the Plattenkalk as a mudflat deposit where carcasses accumulated and were mummified before being covered by fine-grained sediment. Evidence for permanent or temporary subaerial exposure was seen in the presence of tracks, which were then believed to have been produced by terrestrial animals, but are now known to have been made by limulids (Barthel 1978) (Figure 18.4).

Although no general consensus has yet been achieved, the most widely accepted depositional model is the one that has been developed and refined over the years by Barthel (1964, 1970, 1972, 1978; Barthel, Swinburne, and Conway Morris 1990). According to this model, the Plattenkalk micrite is basically carbonate ooze, which was produced by, and had accumulated on, the carbonate platform, with scattered coral patch-reefs to the south. During storm events, this ooze was stirred up and became suspended in the water. Heavier particles settled very quickly, and only the micrite particles were carried over the topographic highs to the basins of the Solnhofen area, where they settled out from suspension. During exceptional storms (monsoons), organisms living on and around

FIGURE 18.4 The limulid *Mesolimulus walchi* at the end of death track. Body length of specimen is 15 cm. (Photo courtesy of G. Viohl, Jura-Museum, Eichstätt)

the sponge bioherms were also carried into the basins. In this hostile environment they could not survive, and after death they became covered by the fine, settling sediment. Deposition of the Plattenkalk was obviously not accompanied by bottom-water currents, since almost all the aptychi and rare bivalves are oriented convex down, and some ammonites are embedded vertically.

Barthel (1964, 1970, 1972, 1978) suggested that there is no essential difference between Flinz and Fäule beds. The Flinz is an accumulation of concentrated carbonate laminae that settled after heavy storms, while the Fäule carbonate was brought into the basins in lower concentrations by minor storms and gentle currents. With the background sedimentation of land-derived clay being more or less constant, the carbonate of the slower-accumulating Fäule beds was more diluted than the Flinz carbonate. Salinity in the lagoon was elevated throughout deposition of the Plattenkalk, and a stable halocline divided the strongly hypersaline and somewhat oxygen-depleted bottom-waters in the basins from the only slightly hypersaline surface waters. Occasional mixing of the entire water column prevented salt precipitation and allowed brief colonization of the lagoon by planktonic coccoliths and benthic foraminfers.

Many variants of this model have been proposed. While assuming sedimentological mechanisms similar to those offered by Barthel, van Straaten (1971) suggested that oxygen-depleted oceanic waters entered the lagoon. However, considering the paleogeographic situation, with a reef barrier in the south, this seems unlikely (Barthel 1978). An allochthonous origin of the Flinz beds has also been proposed by Goldring and Seilacher (1971), but these authors think that the carbonate was brought into the basins by turbidity currents (see also Brett and Seilacher 1991). This may be true for Flinz beds in the Kehlheim–Painten region, but the large areal extent of individual Flinz beds in the western basins cannot be explained by turbiditic sedimentation (Barthel, Swinburne, and Conway Morris 1990). Still another view maintains that the Flinz beds represent allochthonous reef debris, but the Fäule carbonate was produced in the lagoon by coccoliths and benthic foraminifers (Hemleben 1977). Fäule accumulation should thus represent times with normal marine to only slightly hypersaline surface and bottom-waters. The absence of macrofauna is explained by sediment texture, which was too soupy for macrobenthic colonization (Hemleben 1977). This does not, however, explain the rarity of pelagic animals in the Fäule layers (Barthel, Swinburne, and Conway Morris 1990). A radically different theory on Plattenkalk accumulation was developed by Buisonjé (1972, 1985), who suggests that Plattenkalk micrite was derived from coccoliths living together with other planktonic algae in the lagoon and producing toxins during seasonal blooms, poisoning the water. This theory has been rejected mainly because the organic matter content of the

Solnhofen Plattenkalk indicates low productivity in the surface waters (Barthel, Swinburne, and Conway Morris 1990).

Probably the most serious challenge to Barthel's depositional model was presented by Keupp (1977a, 1977b), based on a detailed SEM investigation of the Plattenkalk. According to Keupp, the bulk of the Plattenkalk carbonate was produced *in situ* by cyanobacterial mats growing at the bottom of the basins. These organisms were the only ones that could have lived in these strongly hypersaline waters. Hypersalinity reached up to the surface, and only after mixing of the surface waters were coccoliths and other planktonic organisms able to live for a short time in the lagoon, where their skeletons accumulated in thin bands within Flinz and Fäule beds. Severe storms led to the mixing of the entire water column, killing the cyanobacterial mat and halting carbonate production. During these periods of interrupted carbonate production, the clay-rich bands in the Flinz and Fäule beds would have formed, due to continuous background sedimentation of detrital clay, before the salinity stratification was established again and a new cyanobacterial mat formed at the sediment surface (Keupp 1977a, 1977b). Of course, there is good evidence for the temporary existence of microbial mats at the sediment surface, including both microfacies data and sedimentologic/taphonomic features. But it is difficult to imagine how such regular and tabular limestone beds, which can be traced over very long distances, could have a predominantly stromatolitic origin (Barthel, Swinburne, and Conway Morris 1990). Even more contradictory to Keupp's depositional model is the low organic content of the Solnhofen Plattenkalk; modern microbial mats are known to be sites of high organic productivity (Barthel, Swinburne, and Conway Morris 1990). There are also other patterns that contradict Keupp's model. Taphonomic studies clearly show that the sediment must have formed by suspension fallout, not by the growth of cyanobacteria (Viohl 1994). The same conclusion can be drawn from geochemical studies (Bausch et al. 1994). Furthermore, it must be borne in mind that carbonate-producing cyanobacteria were very common in the adjacent sponge–microbial reefs throughout the Late Jurassic (Meyer and Schmidt-Kaler 1984). Therefore, even if cyanobacteria were major contributors to the Plattenkalk micrite, this does not automatically imply *in situ* carbonate production at the bottom of the basins. It is thus the model of Barthel that is favored by most workers today (Barthel 1978; Barthel, Swinburne, and Conway Morris 1990; Viohl 1996).

TAPHONOMY

In the Solnhofen Plattenkalk, macrofossils occur mainly on the undersides of Flinz beds. Occasionally, they are found inside Flinz beds, and show excellent preservation (Barthel, Swinburne, and Conway Morris 1990). Fossils are known also from Fäule beds, but they are of a lesser quality and are easily destroyed when the shaley layers peel away. Of course, the Solnhofen Plattenkalk is world famous because of the occurrence of preserved soft parts. There are many spectacular examples of soft-part preservation, including (1) different species of jellyfish (Figure 18.5); (2) squid with preserved tentacles, ink sacs, fins, and body outline (sometimes even with phosphatized muscle fibers); (3) polychaete annelids; (4) insects with every detail of their wing venation (Figure 18.6); (5) fully articulated sharks, rays, and chimeras (holocephalians) with preserved calcified cartilage and skin; (6) bony fishes, which sometimes show remains of the skin, digestive tract (Janicke and Schairer 1970), and phosphatized muscular tissue; (7) pterosaur wing membranes with visible hairlike cover; (8) impressions of turtle skin (Joyce 2001); and (9) the feathers of *Archaeopteryx* (Barthel, Swinburne, and Conway Morris 1990).

Sometimes soft parts are preserved as residues of the original organic material—for example, in the ink sacs of squid and the isolated feather of *Archaeopteryx* (Barthel, Swinburne, and Conway Morris 1990). However, this type of preservation is very rare. Equally uncommon is early diagenetic mineralization of soft parts—for example, mineralized digestive tracts of fish and phosphatized muscular fibers found in squid,

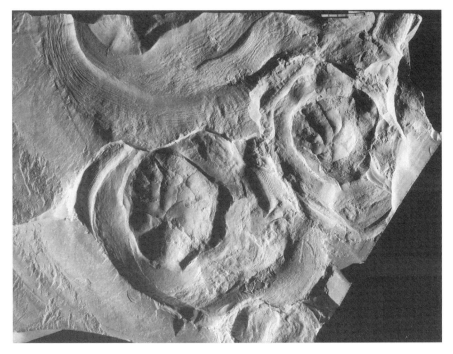

FIGURE 18.5 The jellyfish *Rhizostomites admirandus.* Diameter of specimen in the middle is 23 cm. (Photo courtesy of G. Viohl, Jura-Museum, Eichstätt)

FIGURE 18.6 The dragonfly *Aeschnogomphus intermedius*. Wingspan is 17.5 cm. (Photo courtesy of G. Viohl, Jura-Museum, Eichstätt)

insects, and fish (Barthel, Swinburne, and Conway Morris 1990). Most of the soft-part preservation is in the form of impressions. Surprisingly, it is only the upper surface of the soft parts that has been preserved. This is shown in the Berlin specimen of *Archaeopteryx,* where the feathers are preserved as impressions on the overlying slab, whereas the underlying slab formed a cast (Rietschel 1985). Obviously, only fine mud settling from suspension could produce a cast, whereas the bedding plane on which the fossil came to rest was already partially cemented by bacteria (Rietschel 1985; Barthel, Swinburne, and Conway Morris 1990).

Although soft-part preservation is not the rule, but the exception, almost all the fossils have been preserved fully articulated (Barthel 1978). This implies very quiet bottom conditions, lack of scavenging, and probably fast sediment blanketing or overgrowth by a bacterial mat. Many fossils show signs of dehydration (Seilacher, Reif, and Westphal 1985), most notably the shrinkage patterns in jellyfish, the arched vertebral columns of fish, and the bent neck of *Archaeopteryx* (Figure 18.7). In addition, most of the crustaceans show an arched abdomen, and the arms of the crinoids *Saccocoma* and *Pterocoma* are strongly coiled (Seilacher, Reif, and Westphal 1985) (Figure 18.8). Sometimes the bending has been so strong that the surface of the sediment has been scraped off (Mayr 1967; Barthel, Swinburne, and Conway Morris 1990). Earlier authors interpreted this pattern as desiccation and mummification under subaerial

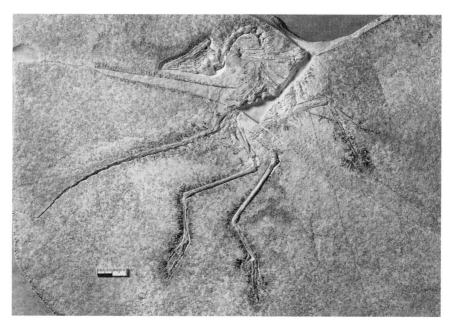

FIGURE 18.7 *Archaeopteryx lithographica,* Eichstätt specimen. Scale bar is 2 cm. (Photo courtesy of G. Viohl, Jura-Museum, Eichstätt)

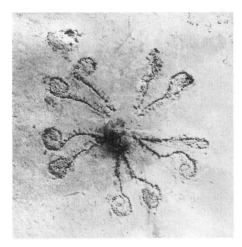

FIGURE 18.8 The crinoid *Saccocoma tenella.* Diameter of specimen is 5 cm. (Photo courtesy of G. Viohl, Jura-Museum, Eichstätt)

conditions, but it is now believed to be the result of dehydration in hypersaline waters (Seilacher, Reif, and Westphal 1985; Barthel, Swinburne, and Conway Morris 1990).

Many of the more common invertebrates of the Solnhofen Plattenkalk are not very aesthetically appealing. Aragonitic shells have invariably been dissolved, and therefore ammonites, gastropods, and bivalves are present only as steinkerns or as flattened impressions. Sometimes, however, the organic periostracum has been preserved (Seilacher et al. 1976). High-Mg calcite, the constituent of echinoderm ossicles, was recrystallized during diagenesis, but in general the internal structure is still visible (Barthel, Swinburne, and Conway Morris 1990). Skeletal parts originally made of low-Mg calcite, like the ammonite aptychi and the belemnite rostra, remained basically unaltered. Chitin, although this is not yet proved, has probably also been preserved in the exoskeleton of arthropods, especially where they show colors (Barthel, Swinburne, and Conway Morris 1990). Cellulose, on the contrary, is not preserved, and most plant remains are therefore preserved only as impressions (Barthel, Swinburne, and Conway Morris 1990).

The fossils of the Solnhofen Plattenkalk show some other remarkable features, most notably the so-called pedestal preservation (from the German *Sockel-Erhaltung*). In this type of preservation, known primarily from flattened ammonites and crustaceans, the fossil stands elevated from the surrounding bedding plane on a pedestal on the lower slab, whereas at the bottom of the overlying slab a corresponding depression is present (Barthel, Swinburne, and Conway Morris 1990; Viohl 1996). Positive buoyancy and upward movement of the embedded fossil in the still-soft sediment were thought to be the cause for pedestal preservation (Seilacher et al. 1976). However, it can be shown that the early cementation of the sediment surface, which prevented the fossil from sinking into the underlying ooze, and a subsequent collapse of the fossil were alone responsible for this kind of preservation (Viohl 1994). Pedestal preservation is such a characteristic feature of the Solnhofen Plattenkalk that it can be used to distinguish between lower and upper surfaces of Flinz beds.

Generally, fragments of ammonite shells, aptychi, and rare bivalves are oriented with their concave side upward, indicating no or only very weak bottom currents at the time of deposition (Barthel, Swinburne, and Conway Morris 1990). However, temporarily weak unidirectional or oscillatory currents were responsible for the orientation of *Leptolepis,* of which many specimens in the so-called *Fischleflinz* appear to be roughly parallel. A few belemnite rostra and some ammonites have been vertically embedded (Seilacher et al. 1976; Barthel 1978), which again indicates very quiet bottom-waters as well as a very soft substratum and perhaps rapid burial. The soupy nature of the sediment was not a constant feature, however. Preservation of landing marks of dead bodies that sank to the seafloor as well as the excellent preservation of limulid and crustacean tracks (Malz 1976; Barthel, Swinburne, and Conway Morris 1990) (Figure 18.4) were possible only if the surficial layer was somewhat stiffened, perhaps by a cyanobacterial mat.

There is ample evidence that throughout deposition of the upper Solnhofen Plattenkalk no living macrobenthos was present at the bottom of the basins. Most of the animals brought into the lagoon were already dead when they reached the bottom. Only some of the more tolerant species, like the gastropod *Rissoa,* the bivalve *Solemya,* and some crustaceans and limulids, were able to survive for a short period of time and produce tracks, at the end of which they are sometimes found (Barthel, Swinburne, and Conway Morris 1990) (Figure 18.4).

Further evidence that all the macrofaunal specimens were displaced from their original habitat into the depositional environment, most likely by storms, comes from the relative frequency of different ecological groups found in the Solnhofen Plattenkalk (Barthel, Swinburne, and Conway Morris 1990). Among marine animals, the following categories are observed (in order of decreasing abundance): planktonic species (jellyfish, crinoids), weak swimmers (ammonites, small fish), nonanchored epibenthos (crustaceans, limulids), strong swimmers (squid and cuttlefish, larger fish, marine reptiles), weakly attached and burrowing benthos (brachiopods, bivalves, polychaetes), and strongly attached reefal organisms (sponges, corals). In some layers, species of the first two categories can occur in high densities, especially the small floating crinoid *Saccocoma* (Figure 18.8)—although for this genus, a benthic mode of life within the lagoon has been proposed (Milson 1994; Manni, Nicosia, and Tagliacozza 1997)—and the fish *Leptolepis.* Some jellyfish also occur in the Gungolding–Pfalzpaint region (Barthel 1978; Barthel,

Swinburne, and Conway Morris 1990). It is believed that these organisms could have colonized the lagoon during brief periods of near-normal salinity of the surface waters but subsequently became the victims of a mass kill (Barthel, Swinburne, and Conway Morris 1990).

Terrestrial organisms are much rarer and are represented mainly by species capable of flight, including insects (only flying forms in the Solnhofen Plattenkalk), pterosaurs, and *Archaeopteryx* (Barthel, Swinburne, and Conway Morris 1990). These species were most likely carried into the lagoon by strong offshore winds (Barthel, Swinburne, and Conway Morris 1990). Ground-dwelling reptiles are exceedingly rare and are known from only a few specimens, and terrestrial plants are equally uncommon.

Once the organisms were swept into the lagoon, they quickly died, most of them in the surface waters. The most common cause of death was probably the hypersaline and/or oxygen-deficient water (Barthel, Swinburne, and Conway Morris 1990). After reaching basin seafloors, carcasses were left undisturbed because no scavengers could exist in this hostile environment (Davis and Briggs 1998). Decay in the strongly hypersaline and possibly oxygen-deficient bottom-water was very slow, so soft parts could be preserved long enough to become buried by the very fine-grained sediment settling from suspension. This fine ooze was a perfect casting medium that preserved soft parts as impressions even if they later decayed. On occasion, cyanobacterial mats growing over the carcasses seem to have been involved in this casting process (Seilacher, Reif, and Westphal 1985; Barthel, Swinburne, and Conway Morris 1990).

PALEOBIOLOGY AND PALEOECOLOGY

Over 750 plant and animal species have been described from the Solnhofen Plattenkalk (Wellnhofer 1990; for a full list down to the species level, see Barthel, Swinburne, and Conway Morris 1990). However, this may not be a very accurate number because many groups are in need of taxonomic revision (Barthel, Swinburne, and Conway Morris 1990). As has been pointed out, the lagoon and especially the bottoms of the Plattenkalk basins were basically devoid of life. The few exceptions were benthic organisms such as cyanobacteria, foraminifers, and rare ostracodes; planktonic organisms such as coccoliths, jellyfish, and floating crinoids; and nektonic animals such as small fish and squid. Some indistinct disruptions of laminae and the occurrence of the putative trace fossil, *Chondrites*, have also been reported (Hemleben 1977), indicating rare bioturbation by small worm-like organisms. All the other fossils found in the Plattenkalk seem to have been swept into the hypersaline and somewhat oxygen-depleted environment. The fossil assemblage of the Solnhofen Plattenkalk is, therefore, a mixture of plants and animals that inhabited the surrounding habitats, including open oceanic waters, the carbonate platform with coral patch-reefs to the south and east, the coast, and the terrestrial environment to the northwest (Viohl 1996).

The vast majority of the preserved organisms are marine animals that lived on and around the reefs bordering the Solnhofen lagoon (Barthel, Swinburne, and Conway Morris 1990). Actual reef builders are, however, very rare (see "Taphonomy'), and only a few specimens of siliceous sponges and bryozoans have been found in the Plattenkalk (Barthel 1978). Corals might be present as fine debris, but unquestionable remains have hitherto not been reported.

Among the organisms colonizing the firm and hard bottoms of the reef environments, and perhaps the subtidal bottoms along the coast, were some brown algae placed into the genus *Phyllothallus.* Most commonly they were encrusted by foraminifers and bryozoans, but sometimes small oysters and inoceramid bivalves were also attached to these plants (Barthel, Swinburne, and Conway Morris 1990). Other hard-bottom dwellers found in the Plattenkalk include brachiopods, bivalves (oysters, "*Inoceramus,*" *Arcomytilus, Buchia*), gastropods (limpets, *Rissoa,* some other genera), serpulid worms, regular sea urchins, and rare benthic crinoids (*Millericrinus*). Bivalves sometimes occur in clusters around ammonite shells, and they have been interpreted as short-lived benthic communities in the Plattenkalk basins (Kauffman 1978). However, these bivalves were most certainly attached to the shell while the ammonite was still floating in the water (Barthel, Swinburne, and Conway Morris 1990).

Some of the bivalves like *Pinna* and *Solemya* were inhabitants of the soft bottoms around the reefs, as were a few gastropods and the rare polychaetes *Ctenoscolex* and *Eunicites.* The same habitat was occupied by most of the numerous crustaceans (mysidaceans, isopods, stomatopods, decapods) (Figure 18.9), horseshoe crabs (*Mesolimulus*) (Figure 18.4), irregular sea urchins, starfish, ophiuroids (Kutscher 1997), and sea cucumbers (Barthel, Swinburne, and Conway Morris 1990).

Many of the animals with swimming capabilities have to be considered as reef and bottom dwellers rather than as open-water forms. In this category, cuttlefish, nautilids, ammonites, some of the sharks (*Pseudorhina, Protospinax*), rays and chimeras, and many of the bony fishes may be included (Viohl 1996). In particular, among the osteichthyes from the Solnhofen Plattenkalk are many species with a body form typical for modern coral fishes (e.g., *Gyronchus*), and several of them possess a strong dentition adapted for crushing shells and coral branches (e.g., *Lepidotes, Gyronchus, Gyrodus*) (Barthel, Swinburne, and Conway Morris 1990). Some aquatic reptiles seem to have occupied coastal waters rather than the vicinity of the reefs (Barthel, Swinburne, and Conway Morris 1990), including sea turtles, some crocodiles (*Steneosaurus*), and rhynchocephalians (*Pleurosaurus*).

FIGURE 18.9 The decapod crustacean *Cycleryon propinquus*. Length of specimen is 25 cm. (Photo courtesy of G. Viohl, Jura-Museum, Eichstätt)

Unquestionable open-water swimmers include squid, belemnites, some of the sharks, small (*Leptolepides*) and large bony fish (*Caturus, Pachythrissops*), ichthyosaurs, plesiosaurs, and crocodiles of the genus *Geosaurus*. With the exception of *Leptolepides,* animals of this category are rare in the Solnhofen Plattenkalk (Barthel, Swinburne, and Conway Morris 1990). The presence of numerous worm-like feces, which can be attributed to squid (*Lumbricaria intestinum*), seems to indicate that these cephalopods lived temporarily in the surface waters of the lagoon (Janicke 1970; Malz 1976). Among planktonic animals, jellyfish (Figure 18.5) and floating crinoids (*Saccocoma* [Figure 18.7], *Pterocoma*) have already been mentioned. Others were probably pseudoplanktonic, including the cirripedian crustacean *Archaeolepas* and some bivalves (small oysters, *"Inoceramus"*), although the latter could probably also have lived as attached epibenthos.

Terrestrial organisms are far less abundant in the Solnhofen Plattenkalk than marine species (Barthel, Swinburne, and Conway Morris 1990). Among the plants, seed ferns (pteridosperms), bennetitales, ginkgos, and conifers are represented. The record of terrestrial invertebrates is completely dominated by insects, in both species numbers and abundances (Barthel, Swinburne, and Conway Morris 1990). They include mayflies (Ephemeroptera), dragonflies (Odonata), cockroaches (Blattoidea), water skaters (Phasmida), locusts and crickets (Ensifera), bugs and water scorpions (Heteroptera), cicadas (Auchenorrhyncha), lacewings (Neuroptera), beetles (Coleoptera), wasps (Hymenoptera), caddis flies (Trichoptera), and flies (Diptera). Apart from the insects, a questionable spider species is the only other known terrestrial invertebrate from the Solnhofen Plattenkalk (Barthel, Swinburne, and Conway Morris 1990). The terrestrial vertebrates, which are exceedingly rare, include various reptiles and *Archaeopteryx* (Figure 18.7). Freshwater fish and mammals have not been found. Among the reptiles, a few lizards (*Eichstaettisaurus, Ardeosaurus*), rhynchocephalians (*Kallimodon, Homoeosaurus*), a small crocodile (*Alligatorellus*), a small dinosaur (*Compsognathus*, known from only one specimen), and seven genera of pterosaurs (*Anurognathus, Ctenochasma, Germanodactylus, Gnathosaurus, Pterodactylus, Rhamphorhynchus, Scaphognathus*) are known. Some of the pterosaurs may have spent part of their lives above and on the water, since fish remains have been found in their digestive tracts (Barthel, Swinburne, and Conway Morris 1990). There is still no consensus about the life habit of *Archaeopteryx*, but it is now generally accepted that this bird was capable of active flight (Padian and Chiappe 1998). Whether it was mainly a ground- or a tree-dwelling animal awaits further clarification.

CONCLUSIONS

The Solnhofen Plattenkalk is certainly the most famous Plattenkalk deposit in the world, and has yielded one of the most spectacular fossil assemblages. The biota that made Solnhofen famous, even outside the scientific community, includes the earliest known bird, *Archaeopteryx*, with preserved feathers. Astonishing soft-part preservation is, however, not restricted to this most famous fossil, but is found as well in jellyfish, squids, polychaete worms, insects, fishes, and pterosaurs (Barthel, Swinburne, and Conway Morris 1990). Like all the other Plattenkalk occurrences (e.g., Bear Gulch [Chapter 9], Monte Bolca [Chapter 20]) (Hemleben and Swinburne, 1991), the Solnhofen Limestone was deposited in a very localized area and in a quite restricted stratigraphic interval.

The so-called Solnhofen lagoon was situated between a low-lying land-mass to the northwest and an extensive carbonate platform with coral patch-reefs to the south. Under subtropical dry climatic conditions, evaporation in the lagoon far exceeded precipitation and freshwater input, and the lagoonal waters became hypersaline (Barthel, Swinburne, and Conway Morris 1990). Within the lagoon, the fine-grained Plattenkalk accumulated only in several small basins, which were surrounded by sponge–microbial reef complexes. In these basins, strongly hypersaline and somewhat oxygen-depleted stagnant bottom-waters collected, and they were inhospitable for metazoans (Barthel, Swinburne, and Conway Morris 1990). The only life-forms that could temporarily colonize the sediment surface were cyanobacteria, rare foraminifers, and even rarer ostracodes. But this hostile environment was very favorable for exceptional fossil preservation, and, indeed, many fossils show preservation of soft parts. However, because the surface waters were inhospitable as well, only a few animal species were able to live temporarily in the lagoon. Fossils are, therefore, exceedingly rare in the Solnhofen Plattenkalk, and the macrofossil assemblage consists—apart from planktonic jellyfish and crinoids—mainly of plants and animals that were swept into the lagoon from the reef area in the south and from the coast and land in the northwest (Barthel, Swinburne, and Conway Morris 1990).

Although the paleogeographic situation is known in great detail, a variety of different depositional models for the Solnhofen Plattenkalk still exist. The opinions differ mainly about the origin of the Plattenkalk micrite. At present, it seems most likely that the bulk of the sediment came from the reefs to the south, where large amounts of debris accumulated. During storms, this reef debris was stirred up and transported to the north, and only the finest fraction reached the Solnhofen area where the micrite settled out from suspension (Barthel 1978; Barthel, Swinburne, and Conway Morris 1990; Viohl 1996). A smaller part of the carbonate was produced *in situ* by cyanobacteria at the bottom of the small basins, whereas coccoliths were only minor contributors to the sediment carbonate. Although the Solnhofen Plattenkalk is probably one of the best studied Lagerstätten, a modern geochemical study of both inorganic materials and organic compounds could possibly help clarify the remaining controversial aspects of the Plattenkalk deposition.

ACKNOWLEDGMENTS

The chapter was critically reviewed by G. Viohl, who helped clarify some key concepts.

REFERENCES

Abbott, A. 1992. *Archaeopteryx* disappears from private collection. *Nature* 357:6.

Abel, O. 1927. *Lebensbilder aus der Tierwelt der Vorzeit.* 2nd ed. Jena: Fischer Verlag.

Barthel, K. W. 1964. Zur Entstehung der Solnhofener Plattenkalke (unteres Untertithon). *Mitteilungen der bayrischen Staatssammlung für Paläontologie und historische Geologie* 4:37–69.

Barthel, K. W. 1970. On the deposition of the Solnhofen lithographic limestone (lower Tithonian, Bavaria, Germany). *Neues Jahrbuch für Geologie und Paläontologie, Abhandlungen* 135:1–18.

Barthel, K. W. 1972. The genesis of the Solnhofen lithographic limestone (low. Tithonian): Further data and comments. *Neues Jahrbuch für Geologie und Paläontologie, Monatshefte* 1972:133–145.

Barthel, K. W. 1978. *Solnhofen: Ein Blick in die Erdgeschichte.* Thun: Ott Verlag.

Barthel, K. W., N. H. M. Swinburne, and S. Conway Morris. 1990. *Solnhofen: A Study in Mesozoic Palaeontology.* Cambridge: Cambridge University Press.

Bausch, W. M., G. Viohl, P. Bernier, G. Barale, J.-P. Bourseau, E. Buffetaut, C. Gaillard, J.-C. Gall, and S. Wenz. 1994. Eichstätt and Cerin: Geochemical comparison and definition of two different Plattenkalk types. *Geobios Mémoire Special* 16:107–125.

Brett, C. E., and A. Seilacher. 1991. Fossil Lagerstätten: A taphonomic consequence of event sedimentation. In G. Einsele, W. Ricken, and A. Seilacher, eds., *Cycles and Events in Stratigraphy,* pp. 283–297. Heidelberg: Springer-Verlag.

Buisonjé, P. H. de. 1972. Recurrent red tides, a possible origin of the Solnhofen limestone (I/II). *Koninklijk Nederlandse Akademie van Wetenschappen, Proceedings* 75:152–177.

Buisonjé, P. H. de. 1985. Climatological conditions during deposition of the Solnhofen limestones. In M. K. Hecht, J. H. Ostrom, G. Viohl, and P. Wellnhofer, eds., *The Beginnings of Birds,* pp. 45–65. Proceedings of the International *Archaeopteryx* Conference, 1984. Eichstätt: Freunde des Jura-Museums.

Davis, P. G., and D. E. G. Briggs. 1998. The impact of decay and disarticulation on the preservation of fossil birds. *Palaios* 13:3–13.

Flügel, E., and H. E. Franz. 1967. Elektronenmikroskopischer Nachweis von Coccolithen im Solnhofener Plattenkalk (Oberer Jura). *Neues Jahrbuch für Geologie und Paläontologie, Abhandlungen* 127:245–263.

Gocht, H. 1973. Einbettungslage und Erhaltung von Ostracoden-Gehäusen im Solnhofener Plattenkalk (Unter-Tithon, SW-Deutschland). *Neues Jahrbuch für Geologie und Paläontologie, Monatshefte* 1973:189–206.

Goldring, R., and A. Seilacher. 1971. Limulid undertracks and their sedimentological implications. *Neues Jahrbuch für Geologie und Paläontologie, Abhandlungen* 137:422–442.

Groiss, J. T. 1967. Mikropaläontologische Untersuchungen der Solnhofener Schichten im Gebiet um Eichstätt (Südliche Frankenalb). *Erlanger geologische Abhandlungen* 66:75–96.

Gümbel, C. W. von. 1891. *Geognostische Beschreibung des Königreichs Bayern.* Section 4, *Geognostische Beschreibung der Fränkischen Alb (Frankenjura) mit dem anstossenden fränkischen Keupergebiete.* Kassel: Fischer Verlag.

Hecht, M. K., J. H. Ostrom, G. Viohl, and P. Wellnhofer, eds. 1985. *The Beginnings of Birds.* Proceedings of the International *Archaeopteryx* Conference, 1984. Eichstätt: Freunde des Jura-Museums.

Hemleben, C. 1977. Autochthone und allochthone Sedimentanteile in den Solnhofener Plattenkalken. *Neues Jahrbuch für Geologie und Paläontologie, Monatshefte* 1977:257–271.

Hemleben, C., and N. H. M. Swinburne. 1991. Cyclic deposition of the plattenkalk facies. In G. Einsele, W. Ricken, and A. Seilacher, eds., *Cycles and Events in Stratigraphy*, pp. 572–591. Heidelberg: Springer-Verlag.

Hückel, U. 1974a. Vergleich des Mineralbestandes der Plattenkalke Solnhofens und des Libanon mit anderen Kalken. *Neues Jahrbuch für Geologie und Paläontologie, Abhandlungen* 145:155–182.

Hückel, U. 1974b. Geochemischer Vergleich der Plattenkalke Solnhofens und des Libanon mit anderen Kalken. *Neues Jahrbuch für Geologie und Paläontologie, Abhandlungen* 145:279–305.

Janicke, V. 1970. *Lumbricaria*—Ein Cephalopoden-Koprolith. *Neues Jahrbuch für Geologie und Paläontologie, Monatshefte* 1970:50–60.

Janicke, V., and G. Schairer. 1970. Fossilerhaltung und Problematica aus den Solnhofener Plattenkalken. *Neues Jahrbuch für Geologie und Paläontologie, Monatshefte* 1970:452–464.

Joyce, W. G. 2001. First evidence of fossil turtle skin impressions recovered from the Upper Jurassic of Germany. *Journal of Vertebrate Paleontology, Supplement* 21:66A.

Kauffman, E. G. 1978. Short-lived benthic communities in the Solnhofen and Nusplingen limestones. *Neues Jahrbuch für Geologie und Paläontologie, Monatshefte* 1978:712–724.

Keupp, H. 1977a. *Ultrafazies und Genese der Solnhofener Plattenkalke (Oberer Malm, Südliche Frankenalb).* Abhandlungen der naturhistorischen Gesellschaft Nürnberg e.V., no. 37. Nuremberg: Naturhistorische Gesellschaft Nürnberg.

Keupp, H. 1977b. Der Solnhofener Plattenkalk—Ein Blaugrünalgen-Laminit. *Paläontologische Zeitschrift* 51:102–116.

Kutscher, M. 1997. Bemerkungen zu den Plattenkalk-Ophiuren, insbesondere *Geocoma carinata* (v. Münster, 1826). *Archaeopteryx* 15:1–10.

Malz, H. 1976. *Solnhofener Plattenkalk: Eine Welt in Stein.* Ein Führer durch das Museum des Solnhofener Aktien-Vereins. Solnhofen: Freunde des Museums beim Aktien-Verein, Maxberg.

Manni, R., U. Nicosia, and L. Tagliacozza. 1997. *Saccocoma* as a normal benthonic stemless crinoid: An opportunistic reply within mud-dominated facies. *Paleopelagos* 7:121–132.

Mayr, F. X. 1967. Paläobiologie und Stratinomie der Plattenkalke der Altmühlalb. *Erlanger geologische Abhandlungen* 67:1–40.

Meyer, R., and H. Schmidt-Kaler. 1984. *Erdgeschichte sichtbar gemacht: Ein Führer durch die Altmühlalb.* 2nd ed. Munich: Bayrisches geologisches Landesamt.

Meyer, R., and H. Schmidt-Kaler. 1989. Paläogeographischer Atlas des süddeutschen Oberjura (Malm). *Geologisches Jahrbuch, A* 115:3–77.

Meyer, R. F. K. 1974. Landpflanzen aus den Plattenkalken von Kehlheim (Malm). *Geologische Blätter für Nordost-Bayern* 24:200–210.

Milson, C. V. 1994. *Saccocoma*: A benthic crinoid from the Jurassic Solnhofen limestone, Germany. *Palaeontology* 37:121–129.

Padian, K., and L. M. Chiappe. 1998. The origin of birds and their flight. *Scientific American* 278: 28–37.

Rietschel, S. 1985. False forgery. In M. K. Hecht, J. H. Ostrom, G. Viohl, and P. Wellnhofer, eds., *The Beginnings of Birds,* pp. 371–376. Proceedings of the International *Archaeopteryx* Conference, 1984. Eichstätt: Freunde des Jura-Museums.

Röper, M., and M. Rothgaenger. 1996. *Die Plattenkalke von Brunn (Landkreis Regensburg): Sensationelle Fossilien aus dem Oberpfälzer Jura.* Eichendorf: Eichendorf Verlag.

Röper, M., and M. Rothgaenger. 1998a. *Die Plattenkalke von Hienheim (Landkreis Kehlheim): Echinodermen-Biotope im Südfränkischen Jura.* Eichendorf: Eichendorf Verlag.

Röper, M., and M. Rothgaenger. 1998b. *Die Plattenkalke von Solnhofen, Mörnsheim, Langenaltheim: Ein Blick in die Welt in Stein.* Treuchtlingen: Verlag Keller.

Rothpletz, A. 1909. Über die Einbettung der Ammoniten in den Solnhofener Schichten. *Abhandlung—Königlich bayrische Akademie der Wissenschaften (II. Abteilung)* 24:311–337.

Seilacher, A., G. Andalib, F. Dietl, and H. Gocht. 1976. Preservational history of compressed Jurassic ammonites from southern Germany. *Neues Jahrbuch für Geologie und Paläontologie, Abhandlungen* 152:302–356.

Seilacher, A., W. E. Reif, and F. Westphal. 1985. Sedimentological, ecological and temporal patterns of fossil Lagerstätten. *Philosophical Transactions of the Royal Society of London, B* 311:5–23.

van Straaten, L. M. J. U. 1971. Origin of Solnhofen limestone. *Geologie en Mijnbouw* 50:3–8.

Veizer, J. 1977. Geochemistry of lithographic limestones and dark marls from the Jurassic of southern Germany. *Neues Jahrbuch für Geologie und Paläontologie, Abhandlungen* 153:129–146.

Viohl, G. 1985. Geology of the Solnhofen lithographic limestone and the habitat of *Archaeopteryx.* In M. K. Hecht, J. H. Ostrom, G. Viohl, and P. Wellnhofer, eds., *The Beginnings of Birds,* pp. 31–44. Proceedings of the International *Archaeopteryx* Conference, 1984. Eichstätt: Freunde des Jura-Museums.

Viohl, G. 1994. Fish taphonomy of the Solnhofen Plattenkalk—An approach to the reconstruction of the palaeoenvironment. *Geobios Mémoire Special* 16:81–90.

Viohl, G. 1996. The paleoenvironment of the Late Jurassic fishes from the southern Franconian Alb (Bavaria, Germany). In G. Arratia and G. Viohl, eds., *Mesozoic Fishes—Systematics and Paleoecology,* pp. 513–528. Munich: Verlag Pfeil.

Wellnhofer, P. 1988. A new specimen of *Archaeopteryx. Science* 240:1790–1792.

Wellnhofer, P. 1990. Die Solnhofener Plattenkalke. In W. K. Weidert, ed., *Klassische Fundstellen der Paläontologie,* vol. 1, pp. 85–97. Korb: Goldschneck-Verlag.

Wellnhofer, P. 1993. Das siebte Exemplar von *Archaeopteryx* aus den Solnhofener Schichten. *Archaeopteryx* 11:1–47.

19

Smoky Hill Chalk: Spectacular Cretaceous Marine Fauna

David J. Bottjer

CHALK, A ROCK MADE PRIMARILY OF THE FOSSILS OF microscopic marine algae (coccoliths), is typically deposited in the deep sea. But in the Cretaceous, it accumulated extensively in shallow epicontinental seas, which covered large portions of many of the continents during that time (Bottjer 1986). Even though it was deposited in the Cretaceous at relatively shallow depths, chalk as a sediment typically accumulates in marine environments with little wave or current action. The chalk cliffs along the northern part of the English Channel—on the English coast in Dover (White Cliffs of Dover) and on the French coast in Normandy—are arguably the most prominent and famous of Cretaceous chalk exposures worldwide. However, this typical Cretaceous rock (the time period takes its name from the Latin word for "chalk": *creta*) is also found along the Gulf Coast (Texas, Arkansas, Mississippi, Alabama) and in the western interior (Kansas, Colorado) of North America (Bottjer 1986).

Chalks found in the western interior not only were deposited in marine environments with relatively little wave or current energy, but also typically formed on seafloors where dissolved oxygen concentrations were low. This led to conditions for the development of a conservation Lagerstätte that is a stagnation deposit. Indeed, the chalk beds of the western interior, particularly the Smoky Hill Chalk Member of the Niobrara Chalk (western Kansas) (Figures 19.1 and 19.2), are renowned for their preservation of articulated Cretaceous marine reptiles, such as mosasaurs, plesiosaurs, pliosaurs, and large turtles, as well as pterodactyls, enormous fish, and the flightless marine bird *Hesperornis*. While

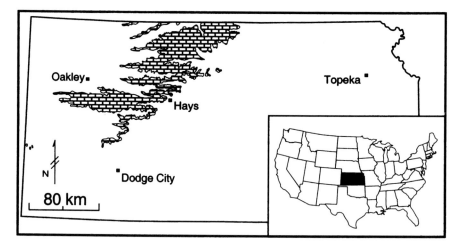

FIGURE 19.1 Outcrop distribution of Niobrara Chalk in Kansas. (Modified from Hattin 1982)

these spectacular vertebrate fossils can be found in museums worldwide, the marine invertebrate fauna, including huge specimens (approaching 2 m in greatest dimension) of the Cretaceous inoceramid bivalve *Inoceramus* (*Platyceramus*) *platinus*, as well as dense bedding-plane accumulations of the strange Cretaceous crinoid *Uintacrinus socialis*, are equally well known.

The fossils of the Smoky Hill Chalk received their earliest attention in the late nineteenth century during the "dinosaur wars" of the American West, due primarily to work of the fossil collector Charles H. Sternberg. Sternberg, later joined by his sons (particularly George), collected the Smoky Hill Chalk beds extensively, and their work provided the bulk of the magnificent articulated vertebrate specimens that can be found in many natural history museums (Rogers 1991). Beginning in the 1920s, George Sternberg participated in developing the fossil collections at a new museum associated with a teacher's college in Hays, Kansas. This museum, now named The Sternberg Memorial Museum, as part of Fort Hays State University, contains outstanding displays of Niobrara fossils collected by the Sternbergs, including large slabs with *Uintacrinus socialis*, a famous *Xiphactinus* "fish-within-a-fish" specimen, as well as a mounted mosasaur and a large *Pteranodon*. There also are excellent collections of Smoky Hill fossils at the Museum of Natural History at the University of Kansas.

GEOLOGICAL CONTEXT

The modern study of chalk strata in western Kansas is based primarily on the work of Donald E. Hattin, a stratigrapher who placed much of the Cretaceous stratigraphy of the western interior on a sound footing.

Despite the long-term popularity of Smoky Hill Chalk as a place to collect spectacular Cretaceous fossils, little was known about its stratigraphy until Hattin's (1982) detailed work was published. This was because much of the chalk on outcrop appears as monotonous strata with few distinguishing features. Since existing outcrops exposed only portions of the entire member, correlations between sections were virtually impossible, and many of the treasured specimens excavated from the Smoky Hill Chalk by early collectors have little or no meaningful stratigraphic data accompanying them.

The Smoky Hill Chalk Member of the Niobrara Chalk is part of a thick sequence of Cretaceous strata exposed in Kansas (Figure 19.2). Hattin's (1982) work showed that this unit consists of approximately 180 m of "mainly olive gray, well-laminated to nonlaminated, flaky-weathering, fecal-pellet-speckled, impure chalk." This chalk outcrops typically in a badlands topography, forming many scenic towers and pinnacles, which is particularly well developed in the western outcrop area in Gove and Logan Counties (Hattin, Siemers, and Stewart 1978). The Smoky Hill contains more than 100 bentonite beds, which range to as much as 11.3 cm in thickness and which were a key component in piecing together a composite stratigraphic section (Hattin 1982). Ammonites are relatively rare in the Smoky Hill, so an inoceramid bivalve biostratigraphy has proved to be most useful for these Coniacean–Campanian strata (Hattin 1982; Kauffman et al. 1993).

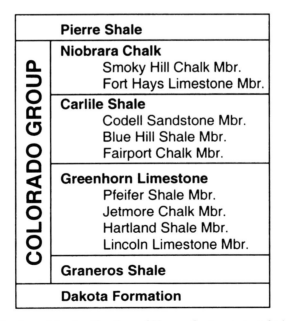

FIGURE 19.2 Stratigraphic classification of Upper Cretaceous rocks in west-central Kansas. (Modified from Hattin 1982)

Paleoenvironmental Setting

During deposition of the Smoky Hill Chalk, Kansas was in a warm to mild temperate climate zone in a relatively warm Cretaceous world (Hattin 1982). At that time, global sea level was as high as it has perhaps ever been, while elevations of the central part of the continent were much reduced from present-day values. Consequently, a North American western interior seaway spread north–south from the present-day Gulf Coast across the continent to the Arctic.

Along with his stratigraphic work, Hattin (1982) has provided the most detailed paleoenvironmental study to date of the Smoky Hill Chalk. Depth of the seaway where the Smoky Hill was being deposited ranged from 150 to 300 m. The coccolith-rich muds that were deposited in this seaway usually provided a soft seafloor, with poorly circulated bottomwaters. Interstitial waters within the sediment were typically extremely low in oxygen content or even anoxic, so bioturbation enhanced the soupy nature of the surface mixed layer on the seafloor, with no deep burrowing. The wide lateral persistence of marker beds, such as the bentonites, indicates an extremely flat depositional topography.

Taphonomy

By far and away, the greatest number of articulated vertebrate specimens retrieved from the Smoky Hill Chalk were collected at the end of the nineteenth and in the early decades of the twentieth century, before the science of taphonomy was developed. However, this is a spectacular stagnation deposit with articulated remains of vertebrates, crinoids, and bivalves. These fossils were typically preserved in an articulated state because the low oxygen concentrations of seawater near the ocean floor precluded the presence of larger scavengers, which would have scattered skeletal elements. Of particular interest is that a new examination of the crinoid *Uintacrinus socialis* has revealed the preservation of soft tissues, which are interpreted to have been preserved as a result of the sealing effects of a microbial mat that covered these crinoids on the seafloor after death (Meyer, Milsom, and Webber 1999; Meyer and Milsom 2001).

Paleobiology and Paleoecology

The Smoky Hill Chalk fauna is particularly well known for its abundant and articulated vertebrate remains, with the occurrence of as many as 135 species (Williston 1898; Frey 1972). Perhaps the most spectacular of these vertebrate remains are those of the mosasaurs. Massare (1997) states that in the study of Mesozoic marine reptiles, six Lagerstätten are particularly important in understanding their ecology and evolution. Of these, one is the Smoky Hill Chalk, three have already been discussed in this book—Monte San Giorgio (Chapter 12), Posidonia Shale (Chapter 15), and Oxford Clay (Chapter 17)—and the remaining two are the Lower Jurassic Blue Lias in England and the Upper Cretaceous Sharon Springs Member of the Pierre Shale, which overlies the Niobrara Chalk in the western interior of North America (Figure 19.2).

In comparison with the other Mesozoic marine reptile Lagerstätten, according to Massare (1997), the fauna of the Smoky Hill Chalk represents a major reorganization of ecological guilds. This is due to the Late Cretaceous rise to dominance of mosasaurs, following the extinction of the ichthyosaurs (Bell 1997). These huge marine lizards, the largest specimens of which exceed 10 m in length, are the most common reptiles found in the Smoky Hill Chalk, particularly the genera *Platecarpus, Tylosaurus* (Figure 19.3), and *Clidastes* (Russell 1967; Carpenter 1990). Mosasaurs were ambush predators that may have been opportunistic feeders and generalists at the top of the trophic pyramid (Massare 1997). The mosasaurs were joined by at least two plesiosaurs that were also ambush predators (*Styxosaurus, Elasmosaurus*), as well as the pliosaur pursuit predators *Dolichorhyncops* (Figure 19.4) and *Polycotylus* (Massare 1997). Massare (1997) concludes that the Late Cretaceous marine reptile record is dominated by mosasaur ambush predators.

In the Smoky Hill (as well as the Sharon Springs Member), large fish and large sharks are also very abundant and appear to have been a much more important component of the marine large predator community than earlier in the Mesozoic. This may indicate that these fish and sharks ecologically replaced some of the earlier reptiles, particularly in the pursuit guilds (Massare 1997). Ample evidence for fish predation exists, as demonstrated by specimens of the bulldog tarpon *Xiphactinus* that are as much as 6 m long. The fish-within-a-fish specimens (Figure 19.5) can have as many as 10 recognizable fish skeletons inside (Benton 1997).

Articulated remains of the bird *Hesperornis,* which was more than 1 m tall and had a long neck, as well as the similar but smaller *Baptornis* (both members of the flightless order Hesperornithiformes), are common in the Smoky Hill Chalk (Martin and Tate 1976; Benton 1997). These birds are thought to have been divers that swam rapidly by kicking their webbed feet and steered by using their tiny wing stumps (Benton 1997). Their jaws have small pointed teeth, and their coprolites indicate that they ate fish (Benton 1997). *Ichthyornis,* a member of the order Ichthyornithiformes that has fully developed wings, is also found in the Smoky Hill Chalk. This smaller bird also has pointed teeth and very likely lived by diving while flying, in a manner similar to that of modern terns (Benton 1997).

Joining *Ichthyornis* in flying over the Smoky Hill seaway was the pterosaur *Pteranodon,* one of the best known and largest of the pterosaurs, which had a wingspan of 5 to 8 m (Eaton 1910; Benton 1997). Pterosaurs are characterized by a wide diversity in skull structure and function, typically related to food preference and aerodynamics, and

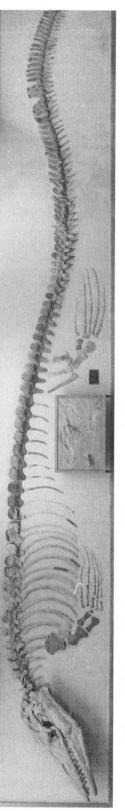

FIGURE 19.3 Mounted skeleton of the large mosasaur *Tylosaurus prorigor* in the Sternberg Museum. Field of view is approximately 12 m. (Photo provided courtesy of Sternberg Museum of Natural History, Fort Hays State University, Hays, Kansas)

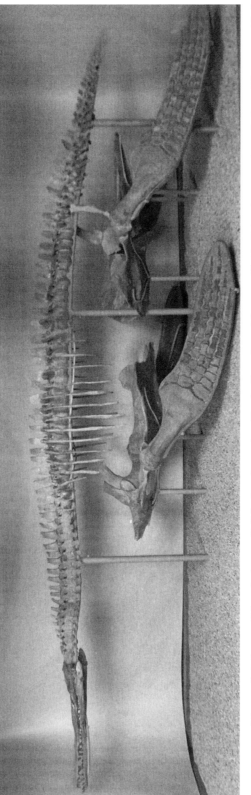

FIGURE 19.4 Mounted skeleton of the plesiosaur *Dolichorhynchops osborni* in the Sternberg Museum. Field of view is approximately 3 m. (Photo provided courtesy of Sternberg Museum of Natural History, Fort Hays State University, Hays, Kansas)

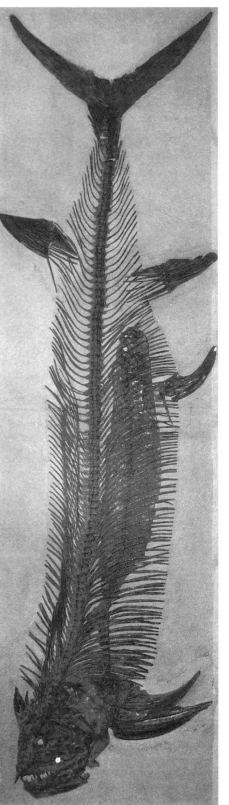

FIGURE 19.5 Mounted fish-within-a-fish skeletons in the Sternberg Museum. The large fish is *Xiphactinus audax*; the small fish is *Gillicus arcuatus*. Field of view is approximately 4 m. (Photo provided courtesy of Sternberg Museum of Natural History, Fort Hays State University, Hays, Kansas)

Pteranodon, with its large pointed crest that is as long as the skull itself, is no exception (Benton 1997). *Pteranodon* had a low takeoff speed, and probably lived similarly to the modern albatross or pelican, soaring over the ocean in search of prey such as cephalopods or fish (Benton 1997).

Diversity of macroinvertebrates on the Smoky Hill seafloor was not high (Hattin 1982). Inoceramid bivalves are probably the most common benthic macroinvertebrate, with remarkable specimens of *Inoceramus (Platyceramus) platinus* approaching breadths of 2 m (Hattin 1982, 1986) (Figure 19.6), thus qualifying as some of the largest bivalves to live in Phanerozoic oceans. Traditionally, it has been thought that these extinct bivalves most likely rested on the seafloor on their left valves. Intriguingly, though, these unusual bivalves show common encrustation by the

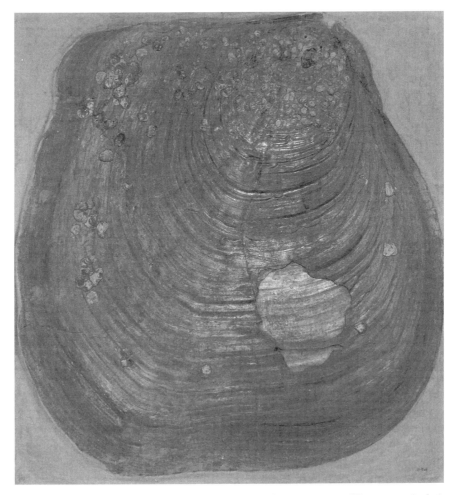

FIGURE 19.6 Mounted specimen of the giant bivalve *Inoceramus (Platyceramus) platinus* encrusted by the oyster *Pseudoperna congesta* in the Sternberg Museum. The height of the specimen is approximately 1.1 m. (Photo provided courtesy of Sternberg Museum of Natural History, Fort Hays State University, Hays, Kansas)

oyster *Pseudoperna congesta,* sometimes on both valve exteriors (Hattin 1986). If the left valve had been resting on the soft substrate, it would not be expected that encrusters such as these oysters would have colonized it. This has led to the hypothesis that these inoceramids potentially had a pseudoplanktonic life mode, attached to floating logs or algae, or that valves were flipped over by large predators such as mosasaurs in search of food, which would have allowed colonization of both valves by encrusters (Hattin 1982, 1986). Hattin (1982, 1986), however, finds that the preponderance of evidence supports the original hypothesis that these enormous bivalves reclined on the seafloor during life. Along with the oysters, epizoans on *I. platinus* also include locally abundant stalked cirripeds, sparse boring sponges, rare serpulids, and fairly common boring cirripeds (Hattin 1982, 1986). Within articulated inoceramid valves are fossils of fish that lived in the mantle cavity (Stewart 1990a, 1990b). Also found in the Smoky Hill are fairly common individuals, or clusters of a few individuals, of the rudist *Durania maxima,* commonly up to 30 cm long, which lived on the seafloor and also have oysters and cirripeds as epizoans (Hattin 1982, 1986). These inoceramids and rudists thus served as islands for epizoans needing a hard substrate on which to live, in a paleoenvironment dominated by a soft, soupy chalk seafloor.

Within a narrow stratigraphic interval of the Smoky Hill are found thin extensive lenses of well-cemented limestone composed primarily of

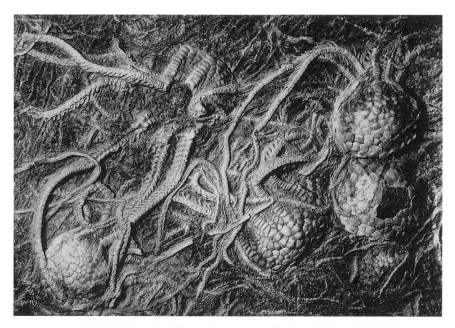

FIGURE 19.7 Typical dense bedding-plane accumulation of the crinoid *Uintacrinus socialis.* Field of view is approximately 29 cm. (Photo courtesy of D. L. Meyer, University of Cincinnati)

extremely well-preserved specimens of the unusual crinoid *Uintacrinus socialis* (Hattin 1982) (Figure 19.7). Springer (1901) interpreted these crinoids as swimmers, like the living comatulid crinoids, and postulated that their occurrence in huge numbers forming thin beds indicates that they died during mass-mortality events. However, recent work indicates that these crinoids, with calyces about 5 cm in diameter, lived as aggregations on the seafloor much like modern brittle stars, suspension feeding with arms reaching up to 1.25 m in length (Meyer, Milsom, and Webber 1999; Webber, Meyer, and Milsom 1999; Meyer and Milsom 2001).

CONCLUSIONS

The Smoky Hill Chalk has a long and storied history as one of the best Lagerstätten in which to find Cretaceous vertebrates that lived in and adjacent to the ocean. This reputation began with the Sternbergs, and many Smoky Hill specimens are prominently displayed in museums. The Smoky Hill Chalk fauna thus provides a marine counterbalance to the extensive information that exists on Cretaceous terrestrial environments, which are characterized by dinosaurs. Relatively little detailed work has been done on the taphonomy of fossils from this unit, but the future potential for such studies would seem to be great, given the spectacular preservation of a wide variety of organisms found in this Lagerstätte.

ACKNOWLEDGMENTS

Many thanks are offered to D. E. Hattin and N. G. Lane, who provided an introduction to the numerous wonders of the Kansas chalks.

REFERENCES

Bell, G. L., Jr. 1997. Mosasauridae: Introduction. In J. M. Callaway and E. L. Nicholls, eds., *Ancient Marine Reptiles*, pp. 281–292. San Diego: Academic Press.

Benton, M. J. 1997. *Vertebrate Palaeontology*. London: Chapman and Hall.

Bottjer, D. J. 1986. Campanian–Maastrichtian chalks of southwestern Arkansas: Petrology, paleoenvironments and comparison with other North American and European chalks. *Cretaceous Research* 7:161–196.

Carpenter, K. 1990. Upward continuity of the Niobrara fauna with the Pierre Shale fauna. In S. C. Bennett, ed., *Niobrara Chalk Excursion Guidebook*, pp. 73–81. Lawrence: University of Kansas Museum of Natural History and Kansas Geological Survey.

Eaton, C. F. 1910. *Osteology of Pteranodon*. Memoirs of the Connecticut Academy of Arts and Sciences, vol. 2. New Haven: Yale University.

Frey, R. W. 1972. *Paleoecology and Depositional Environment of Fort Hays Limestone Member, Niobrara Chalk (Upper Cretaceous), West-central Kansas*. University of

Kansas Paleontological Contributions, no. 58. Lawrence: University of Kansas Paleontological Institute.

Hattin, D. E. 1982. *Stratigraphy and Depositional Environment of Smoky Hill Chalk Member, Niobrara Chalk (Upper Cretaceous), of the Type Area, Western Kansas.* Kansas Geological Survey Bulletin, vol. 225. Lawrence: Kansas Geological Survey, University of Kansas.

Hattin, D. E. 1986. Carbonate substrates of the Late Cretaceous sea, central Great Plains and southern Rocky Mountains. *Palaios* 1:347–367.

Hattin, D. E., C. T. Siemers, and G. F. Stewart. 1978. *Guidebook: Upper Cretaceous Stratigraphy and Depositional Environments of Western Kansas.* Kansas Geological Survey Guidebook Series, no. 3. Lawrence: Kansas Geological Survey, University of Kansas.

Kauffman, E. G., B. B. Sageman, J. I. Kirkland, W. P. Elder, P. J. Harries, and T. Villamil. 1993. Molluscan biostratigraphy of the Cretaceous western interior basin, North America. In W. G. E. Caldwell and E. G. Kauffman, eds., *Evolution of the Western Interior Basin,* pp. 397–434. Geological Association of Canada Special Paper, no. 39. St. John's, Newfoundland: Geological Association of Canada.

Martin, L. D., and J. Tate, Jr. 1976. The skeleton of *Baptornis advenus* (Aves: Hesperornithiformes). *Smithsonian Contributions to Paleobiology* 27:35–66.

Massare, J. A. 1997. Faunas, behavior, and evolution: Introduction. In J. M. Callaway and E. L. Nicholls, eds., *Ancient Marine Reptiles,* pp. 401–421. San Diego: Academic Press.

Meyer, D. L., and C. V. Milsom. 2001. Microbial sealing in the biostratinomy of *Uintacrinus* Lagerstätten in the Upper Cretaceous of Kansas and Colorado, USA. *Palaios* 16:535–546.

Meyer, D. L., C. V. Milsom, and A. J. Webber. 1999. A crinoid conundrum. *Geotimes* 44:14–16.

Rogers, K. 1991. *The Sternberg Fossil Hunters: A Dinosaur Dynasty.* Missoula: Montana Press.

Russell, D. 1967. Systematics and morphology of North American mosasaurs. *Peabody Museum of Natural History Bulletin* 23:1–240.

Springer, F. 1901. *Uintacrinus:* Its structure and relations. *Harvard College, Museum of Comparative Zoology Memoir* 25:1–89.

Stewart, J. D. 1990a. Preliminary account of halecostone–inoceramid commensalism in the Upper Cretaceous of Kansas. In A. J. Boucot, ed., *Evolutionary Paleobiology of Behavior and Coevolution,* pp. 51–57. Amsterdam: Elsevier.

Stewart, J. D. 1990b. Niobrara Formation symbiotic fish in inoceramid bivalves. In S. C. Bennett, ed., *Niobrara Chalk Excursion Guidebook,* pp. 31–41. Lawrence: University of Kansas Museum of Natural History and Kansas Geological Survey.

Webber, A. J., D. L. Meyer, and C. V. Milsom. 1999. New findings on the stratigraphic position of *Uintacrinus socialis* (Cretaceous, Santonian). *Geological Society of America Abstracts with Programs* 31:A-104.

Williston, S. W. 1898. Birds, dinosaurs, crocodiles, mosasaurs, turtles (*Toxochelys* by E. C. Case): The Kansas Niobrara Cretaceous. *Kansas Geological Survey* 4:41–411.

20
Monte Bolca: An Eocene Fishbowl

Carol M. Tang

FOR FOUR CENTURIES, NOBLEMEN, CARDINALS, AND EVEN emperors are said to have coveted the fossil fish from the Pesciara (Italian for "fishbowl") of Monte Bolca in northern Italy. This nickname is based on the extraordinarily preserved fish of Eocene age that have been excavated since the sixteenth century, but the deposit also contains well-preserved remains of reptiles, crustaceans, molluscs, plants, cephalopods, and jellyfish. More than 500 species of terrestrial and marine vertebrates, invertebrates, and plants are represented in this Lagerstätte, including 90 families of fish alone. The first printed monograph on paleoichthyology, Giovanni Serafino Volta's *l'Ittiolitologia veronese* (1796–1808), and some of Louis Agassiz's pioneering work on comparative zoology were based largely on fossils from Monte Bolca. The diversity, abundance, and excellent preservation of these fossils have greatly influenced systematic, paleoecologic, and evolutionary studies in ichthyology. But much less work has been conducted on the other fossils or on the geologic, paleoenvironmental, and taphonomic aspects of the Monte Bolca Lagerstätte, and unlike that for the fish specimens, which have been widely disseminated in museums around the world, most of the information we do have about this deposit is not widely known.

HISTORICAL BACKGROUND

There is a very rich history on the collection and study of Monte Bolca fossils (Gaudant 1997). Accounts of the extraordinary fish from Bolca go back to the mid-sixteenth century. The first written description of Monte Bolca fossil material came from Andrea Mattioli, a doctor from Siena, who in his 1555 book reported seeing "some slabs of stone which, on being split in half, revealed the shapes of various species of fish, every detail of which had been transformed into stone" (Sorbini 1972).

In 1571, Francesco Calceolari of Verona had a collection of Monte Bolca fossils in his museum of natural history, which is believed to have been one of the first museums, if not *the* first museum, of its kind in the modern era (Sorbini 1972). In 1796, Abbot Serafino Volta began publishing his monograph on fossil fish–the first major work ever printed on the Monte Bolca fossils and on paleoichthyology (Sorbini 1983). This treatise includes beautifully illustrated examples of fossil fish collected from the Bolca area.

In the mid-nineteenth century, the Cerato family began excavating the Pesciara on a regular basis (Sorbini 1972). Working with only hand tools, such as hammers, picks, and chisels, generation after generation of Ceratos have carved out a latticework of tunnels running alongside fossiliferous beds. The family home, with its collection of fossils, was the destination of many scientists, tourists, and dignitaries for centuries. One local story claims that Emperor Franz Josef once extended his visit to Bolca because he was so intrigued by the fossils (Stanghellini 1979). Even though the excavation now occurs aboveground in open quarries, the Ceratos are still carrying out this family tradition. It is estimated that the family has extracted more than 100,000 fossils, which are now distributed in museums and private collections around the world (Blot 1969).

Fossils collected by the first generation of Cerato "fishermen" were obtained by Count G. B. Gazola, an Italian naturalist of the eighteenth century (Sorbini 1972). Although several noblemen had significant collections of Monte Bolca fossils at this time, by the end of the eighteenth century Gazola had purchased most major collections and exhibited his fossils in two halls in his palace (Sorbini 1983).

During Napoleonic times, Count Gazola may have sold many pieces from his fossil collection to the French. However, he maintained his own collection by keeping duplicates of many of the pieces and saving some of the best. What remains of the Gazola collection is now housed in the Natural History Museum of Verona, along with more recently excavated specimens (Sorbini 1983).

The French shipped their share of the fossils to Paris. Today, these original purchases account for the large collection of Monte Bolca fossils in the Paris Natural History Museum, which is surpassed only by collections at the Verona museum and the museum at Padua University (Blot 1969). In Paris, Louis Agassiz, a founder of comparative zoology, studied the Monte Bolca collection. The excellent preservation of the fish allowed him to make morphological comparisons between the Monte Bolca specimens and other fossil and living fish species. His five-volume monograph, published between 1833 and 1843, is still used as a reference for Monte Bolca fish species. Thus, the Monte Bolca fossils have exerted much influence in the fields of paleontology and zoology.

GEOLOGICAL CONTEXT

The geological and paleoenvironmental aspects of the Eocene fossiliferous deposits of Monte Bolca were first described by Fabiani (1914), followed by Sorbini (1968) and Massari and Sorbini (1975), and summarized by Sorbini (1983). The following discussion is based on these studies.

The original and most famous collecting locality, Pesciara, is about 2 km northeast of the village of Bolca, near Verona in northern Italy (Figure 20.1). Other localities in the region have also yielded fossils of the same age, including Monte Postale, which contains fish and molluscs; Monte Vegroni, which is famous for palms; and Praticini, Loschi, Le Pozzette, and Zovo e Valleco, which have yielded crocodiles and tortoises. These sites have been incorporated into a fossil park where tourists may view the quarries.

Although the two best-known and best-studied localities, Pesciara and Monte Postale, are only about 100 m apart, the stratigraphic relationships between them are unclear (Figure 20.2). Although the fish fauna are

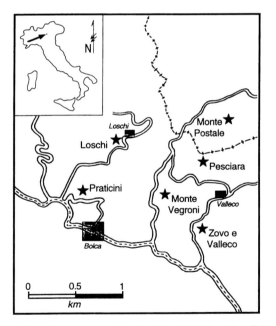

FIGURE 20.1 Several of the most productive fossil-yielding localities in the region around Monte Bolca in northern Italy. (Modified from Stanghellini 1979 and Sorbini 1983)

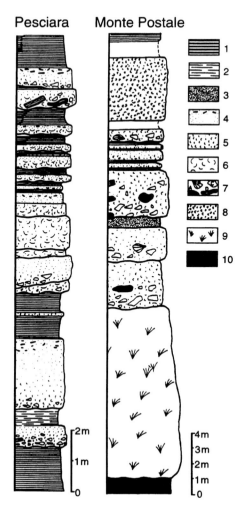

FIGURE 20.2 Stratigraphic section of the Middle Eocene fossil-bearing limestone units. Lithologies include (1) laminated carbonate mudstone; (2) partially laminated carbonate mudstone; (3) bioturbated carbonate mudstone; (4) massive calcarenite; (5) carbonate grainstone; (6) molluscan limestone; (7) coarse breccias with variable clasts; (8) foraminiferal carbonate rainstones; (9) coral–algal biolithite; and (10) hyalo-clastite. The well-preserved fish and plant fossils are contained within the laminated carbonate mudstones (1), although invertebrate fossils and microfossils are found in many of the units. (Modified from Massari and Sorbini 1975 and Sorbini 1983)

generally similar between the two localities (Sorbini 1983), the specific ecological structure (Landini and Sorbini 1996), paleoenvironmental interpretation, and stratigraphic sequence differ (Massari and Sorbini 1975). The Pesciara deposit consists of a cliff-forming block of limestone surrounded by volcanic deposits. The unit is about 19 m thick and extends for only a few 100 m². The dip and folds present along one side of the block suggest that the whole block was transported and deposited while the sediments were still unlithified.

The fossil-bearing section consists of several calcareous units (Figure 20.2). Exceptionally well preserved fossils are derived from units of laminated sheet limestones underlain by reefal limestones. Fish and plant fossils can be found at five levels. The beds containing fish and plant fossils show a varved structure in which relatively thick biomicritic laminae alternate in a regular pattern with very thin micritic laminae, which are homogenous and slightly clayey. The thicker biomicritic laminae contain silt-size particles and foraminifera, ostracodes, molluscs, and small thin-shelled gastropods. These biomicritic laminae are gradational with the micritic laminae, producing a graded bedding texture for each pair of laminae. Finely disseminated pyrite and bitumen can be found in these laminated beds.

The five fish-bearing units are separated from one another by coarse limestone debris beds containing clasts as large as 60 cm in diameter. The clasts can be foraminiferal biomicrites, biosparites containing alveoline and miliolid benthic foraminifera, coral- and algal-bound bioliths, and laminated biomicrites. These clasts are sometimes rounded and show evidence of boring prior to their transport and final deposition.

The Monte Bolca fossils are considered to be Middle Eocene in age (*NP 14* zone) based on the calcareous nannofossil *Discoaster sublodensis* found in one clay horizon (Medizza 1975). Using the Harland et al. (1989) time scale, this would place deposition of the Monte Bolca Lagerstätte at about 49 Ma ago.

PALEOENVIRONMENTAL SETTING

The organisms preserved in the Monte Bolca Lagerstätte represent a number of ecosystems, including pelagic marine, coral reef, seagrass beds, brackish lagoons, fluvial systems, coastal ponds, and terrestrial habitats (review in Landini and Sorbini 1996). The fauna and flora are indicative of a tropical or subtropical region. Lithologic components within the Monte Bolca deposits were derived from a number of areas, including reefal, littoral, and lagoonal (Massari and Sorbini 1975). The overall environment is that of a tropical coastal region in close proximity to a coral reef and a large landmass (Sorbini 1968, 1983; Massari and Sorbini 1975).

In the past, it was suggested that the Pesciara Lagerstätte was deposited within a lagoon that occasionally became isolated from the open ocean due to barrier bar migration (Sorbini 1983). During times of isolation, there would have been no water circulation, and changes in the water column could have led to anoxia and deposition of laminated sediments.

More recently, however, Landini and Sorbini (1996) hypothesized that the fish-bearing strata accumulated parallel to the coast in a silled depression on the seafloor landward of a barrier reef. In the area of the depression, the overlying water column was open to pelagic, seagrass, and coral reef influences. But at the bottom of the basin, circulation could have become restricted, resulting in stagnation near the seafloor and the accumulation of organic-rich sediments containing exceptionally preserved fossils.

In either case, the nonlaminated layers containing normal marine invertebrates represent intervals when normal oceanic conditions prevailed. Benthic organisms colonized the seafloor, and bioturbation was present. Storm events are evidenced by the large limestone blocks ripped from the nearby reef and redeposited in tempestites. The faunal components, borings, and abrasion seen in limestone clasts suggest that they originated in a shallow-water environment, possibly the littoral zone (Massari and Sorbini 1975). Thus, at times the region experienced normal marine circulation, was connected with the reef and pelagic realms, received sediments from littoral and coral areas, and was occasionally disturbed by storm events.

The deposition of laminated sediments and fully articulated fossils suggests that, at other times, the area became restricted and stagnant (Massari and Sorbini 1975). It is during these intervals that deposition of abundant, articulated, and exceptionally well preserved fossils occurred. Anoxic or low oxygen conditions on the seafloor are indicated by the lack of bioturbation, general dearth of benthic organisms, absence of scavenging and decomposition of organic remains, presence of organic carbon, and presence of pyrite.

Because of the abundant volcanoclastic material surrounding the carbonate strata and the gaping mouths of the fish, early studies concluded that these organisms were killed by volcanic gases and sediment before being buried (Blot 1969). It is now thought that both the mass kills and the exceptional preservation are related to the cyclic, varve-like nature of the laminations. Sorbini (1972) hypothesized that the anoxic episodes may have been triggered by dinoflagellate blooms within the lagoon. The plankton may have been toxic to the fish, and the large amount of decomposing organic matter could have caused anoxic conditions in both the water column and the seafloor sediment. Modern mass fish kills due to plankton blooms have been well documented in a number of cases. Landini and Sorbini (1996) have proposed that the water column above the depression could have undergone eutrophication due to the influx of organic matter from seasonal terrestrial runoff. Similar to the processes invoked in the plankton bloom scenario, the decomposition of large amounts of organic matter would deplete dissolved oxygen concentrations and could lead to anoxic conditions in the water column and the seafloor.

TAPHONOMY

Very little taphonomic work has been conducted on Monte Bolca deposits, and, except for the collection of very well preserved articulated fish, even basic descriptions of the preservation quality of the fossils are generally difficult to find. Fossils preserved in the five Lagerstätten layers in the laminated section can be fully articulated (Figures 20.3–20.5), and organic matter such as tissue, skin, and scales, along with pigmentation patterns, can be preserved. A popular book about the Monte Bolca fossils states that the fish "not only maintain their natural form and shape, but even their amber, brown or greenish colour" (Stanghellini 1979). The presence of laminations and pyrite indicates the presence of low oxygen conditions on the seafloor, which would have

FIGURE 20.3 Example of an extremely well preserved *Exellis velifer.* Length of specimen is 13.5 cm. (Photo courtesy of Museo Civico di Storia Naturale di Verona)

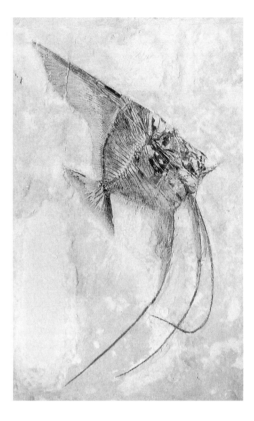

FIGURE 20.4 Specimen of an angelfish, for which Monte Bolca is famous. Even delicate skeletal structures are preserved in this *Ceratoicthys pinnatiformis*. Length of specimen is 17.5 cm. (Photo courtesy of Museo Civico di Storia Naturale di Verona)

prevented the breakdown of organic material and the scavenging of carcasses.

PALEOBIOLOGY AND PALEOECOLOGY

The paleobiological and paleoecological significance of the Monte Bolca Lagerstätte lies not only in the spectacular preservation of fossils, but also in the great abundance and diversity of the fauna and flora preserved.

Marine Invertebrates

Throughout the Monte Bolca deposit, corals, molluscs, and foraminifera typical of a coral reef environment can be found. They are usually deposited in the form of transported debris.

In the laminated layers, polychaete worms, jellyfish (new genus *Semplicibrachia*), and a cephalopod (probably suborder Metateuthoidea) have been found (Sorbini 1983). Secretan (1975) has described several crustaceans from Monte Bolca.

Marine Vertebrates

The fossil fish of Monte Bolca are by far the best-studied component of the fauna. As discussed earlier, systematic comparative anatomy was conducted on Monte Bolca fish in the nineteenth century by Agassiz. Since then, Blot (1969, 1980) has undertaken many detailed systematic revisions, and Landini and Sorbini (1996) have begun paleoecological comparisons with modern reef fish faunas.

Both elasmobranch and teleost fish are present, including angelfish, moray eels, mackerels, sea pikes, rays, and sharks (Figures 20.3–20.5). Thus far, more than 250 species have been identified from Monte Bolca, representing 140 genera, 90 families, and 19 orders (Landini and Sorbini 1996). Eighteen of the orders are extant, as are 80 percent of the families and 30 percent of the genera. Monte Bolca specimens of pycnodontiformes—the only extinct order—constitute the last stratigraphic record of this order. All species are extinct, although for many there are minimal osteological differences when compared with modern species (Blot 1969; Sorbini 1975). The fish found in this deposit already have

FIGURE 20.5 One of the many rays (*Trypon muricorte*) preserved in Monte Bolca. Length of specimen is 74 cm. (Photo courtesy of Museo Civico di Storia Naturale di Verona)

modern characteristics and are clearly similar ecologically to modern tropical fish assemblages (Landini and Sorbini 1996). However, while it was believed that many species were so similar to living ones that they belonged in modern genera, current workers have documented that many Monte Bolca fish possess primitive characters, warranting the erection of new genera. Detailed morphological studies on these genera as well as on those with both primitive and advanced characters have allowed ichthyologists to study the evolutionary relationships among many of these taxa (Tyler 1998). Some lineages, such as the anguilliforms and bericiforms, appear to have experienced more morphological changes since the Eocene than other groups, such as the perciforms (Sorbini 1984).

Almost all the fish are marine, although some genera have modern relatives that live under brackish conditions (Landini and Sorbini 1996). Many share characteristics with those living today in tropical waters of the Indo-Pacific and Atlantic Oceans. These similarities may be due to the oceanic communication between these bodies of water through the Mediterranean during the Eocene (Sorbini 1983).

A detailed trophic analysis and comparison with modern fish communities indicate that the fish fauna from the Pesciara locality can be assigned to several ecological categories related to modern coral reef ecological zonation and probably represent fauna from a nearshore, peri-reefal area (Landini and Sorbini 1996). The Eocene fossils are clearly analogous to modern fish from (1) the sand and/or seagrass areas around the reef, (2) the reef front, and (3) the peri-reefal and pelagic areas.

All trophic guilds in modern reef environments are represented in the Pesciara fossils, although different distribution patterns seem to exist (Landini and Sorbini 1996). For example, the primary energy flux passes through the planktivore-predator food chain, suggesting that pelagic phytoplankton and zooplankton played a much greater role in the Monte Bolca ecosystem than in modern true reef environments. The number of herbivore and coral-eating taxa is low, and there appear to be a number of more generalist, nonendemic taxa.

Marine Plants

Remains of algae are quite common in Monte Bolca and include *Halocloris* (Landini and Sorbini 1996), red algae (*Pterigophycos*), green algae (*Arystophycos*), and brown algae (*Postelsiopsis caput medusae*) (Sorbini 1983).

Terrestrial Invertebrates

A diverse number of insects have been preserved, including crickets, termites, beetles, water bugs, mosquitoes, and dragonflies (Massalongo 1856; Omboni 1886). Some of the water bugs of the order Hemiptera are similar to such modern genera as *Velia* and *Gerris* (water strider). *Gryllotalpa tridactylina* of the order Orthoptera is the first specimen of this genus in the fossil record (Sorbini 1983).

Terrestrial Vertebrates

Occasional remains of reptiles and impressions of bird feathers can be found within the deposit (Sorbini 1983). Plates of land tortoises have been uncovered at Mount Vegroni and Mount Purga. In the 1940s, the Cerato family excavated a 2 m long crocodile (*Crocodilus vicetinus*), which is now on display in the Natural History Museum of Verona (Sorbini 1972). A snake (*Archaeophis bolcensis*) was described by Massalongo (1859).

Terrestrial Plants

Hundreds of kinds of both freshwater and land plants are represented in the flora of Monte Bolca, some of which were studied by Massalongo (1856, 1859). So many new specimens and taxa have been found, however, that the Monte Bolca flora is in need of modern systematic study. *Eichorniopis* and *Maffeia* are two genera of freshwater plants that were living in fluvial and coastal pond systems close to the shoreline (Massalongo 1856). A diverse assemblage of herbaceous, shrubby plants are present as leaves, fruit, and branches (Massalongo 1859). Coconut trees, *Ficus, Eucalyptus,* and palm trees—some of which are hypothesized to have been 5 m tall—are also found in the various quarry localities (Massalongo 1859).

CONCLUSIONS

Except for the field of ichthyology, which has been greatly influenced by Monte Bolca fossils, geologists and paleontologists have not given this deposit as much attention as other Lagerstätten, probably because few people have studied this deposit and most of the publications resulting from their work have not been published in English. More detailed stratigraphic work, taphonomic study, and geochemical analyses could provide information not only on the paleoenvironment of the region, but also on the paleoceanographic conditions that led to deposition of this Lagerstätte. For example, stable isotopic analyses of the organic carbon within the laminae could indicate whether it is primarily terrestrial or marine in origin and thus differentiate between the two mass-kill scenarios. Oxygen isotopic analyses of carbonate tests could also indicate whether freshwater runoff was a significant factor during deposition of the fish-bearing beds. In addition, the application of cyclostratigraphic techniques may elucidate the apparently cyclic nature of the oceanographic, climatic, and sedimentologic conditions controlling the oxygenation of the seafloor, the lamination of sediments, and the mass killing of fauna.

Surprisingly, much more paleobiological work could also be conducted at Monte Bolca. Even with centuries of study, very little is actually known about many of the hundreds of invertebrate, vertebrate, and plant species that are remarkably preserved at Pesciara and other related localities. Because of the abundance, diversity, and excellent preservation of its fauna and flora, the Monte Bolca deposits represent a rare opportunity to study reef paleoecology and detailed morphological characters. This type of future work could influence the fields of systematics, evolution, and evolutionary paleoecology.

ACKNOWLEDGMENTS

I am indebted to the late L. Sorbini for providing a careful review of this chapter, a preprint of a manuscript on the paleoecology of Monte Bolca fish, a number of photographs, and general support for this work. I also thank H. Rieber for providing relevant literature.

REFERENCES

Agassiz, L. 1833–1843. *Recherches sur les poissons fossiles.* 5 vols. Neuchâtel: Petit-pierre.

Blot, J. 1969. *Les poissons fossiles du Monte Bolca classes jusqu'ici dans les familles des Carangidae, Menidae, Ephippidae, Scatophagidae.* Museo Civico di Storia Naturale di Verona Memorie Fuori, 2nd ser. Verona: Museo Civico di Storia Naturale.

Blot, J. 1980. La faune ichthyologique des gisements du Monte Bolca (Province de Verone, Italie): Catalogue systematique presentant l'état actuel des recherches concernant cette fauna. *Bulletin du Museum National d'Histoire Naturelle, Section C: Sciences de la Terre Paleontologie Géologie Mineralogie* 2:339–396.

Fabiani, R. 1914. La serie stratigrafica del Monte Bolca e dei suoi dintorni. *Memorie Prima Geologia Mineralogia Università Padova* 2:223–235.

Gaudant, J. 1997. Les poissons pétrifiés du Monte Bolca (Italie) et leur influence sur les théories de la Terre au milieu du Siècle des lumières, d'après un ma-nuscript inachevé de Jean-François Seguier (1703–1784). *Bulletin de la Société géologique de France* 168:675–683.

Harland, W. B., R. L. Armstrong, A. V. Cox, L. E. Craig, A. G. Smith, and D. G. Smith. 1989. *A Geologic Time Scale.* Cambridge: Cambridge University Press.

Landini, W., and L. Sorbini. 1996. Ecological and trophic relationships of Eocene Monte Bolca (Pesciara) fish fauna. In "Autecology of Selected Fossil Organisms: Achievements and Problems" [special issue]. *Bolletino della Società Paleontologica Italiana* 3:105–112.

Massalongo, A. 1856. *Studi paleontologici.* Verona: Antonelli.

Massalongo, A. 1859. *Specimen photographicum animalium quorundam plantaumque fossilium agri veronensis.* Verona: Vicentini-Franchini.

Massari, P. F., and L. Sorbini. 1975. Aspects sedimentologiques des couches a poissons de L'Eocene de Boca (Verone–Nord Italie). *IXe Congrès International de Sedimentologie, Nice,* 55–61.

Medizza, F. 1975. Il nannoplancton calcareo della Pesciara di Bolca (Monti Lessini). *Studi e ricerche sui giacimenti terziari di Bolca, Museo Civico di Storia Naturale Verona* 2:433–453.

Omboni, G. 1886. Di alcuni insetti fossili del Veneto. *Atti Istituto Veneto di Scienze, Lettere ed Arti* 4:1–14.

Secretan, S. 1975. Les crustaces du Monte Bolca. *Studi e richerche sui giacimenti terziari di Bolca, Museo Civico di Storia Naturale Verona* 2:315–427.

Sorbini, L. 1968. Contributo alla sedimentologia della Pesciara di Bolca. *Memorie Museo Civico di Storia Naturale Verona* 15:213–221.

Sorbini, L. 1972. *I fossili di Bolca*. Verona: Corev.

Sorbini, L. 1975. Gli Holocentridae di M. Bolca II *Tenuicentrum pattersoni* nov. gen. nov. sp.: Nuovi date a favore dell'origine monofiletica del Bericiformi (Pisces). *Studi e richerche sui giacimenti terziari di Bolca, Museo Civico di Storia Naturale Verona* 2:455–472.

Sorbini, L. 1983. The fossil fish deposit of Bolca (Verona), Italy. In H. Rieber and L. Sorbini, eds., *First International Congress on Paleoecology, Excursion 11A Guidebook*, pp. 23–33.

Sorbini, L. 1984. Resultats de la revision des Berciformes et des Perciformes generalises du Monte Bolca. *Studi e richerche sui giacimenti terziari di Bolca, Museo Civico di Storia Naturale Verona* 4:41–48.

Stanghellini, E. 1979. *Bolca and Its Fossils*. Verona: Espro.

Tyler, J. C. 1998. A new family for a long known but undescribed acanthopterygian fish from the Eocene of Monte Bolca, Italy: *Sorbiniperca scheuchzeri* gen. & sp. nov. *Eclogae Geologicae Helvetiae* 91:521–540.

CONTRIBUTORS

David J. Bottjer
Department of Earth Sciences
University of Southern California
Los Angeles, California 90089

Walter Etter
Naturhistorisches Museum
Augustinergasse 2
4001 Basel
Switzerland

James W. Hagadorn
Department of Geology
Amherst College
Amherst, Massachusetts 01002

Stephen A. Schellenberg
Department of Geological Sciences
San Diego State University
San Diego, California 92182

Carol M. Tang
California Academy of Sciences
Golden Gate Park
San Francisco, California 94118

ADDITIONAL CREDITS

INDEX

Numbers in italics refer to pages on which illustrations appear.